Risk Management and Society

Advances in Natural and Technological Hazards Research

VOLUME 16

The titles published in this series are listed at the end of this volume.

Risk Management and Society

Edited by

EVE COLES and DENIS SMITH
Centre for Risk and Crisis Management,
University of Sheffield, U.K.

and

STEVE TOMBS
Liverpool John Moores University, U.K.

KLUWER ACADEMIC PUBLISHERS
DORDRECHT / BOSTON / LONDON

A C.I.P. Catalogue record for this book is available from the Library of Congress.

ISBN 0-7923-6899-1

Published by Kluwer Academic Publishers,
P.O. Box 17, 3300 AA Dordrecht, The Netherlands.

Sold and distributed in North, Central and South America
by Kluwer Academic Publishers,
101 Philip Drive, Norwell, MA 02061, U.S.A.

In all other countries, sold and distributed
by Kluwer Academic Publishers,
P.O. Box 322, 3300 AH Dordrecht, The Netherlands.

Printed on acid-free paper

Dedication

This book is dedicated to the memory of Professor Barry Turner, in thanks for the inspiration.

CONTENTS

FOREWORD

This collection of essays represents a range of perspectives on the management of risk within the modern age. Our concern in bringing this work together was to provide readers with access to a multi-disciplinary account of the issues and to ensure that critical perspectives on the issues of risk management were encouraged. Three criteria have been employed in selecting contributions to this volume. First, we have made a judgement regarding the quality of respective contributions - thus as editors we have asked, To what extent does this contribution, taken in isolation, represent a contribution to an understanding of the various elements of 'risk management and society'? Second, to what extent do the contributions as whole produce a coherent text. Finally we are concerned to ascertain to what extent do the contributions actually reflect the range of work - in particular, the diversity of methodological approaches, objects of enquiry, and theoretical and political frameworks that constitute the area of risk management?

This collection owes a debt of thanks to the late Barry Turner. The genesis of this collection came at a meeting in Liverpool, an event overshadowed by Barry's death just as the conference started. Barry had been a keynote speaker at the first risk conference, which was held in Bolton, and he had set the scene for the second conference, which was held in Liverpool. The decision to bring together a range of researchers around many of the issues raised by Barry's work came out of discussions at the Liverpool conference and subsequent events held in Bradford and Durham.

Many of the contributors to this volume had known Barry as a generous supporter of new research talent within the field of risk and disaster management. Barry's contribution to the field is immeasurable. His early work on disasters was a landmark study that represented the first major attempt of the social sciences to deal with the issues of disaster management. More importantly, he introduced the concept of crisis/disaster incubation in which the actions of management and those responsible for the control of the system played an active role in creating the preconditions under which that failure took place. For all of us, Barry's death was premature and it deprived the academic community of one of its most innovative thinkers. We all owe him a considerable debt of gratitude. It is to his memory that this book is dedicated.

Denis Smith
Steve Tombs
Eve Coles

Sheffield and Liverpool

NOTES ON CONTRIBUTORS

Matthias Beck formerly lectured at the Universities of St Andrews and Glasgow before taking up his current appointment as a Professor at Glasgow Caledonian University. In collaboration with Charles Woolfson and John Foster he has published extensively on health and safety issues and regulation, both in the UK offshore oil sector and in onshore industry. In addition, he has written a number of articles on financial regulation and dual labour markets.

Richard Booth is Professor of Occupational Safety and Health at Aston University. He is the author of about 175 papers, articles and chapters in edited books on health and safety management and safety technology. He is Chairman of Health and Safety Technology and Management Ltd (HASTAM). He was founder-Chairman of the UK National Examination Board in Occupational Safety and Health from 1979 to 1986 and President of the Institution of Occupational Safety and Health (IOSH) from 1986 to 1988.'

Eve Coles is a research associate within the Centre for Risk and Crisis Management at Sheffield University Management School. She was formerly editor of 'Emergency' and was assistant editor of a nine volume series on Natural Hazards and Disasters with Alf Keller. She is a Fellow of the Institute of Civil Defence and Disaster Studies and of The Royal Society of Arts, Manufacture and Commerce. Her research interests include emergency planning and management, risk and crisis management and public sector management. She has published in various journals including *Local Government Studies, Public Money and Management* and *The Australian Journal of Emergency Management.*

Dominic Elliot is a Senior Lecturer in Strategic Marketing at the Centre for Risk and Crisis Management, University of Sheffield. He regularly speaks on the subjects of Crisis and Strategic Management at national and international conferences and as a visiting lecturer at a number of universities. Reflecting the practical focus of his work he has worked as a consultant and trainer with senior executives from a range of organisations including: Chartered Institute of Marketing, IBM, Philips Electronics, Government of Lesotho, Football Association, Royal Mail, Transco and the Department of Trade and Industry. He has published many articles and books in the fields of Strategic and Crisis Management and Business Continuity. Dominic founded the Centre for Business Continuity Planning at De Montfort University and is a Fellow of the Business Continuity Institute. Dr. Elliott is Review's Editor for Risk Management: An International Journal.

Christopher Gifford is a consultant mining engineer. Until is retirement from full time work he was HM District Inspector of Mines and Quarries in the Health and Safety Executive. He is a regular contributor to seminars and conferences and his interests lie in health and safety and regulation.

Ian Glendon is Associate Professor in Applied Psychology at Griffith University (Gold Coast Campus), Australia. He is a psychologist specialising in organisational issues, has been active in the risk cognition and risk management field since 1975 and has worked on numerous research and consultancy projects. Topics include risk cognition; accident and disease reporting; analysis of accident injury data; risk homeostatsis; employee involvement in health and safety; human error and reliability; occupational stress; health and safety management; safety culture and safety auditing; risk management; human factors; health and safety expertise. He has written over 100 book chapters, journal papers, reports, booklets, manuals, professional articles and other publications as well as presenting over 80 papers at national and international conferences in around a dozen different countries. He has co-authored three books in the field of human safety and risk management.

Heather Höpfl is Professor of Organisational Psychology and Head of the School of Operations Analysis and Human Resource Management at Newcastle Business School, University of Northumbria at Newcastle. She is an Adjunct Professor of the University of South Australia and a Visiting Professor of the Academy of Entrepreneurship in Warsaw and of Bolton Institute, UK. She is a former Chair of the Standing Conference on Organisational Symbolism (SCOS) and a Fellow of the British Academy of Management and is on the editorial boards of a number of major journals. Her research interests are primarily in the field of organisational performance and the tacit and taken for granted aspects of organisational behaviour. From 1990 – 1995 she worked on the development of an interpretative environment around the safety function with British Airways. She is currently working with Prof George Schreyoegg of the Free University of Berlin on aspects of organisational performance and with Prof Hugo Letiche of the Humanistic University of Utrecht on theatre and film in the analysis of organisational behaviour.

Jo McCloskey is Head of the Marketing School at Bristol Business School. She has previously taught at Liverpool John Moores and De Montford Universities and has worked in a range of higher education institutions in Ireland and Africa. She has a wide range of consultancy experience having undertaken projects in North America, Africa and Europe. Her research interests are in Strategic Marketing, Corporate Communications Strategies and Risk and Crisis Management. She has published in several journals such as Public Money and Management, Management Decision and Management Learning and has contributed to a number of books that have been re-printed and translated into European editions.

Bryan O'Loughlin Graduated in Electrical Engineering at Manchester University in 1963 and spent a career in the electricity supply industry. Initial experience as an operational engineer, joining the Safety Branch of the Electricity Council in 1972. He is manager of the Health and Safety Branch and Safety Advisor to National Grid Company until 1994. Now Visiting Fellow at Aston University.

Frank Pearce is a Professor of Sociology at Queens University, Kingston, Ontario. He has published extensively in North America and the UK on social theory and on corporate crime. He is author of *Crimes of the Powerful* (Pluto, 1976) and the *Radical Durkheim* (Unwin Hyman, 1989), and *Toxic Capitalism: corporate crime and the chemical industry*, (with Steve Tombs, Ashgate, 1998, Canadian Scholars' Press, 1999); he also co-edited *Global Crime Connections* (with Mike Woodiwiss, Macmillan, 1993) and *Corporate Crime: contemporary debates* (with Laureen Snider, University of Toronto Press, 1995).

Clive Smallman is Senior Research Associate at the Judge Institute of Management Studies and a Fellow of Hughes Hall, University of Cambridge. He is a corporate member of the British Computer Society (MBCS), a chartered information systems practitioner, a graduate member of the Institute of Mathematics and its Applications (GIMA) and is an affiliate member of the Chartered Institute of Personnel and Development. He has previously worked for Marconi Research, the Prudential, British Gas, General Accident, and Admiral Consulting. He joined the Bradford University Management Centre in 1992 and moved to Cambridge in May 1999. His main research interest is in operational risk management (managing risks in business processes and the people that operate them), and central to this is the management of information and knowledge.

Denis Smith is Professor of Management at the University of Sheffield where he is also head of the Strategy, Systems and Management Division. In addition, he is head of the Centre for Risk and Crisis Management, which has been operating as a research group under his leadership since 1990. Prior to his appointment at Sheffield, he had been Professor of Management at both the University of Durham and Liverpool John Moores University, where he was also head of Liverpool Business School. He has previously held faculty positions at the University of Manchester, Nottingham Trent University and De Montfort University (Leicester). He has also been visiting professor of Human Resource Management at Kobe University in Japan, visiting professor of Strategic Management at the University of Sheffield and a visiting adjunct professor at San Diego State University in the USA. He is a Fellow of the Royal Geographical Society, a Fellow of the Institute of Personnel and Development as well as being a member of a number of professional bodies including, the Ergonomics Society and the Human Factors and Ergonomics Society. His post-doctoral studies were in the area of science and technology policy and his current research interests include: social issues surrounding the use of new technology; technology transfer and innovation; latent error potential in health care systems; crisis incubation and escalation within organisations; risk perception and communication; crisis learning; latent human error in systems failure; and environmental management. Outside of his university activities, Professor Smith is a non-executive director of St Helens Rugby League Football Club and has previously served in a similar capacity for Mersey Regional Ambulance Service. During

1999-2000 he also served as a member of the Ministerial Expert Committee, 'Learning from Experience in the National Health Service', chaired by the Chief Medical Officer.

Walter Stahel is an alumni of ETH, the Swiss Federal Institute of Technology, in Zürich, where he received his diploma in architecture in 1971. He has joined the Geneva Association (International Association for the Study of Insurance Economics) in 1987, as director in charge of risk management research. He is also one of the founder-directors of the Product-Life Institute and was previously a project-manager at the Battelle Geneva Research Centres, Geneva, Centre for applied economics, in the fields of business strategy and feasibility studies. He is been a member of the first Environmental Council of the German Railways, and of the Umweltrat of the UmweltBank, Nürnberg, since its foundation in 1997. He was member of the Jury of the SHE (Safety, Health and Environment) Excellence Awards 1996, 1998 and 1999 of Du Pont de Nemours. He is a member of the Eco-Dream-Team of Interface Inc, Atlanta GA, and of the work group on "Basic Needs" for the world exhibition EXPO 2000 Hannover. He is the author of books and numerous articles on policies, strategies and tools to foster an economic and societal development towards a more sustainable society.

Steve Tombs is a Professor of Sociology in the Centre for Criminal Justice, Liverpool John Moores University. He has a long-standing interest in the incidence, nature and regulation of corporate crime. His main recent publications are *Corporate Crime* (Longman, 1999), with Gary Slapper, and *Toxic Capitalism: corporate crime and the chemical industry*, (Ashgate, 1998, Canadian Scholars' Press, 1999), with Frank Pearce. He is Chair of the Centre for Corporate Accountability.

Dave Whyte is a Lecturer in Criminology at Manchester Metropolitan University , and until recently a researcher at the Centre for Criminal Justice, Liverpool John Moores University. He has written widely on the regulation of worker safety in the UK oil industry, on corporate crime and on the Scottish legal system in various journals and forums, including: *The Big Issue, Blowout, Industrial Relations Journal, Occupational Safety and Health, Public Money and Management, Social and Legal Studies*, and *Studies in Political Economy*.

Charles Woolfson is currently Director of Faculty of Social Sciences Graduate School (Associate Dean), Deputy Director of the Centre for Regulatory Studies and Director of the European Centre of Occupational Health, Safety and the Environment (ECOHSE). In collaboration with Matthias Beck and John Foster he has published extensively on health and safety issues and regulation, both with respect to the UK offshore oil sector and to onshore industry. In addition, with John Foster he has written a number of monographs on industrial conflict.

ACKNOWLEDGEMENTS

In a book of this nature, there are inevitably a number of people who have made a major contribution to bringing the manuscript to a final version. In particular, we would like to thank Petra van Steenbergen and Manja Fredriksz of Kluwer for their support throughout the period at which the book was pulled together and for their patience with three editors who seemed, at times, continually to be moving around the UK's university sector! Alf Keller also deserves our thanks for his initial support and encouragement. Thanks must also go to Christine Jawad, Anne Hardy and Mandy Robertson for their secretarial support throughout the various stages of the book. Chris and Anne deserve a particular thanks for putting the papers into their final camera ready copy. Anne also deserves a special mention for her re-drawing of the diagrams for the 'Loss Prevention' chapter.

Each of the chapters were sent out to scholars and practitioners for review and comment. This is a difficult and time consuming task and we owe each of those individuals a vote of thanks. They include: Mark Butler, Mike Clark, Kevin Daniels, Dominic Elliott, David Evans, Ken Green, Richard Griffiths, Alan Irwin, Jane Jacks, Trevor Sheldon, Paul Sparrow, Brian Wynne

Finally, thanks are also due to our various families for support and patience during the period when we each seemed to be attached to our respective computers. Eve Coles would like to thank all her family for putting up with an inattentive wife and mother. To Jim go particular thanks for the job he did on some of the diagrams and for not complaining when forced to live off ready meals and takeaways! Denis Smith would like to express his thanks to Janice, Rachael and Liam for their understanding and support -yes Liam, it is OK to type on the computer now. Steve Tombs wishes to thank, for their various forms of advice, support and friendship, David Bergman, Alan Dalton, Chris Gifford, Pete Gill, Rory O'Neil, Frank Pearce, Gary Slapper, Joe Sim, Dave Whyte and Charles Woolfson. Particular thanks go to Pam and Patrick, who have carried the burdens which have yet again been forced on them by a partially absent partner and father with smiles and love. All three would like to acknowledge the intellectual debt that they owe to the late Barry Turner to whom this book is dedicated.

OF COURSE IT'S SAFE, TRUST ME!

Conceptualising issues of risk management within the 'Risk Society'

DENIS SMITH
Sheffield University
Management School
9 Mappin Street
Sheffield, S1 4DT, UK

STEVE TOMBS
Liverpool John Moores University
School of Law, Social Work & Social Policy
Josephine Butler House
1, Myrtle Street
Liverpool L7 4DN, UK

Introduction

One might be forgiven for thinking that notions of risk appeared to adopt a higher media perspective during the latter part of the 20th Century than in previous periods of our history. This was due, in part, to the many and varied media conceptualisations of the problems that are allegedly facing society. For many, the media assault on our senses created a feeling that the world was becoming a more dangerous place. Humans seem to be discovering more exotic ways to die or to be severely injured as a result of the activities of modern societies. Even nature appears to be determined to exact a greater price from the human race through El Nino and a range of natural disasters. Of course, such expressions of the risk problem at the start of the second millennium rather mask the complexity of the many problems that face humankind. To an extent, it is this complexity and the uncertainty of our knowledge around the issues of hazard that creates, and sits at the heart of, the risk problem. Thus a key part of contemporary difficulties lies in our (in)abilities to make sense of 'risk'. This is especially so within a highly charged environment in which trust, issues surrounding the burden of proof and the legitimacy of expertise are crucial issues. Indeed, a proliferation of concern with risk seems to have arisen at a time when questions are being asked of both the ability and even the willingness of government to 'protect' society from the complex web of risks that seem to face us (see Rose, 1996). Some recent examples serve to illustrate the nature and extent of the problem.

1

E. Coles, D. Smith and S. Tombs (eds.), Risk Management and Society, 1–30.
© 2000 *Kluwer Academic Publishers. Printed in the Netherlands.*

On one level, there seems little doubt that our societies are now more informed about risk than at any other time in human history. The visual impact of many forms of hazard creates an opportunity for the media to provide graphic images of destruction to the viewing public. For example, the bombing of the City of London and the virtual destruction of the Baltic exchange on 24th April 1993 illustrated to UK industry and the world's viewing public that even the largest and most powerful actors - the corporations that dominate and reside in the City of London - are not immune from the possibility of a major disruption to their business. The subsequent bombing of Manchester town centre on a busy Saturday reinforced the potential for disruption that can face business. The Oklahoma and Atlanta bombings in the USA illustrated to a shocked US public, used to seeing devastation elsewhere, that their own cities were also vulnerable. The bombing of Wall Street in February 2000 served to reinforce this vulnerability. In Japan, the gas attack on Tokyo's underground was a terrible reminder of the potential that exists for both terrorist and accidental toxic and biological contamination of our cities. The difficulties that face government in protecting the populace from such events is all too obvious. The mass evacuation of a large population center within a short period of time creates immense problems for emergency planners. The potential for such a risk was reinforced by the nuclear accident at Tokamura (1999) in the greater Tokyo area and it brought back to our consciousness the specter of widespread nuclear risk, which had been raised by the Windscale, Three Mile Island and Chernobyl accidents. Almost as if to mark the passage into a new century of ever-heightening risk, it was announced in late 1999 that a number of governments were examining the feasibility of deflecting any large asteroid or meteor that threatened to hit the earth and create the potential for the so-called "extinction level event".

Notions of risk also became manifest in new forms through the use of the very technology that was designed to improve our lives. As we entered the new millennium, the City of London, along with almost every other organisation of any size in the UK, was urged by Government to prepare for the risks associated with the use of IT and the so-called millennium bug. Organisationally-based risks were also on the public agenda. The claims by a senior clinician, made in January 2000, that the NHS was not meeting expectations prompted a knee-jerk reaction by government to put yet more money into a health service deemed to be ailing, in an attempt to placate criticism of policy failure around the core of the welfare state. However, within a month, the media announced that a patient undergoing surgery had the wrong kidney removed; he died within weeks. Rumours abounded concerning the accident rates within operating theatres and public confidence in the country's health care system was damaged still further (see Leape and Berwick, 2000). Mounting criticism suggested that health care was in crisis and that the solutions needed to be radical and would require a complete re-think of the system (see, Reason, 2000; Nolan, 2000). On top of these concerns, the highly public conviction of Dr Shipman for murder raised serious concerns over the trust that we place, and the control systems in place, within that most proximate form of healthcare, namely the GP's surgery. These criticisms coincided with a flu outbreak that not only stretched the health service's resources to the limit but also illustrated the risks that we faced from both possible epidemics and pandemics (see Davies, 1999) and the range of viruses that face us on a global scale (Ryan, 1996). The massive advances in transport allow viruses to spread quickly beyond their normal confines. Similar advances in telecommunications ensure that the horror of these events is beamed into our homes. Those charged with managing the problem are faced with questions from a panel of experts drawn from all corners of the world. Matters are made worse by the fact that many of these viruses evolve to defy our attempts at controlling them and some, notably the avian flu virus outbreak in Hong Kong, have even jumped the species barrier (see Centres for Disease Control and Prevention, 1998a; 1998b; 1999; Davies, 1999).

The cumulative effect of such acute incidents should not be under-estimated. Yet their effects on popular notions (indeed fears) of risk are all the greater since they need to be set in a context of a vast array of seemingly intractable chronic risks. For example, in the UK, the spread of Bovine Spongiform Encephalopathy (BSE) amongst cattle, and its transmission to humans as new variant Cruzweldt Jacob Disease (nvCJD) illustrates how problems can emerge which defy the prevailing scientific paradigms operating at the time. The public conflict that resulted from debates amongst "experts" resulted in a loss of consumer confidence in government, a loss of confidence which sits at the heart of current, highly charged, debates on genetically modified foods (see Smith and McCloskey, this volume). These latter concerns are just a sub-set of concerns related to a general sense of widespread environmental damage being caused by human activity, damage that threatens to change the planet on which we live, and the conditions of our very existence. Examples are too numerous to mention, but familiar to all. The fact that the late twentieth century witnessed a global explosion of environmental concern fueled by - and perhaps fueling - the 'discovery' of new environmental hazards only illustrates how, thirty years after the warnings of Rachel Carson (1962) on the use of pesticides (and DDT in particular), societies have either failed to learn, or failed to act upon that learning.

The range of these phenomena captures an element of the range of real and perceived, acute and chronic, localised and widespread hazards within our society. Along with the evidence of an apparently unprecedented number of accidents, disasters and crises (experienced on a global scale during the 1980s and 1990s) they have served to underline the increasing realisation that there is a need to address, subject to critique, and develop what currently represents the sub-discipline of risk management. The events also illustrate the fallacy of control that predominates around issues of risk (Smith, 2000). Some of these events were unforeseen or, in some cases, their probability of occurrence was considered to be so low that they were virtually ignored in terms of preventative action. In some cases, sequences of events bypassed the defences that were put in place to prevent the accidents, or the defences were simply inadequate to cope with the demands placed upon them. Alongside a growing realisation of the extent of the hazards we face, recent years has witnessed the development of burgeoning new industries around issues of risk management. It is clear that many organisations are turning themselves, with greater or lesser efficacy, to the task of developing expertise in risk, whilst Universities and consultants in the UK, Europe and North America have begun to develop programmes of advice, training, research and teaching in all areas of risk management. At the same time, social theorists have sought to develop new, more insightful ways of conceptualising and thus dealing with such issues. It is to a discussion of some of these concepts that we now turn our attention.

Conceptualising Risk

Whilst the notion of the risk society has attracted considerable attention within the academic literature (see Beck, 1992; Giddens, 1990), the conceptualisation of this phenomenon has also been a matter of considerable debate. For Beck (1992) the processes of modernization have given rise to the creation of risks that threaten large numbers of people and yet have become more opaque with the result that,

> "in the course of the exponentially growing production process in the modernization process; hazards and potential treats have been unleashed to an extent previously unknown" (Beck, 1992 p19).

Such concerns about risk have been expressed by a number of writers who have approached the problem from a number of perspectives (see, for example, Shrivastava, 1992; Lupton, 1999; Shrader-Frechette, 1991; 1993; Erikson, 1994; Draper, 1991; Sheldon and Smith, 1992). Despite such concerns, there are those who express strong doubts regarding the true extent of the emergence of higher levels of risk within the modern period (see Cohl, 1997; Furedi, 1997). Popular representations of risk are evident in much of the television and cinema offerings that, many argue, foretell an apocalyptic future for society. Glassner (1999) has labeled this obsession with risk a "pathology of fear". Citing health as one of a number of examples, he observes that,

> "The scope of our health fears seems limitless. Besides worrying disproportionately about legitimate ailments and prematurely about would-be diseases, we continue to fret over already refuted dangers" (Glassner, 1999, p. xii).

Glassner makes the point that our pessimistic views on life are such that if we give ourselves "a happy ending ... we write a new disaster story" (Glassner, 1999, p. xi). Whether our preoccupation with issues of risk is a symptom of the human condition, or a reflection of (and on) the state of the world in which we live, is clearly a matter of some debate. What has become clear, however, is that many of the previous notions of risk have become somewhat flawed as a means of conceptualising the range of problems that have to be dealt with by both society and the range of organisations within it. Whilst Beck has stated that "risks are not an invention of modernity" (Beck, 1992 p21), within the risk literature is clearly the argument that a shift from individually-based risk to a more widespread exposure to hazard differentiates contemporary society from its forebears. The subtle, more opaque relationships that exist between cause and effect creates problems in terms of the knowledge base around risk generation and crisis incubation. There is widespread recognition concerning the limitations of technical expertise in issues of risk (see Beck, 1992; Fischer 1980; 1990; Collingridge and Reeve, 1986; Irwin, 1995; Lasch 1995; Smith, 1990; 2000), and on the need for a role for other forms of non-expert knowledge in risk debates (see Irwin, 1995; Irwin and Wynne, 1996). Much of this tension becomes conceptualised in terms of public-expert conflicts around the nature of the hazard and its associated probabilities. Previous attempts to deal with such issues saw them simply in terms of a deficit model, which saw the problem simply in terms of giving the public the facts about the problem and their concerns will go as a consequence. However, it is now clear that this was a flawed strategy. Information dissemination, without a foundation of trust, is destined to be treated with considerable suspicion. Never before have the public been exposed to so much information, although this also brings with it considerable problems. Lasch (1995), observes that the information given to the American public, for example, "tends not to promote debate but to circumvent it" and that "although Americans are now drowning in information ...... (they) ..... are notoriously ill informed" (Lasch, 1995, p. 11). Herein lies an apparent paradox within risk debates. Whilst publics clearly need and, indeed, have a right to, information, there is also the view that if there is too much information, then the same publics may not be able to make sense of the complex, contradictory information that they are given. Similar problems occur when publics are presented with contradictory expert opinion and this has been held to increase the extent of conflict surrounding issues (see Collingridge and Reeve, 1986). Nevertheless, a consistent though certainly not ubiquitous feature of risk debates is that it is those organisations with the greatest power (expressed in terms of capital, influence and knowledge) whose voices, views, and interpretations of events achieve dominance. The questions that remain centre on the role that 'conventional', scientifically determinist approaches to risk management may have contributed to this situation. It is here that the greatest source of tension exists within the contemporary societies seeking to manage risk, and which provides the challenge to social science research.

Research efforts within the social sciences have, over the last 25 years, sought to deal with the complex array of issues surrounding risk management. Lupton (1999) distinguishes between three main groups of social research that stand in criticism of the techno-rational perspectives on risk management. The first of these deals with the cultural/symbolic approach to risk (Lupton, 1999) which centres around the social anthropological work of Mary Douglas and colleagues (Douglas, 1980; 1985; Douglas and Wildavsky, 1982; Schwartz and Thompson; 1990). This work has focused primarily on the cultural response to issues of hazard. This has been grounded within a socio-cultural context and has adopted a structuralist approach to looking at the relationships between group cohesion and other constraints on social groups (grid-group model). This work raises a number of important issues concerning the hierarchical nature of societies, their approaches to regulation and the role of the group (rather that the individual) in dealing with issues of risk

The second body of work outlined by Lupton (1999) concerns a range of issues that are grouped around the notion of the "risk society" and it is this body of research that has, perhaps, attracted the greatest amount of attention. The risk society approach is dominated by the view that sees risk emerging from the activities of modern society and sees society reflecting upon its activities and the problems that emerge from that process. This notion of "reflexive modernization" can be formed by reference to Beck's (1992) observation that,

> "The concept of risk is directly bound to the concept of reflexive modernization. Risk may be defined as a systematic way for dealing with hazards and, insecurities induced and introduced by modernization itself. Risk as opposed to older dangers, are consequences which related to the threatening force of modernization and to its globalization of doubt. They are politically reflexive" (Beck, 1992 p.21).

Reflexive modernization raises a number of important issues with regard to risk, not least of which is this "globalization of doubt" and the impact that it has on trust, expert knowledge and power. Beck (1992) makes the point that such a process of reflexivity leads both to emergent risks from the process of modernization and also causes society to examine the "problem resulting from techno-economic development itself" (Beck, 1992 p19). Given the dominance of this literature, we shall address some aspects of this in more detail later in this paper.

The final approach outlined by Lupton is concerned with notions of governmentality. This body of work is essentially concerned with the shift in emphasis from social insurance towards self-insurance (see, Lupton, 1999; Rose, 1996). This approach to risk can be framed in terms of the ways in which reality becomes constructed through the use of various expressions of knowledge, the discourse between interested parties and the role of both expertise and institutions within the process (see Lupton, 1999). For these theorists, in the sphere of risk as in other discourses, the key trend marking advanced social orders (typified by neo-liberalism in the later decades of the last century) has been the generalised adoption of technologies of the self, through which risk is individualized. Here responsibilised individuals are being offered, or develop themselves, various ways of managing risk. At the same time, governments and states absolve themselves from various responsibilities of popular welfare. Invariably, this leads to a process of social exclusion around the ability to mitigate the consequences of such risks and results in the development of a class of people for whom hazards are possibly greater than for other members of society – a point that will be returned to later.

For our present purposes, whilst some attention will generally be focused on the work of the "risk society", elements of the other research perspectives will be incorporated into the discussion. There are, however, a number of themes that emerge within the chapters which constitute this volume. The first of these is the role of *quantification and audit* within the processes of risk analysis. This theme is significant within risk management practice and a veritable industry has emerged which seeks to measure the extent of the risks in many of our systems and practices. The second theme that emerges concerns the role of *organisational culture* in shaping the nature of many risk debates and, indeed, incubating the potential for risk in the first place. Culture is a seductive and yet elusive term, which seems to offer considerable potential for addressing many of the issues raised by risk within modern society. It impacts upon such issues as power relationships, the role of knowledge and expertise in shaping debates and the nature of communication around risk issues. Culture can also have a dark side and may contribute to the incubation of failure potential within organisations (Smith, 1995). These issues are touched upon by a number of the contributions to this volume. The final theme that emerges within this book centres on the notion of *management* itself. Despite its widespread use as a concept, "management" can be seen as an abstract term (Lilienthal, 1967), which often defies effective description. In addition, the term often implies that managers have the ability to control the uncertainty surrounding those elements of the system in which they operate. If there was no uncertainty within organizations then there would be little need for decision making and, therefore, one might argue, for managers. If we take this argument to its logical conclusion then we can claim that managers exist to cope with uncertainty. Whilst this is hardly a radical assumption, one might question how many managers would agree with the proposition. Similarly, it could be argued that it is the inability of managers to deal with the demands of emergence (and its associated uncertainty) that creates many of the problems that are discussed in this collection of essays. The key issues center on the role of knowledge and expertise within the function of management and the manner in which uncertainty is communicated both internally and externally. Both of these issues are touched upon by several of the contributors to this volume.

Taken together, these three themes provide us with the opportunity to examine the relationship between risk generators and those elements of society that are exposed to the consequences of that risk. The remainder of this chapter considers the implications of these themes in more detail.

Quantification and Audit

Clearly identified in the chapters which constitute this volume are a whole series of techniques which are commonly used as a means of 'managing' risk. These are, for the most part, dealt with extensively within the literature, though such treatments are predominantly exhortational, prescriptive or normative. What the chapters in this volume seek to do is to explore their application, their limitations and their potential for effective risk management, through a variety of empirical and theoretical considerations. These techniques range from the 'hard' end of the continuum of risk management techniques - utilising engineering and design approaches (themselves based upon science-technology principles and quantification) - through to the 'softer' human factors, which have also adopted a quantitative approach but also have implications in terms of recruitment, training, and culture change. This broad continuum is largely explained by the development of risk management as an activity and sub-discipline. The origins of risk management have been seen to lie in a number of academic disciplines including natural hazard research (Bourriau, 1992; White and Burton, 1980), economics and finance (Bernstein, 1996) and engineering (Rowe, 1977). More latterly, these technocratic approaches to risk have come under considerable scrutiny from the social sciences (see Fischer, 1980; 1990) although some, notably human factors and ergonomics, have themselves sought to emulate the 'scientific' approaches to risk.

Almost irrespective of its origins, risk has tended to assume a quantitative approach and this remains the dominant paradigm within the literature. Thus, as Smallman (this volume) notes,

> "Risk management is almost totally dominated by the 'quantificationists'. Their assumption is that systematic quantification of risk is the only method by which risk may be rationally analysed and measured against pre-determined objectives. Such supposed rationality, coupled to technically sophisticated risk assessment methods which parallel cost benefit analysis sits well with bureaucrats and legislators .. {it} seems to offer a feeling of security" (Smallman, **24-5**, this volume).

Such techniques have, of course, been subject to extended critique - a critique extended by several of the contributions that follow here (see, notably, Beck and Woolfson, Pearce and Tombs, Smallman, Smith and McCloskey, this volume). But the appeal of the rational-scientific paradigms in general, and the vested interests in whose hands they are often used, has served to bolster quantification in the face of its critics. Whether or not such techniques are predominant within all of the various approaches to risk management, quantification continues to play some part in most attempts to manage risk. Given that an understanding of the probabilistic dynamic of hazards requires some basis in historic data, one would expect that some benchmark against which to measure changes in its frequency would exist. However, when such an approach refuses to acknowledge the validity of other forms of knowledge and is based upon poor *a priori* data, then such an approach can invariably be called into question. When the hazards under consideration are 'new' (or emergent) and the evidence is extrapolated from laboratory-based experiments, then there obviously remains a number of problems with the quantified approach.

In the first instance, there may simply be insufficient evidence to justify a probabilistic estimation of the hazard's frequency. Such a problem faced the investigating team at the Canvey Island complex in the late 1970s. Here the solution to the lack of data was to incorporate experts' best estimates of likely failure rates into the risk assessment (see Smith, 1990). To claim that such an analysis is objective is clearly a spurious argument, especially when it is set against the supposed 'subjective' views of local residents who are exposed to the hazards. One might argue that, ultimately, all approaches to risk management are subjective, as even the most quantified approaches are set within dominant academic paradigms of the various technical disciplines involved in the analysis. Added to this is the influence of organisational culture upon the investigating team, which raises further questions concerning the legitimacy of knowledge and the role of power and culture in shaping the social construction of that knowledge. This social legitimacy of knowledge is an important issue within the process of risk management. Lasch (1995), for example, has argued that "knowledge is merely another name for power" (p. 12), a point that has also been made by a number of observers (Collingridge and Reeve, 1986; Smith, 1990, 1991). A third area of concern is that experts often communicate only with other experts (Lasch, 1995) and this process contributes to the social exclusion of potential victims in two distinct ways. The first of these is simply that those affected by the hazards are often deemed not to be part of the "expert" community and are, therefore, either simply ignored or their views are relegated in importance. Evidence for this can be found in a whole range of debates concerning environmental impact (Irwin, 1995; Irwin et al, 1996; Smith, 1991), medicine and public health (Epstein, 1996; Bennett and Calman, 1999) and risk (Smith, 1990; Whyte et al, 1995). The second, and more subtle, form of social exclusion is through the use of a complex language of technocracy which is not generally accessible to those outside of the expert group (see Fischer, 1980; 1990; Porter, 1995; Smith and McCloskey, this volume). In most cases, this language is one of mathematics that, some argue, has become a surrogate for trust (see Porter, 1995). Trust and expertise thus combine to create our final set of concerns with the quantitative approach. Ultimately, the expert community is accountable to those organisations for whom they work and the professional bodies who accredit them. Even the regulatory agencies themselves face this problem and, therefore, one might question the democratic nature of this process. This point is best illustrated by reference to the case of health and safety management.

The UK body charged with regulating workplace health and safety - the Health and Safety Executive - has, following longstanding work by companies, trade associations, consultants and academics, begun to emphasise the use of auditing tools, performance standards, risk assessment, and the practice of benchmarking against best practice for risk management (Health and Safety Executive, 1991, 1997). All of these rely on some form of utilisable measures of health and safety performance. For the purposes of this brief discussion, we shall use the HSE's Successful Health and Safety Management (1991, 1997) (SHSM) to illustrate the main dynamics of this approach.

SHSM advocates the use of internally created and monitored performance standards. But this raises immediate problems. First, we know that there are massive (if variable levels of) under-reporting across all industries, formally revealed by HSE's use of questions in recent Labour Force Surveys (from 1990 onwards). This does not bode well for any performance standards based upon accurate measures of injuries, let alone other types of accidents. If organisations are failing to report externally, despite a legal obligation to do so, and notwithstanding moves by HSE to make such reporting simpler and less time-consuming, then one might at least be sceptical about the propensity of internal reporting systems to operate effectively. Secondly, we have to be sceptical about internal measurement because of what has been labeled the "DuPont effect". This describes the tendency for organisations to under-report incidents, especially where the organisation places great emphasis on, or attaches rewards or sanctions, to the recording of incidents and lost time accidents. Quite perversely, therefore, safety management systems can produce very real pressures not to report injuries and other incidents. The offshore oil industry again provides numerous documented examples of the "creative" approaches, which companies have utilised in order to minimise reported lost time injuries (not least because contract firms may lose lucrative incentive bonuses). Even where incidents are reported internally, it is questionable whether these will then be reported externally if there is a belief that negative consequences may follow. These consequences may include, the raising of insurance premiums, the attention of inspectorial activity, generalised negative publicity, and so on. Indeed, today there is widespread evidence of the falsification of injury data to those 'outside' the organization, which is attributable to such forces. The recent exposure of BNFL regarding the falsification of testing data on exported plutonium (Boggan, 1999) may be rare in terms of the public profile it received, but it is far from unique.

Obviously, any unreliability within incident data is likely to distort attempts at benchmarking. It is here that auditing may come in as a useful tool for the development of more effective safety and health management systems. There are now several well-established auditing systems in use. The best known include the British Safety Council's 5 Star, CHASE, Coursafe, DUPONT, the International Safety Rating System, Management Safety Systems Assessment in the Evaluation of Risk, RoSPA's Quality Safety Audit, the Professional Rating of Implemented Safety Management, and SHARP, not to mention HSE's own Safety Climate Assessment Toolkit. The HSE itself has advocated the use of bespoke auditing systems - now available in electronic rather than simply paper forms. The HSE's view is based on evidence that where auditing systems are "structured in a way and put into operation in a way which fits with the characteristics and needs of user organisations", where they utilise information from a range of sources, and where they are subject to external verification, they can form an element of improved health and safety management.

While auditing may play a useful rule in disclosing relevant safety information, there is some evidence that the use of auditing tools can also be problematic. Where auditing is used simply as means of complying with the demands of insurers, or functions as a superficial checklist, then it may actually obscure the real work practices that are prevalent within an organisation. Perhaps most significant problem of all is that the development and use of audits in an effective manner requires a significant commitment of resources. Where such resources are not committed by an organisation and/or where incentives to underreporting are strong, we cannot expect auditing to present an effective safeguard of safety performance

The weight of this general critique of techniques of quantification and auditing - as well as the more specific considerations in the brief focus on health and safety management - is to cast considerable doubt on the validity and utility of such techniques, a doubt explored in more detail by the various contributors to this volume. At the very least, it is clear that quantification and auditing do not possess the 'hard' characteristics of rationality and objectivity ascribed to them by their proponents. Consequently, we need to explore the "softer" and more abstract dynamics of the risk management process, namely organizational culture and the nature of "management" itself.

Organisational Culture

Culture has emerged as an increasingly fashionable approach to dealing with risk and can be seen to sit at the 'soft' end of the spectrum of risk management techniques (see Pidgeon, 1997; Reason, 1998). Indeed, this is the focus of the chapters here by both Hopfl, and by Smith and Elliott, each of which provide sectorally-based discussions of the issues surrounding culture and its relationships to risk management. Smith and Elliott focus attention upon a particular organisational culture which is based within the fire-fighting services. Their study indicates that the dominant culture of the service can create a bulwark against effective risk management (expressed in terms of occupational stress). Hopfl, by contrast, details the efforts of one specific company, namely British Airways, to institute a general programme of culture change - launched, interestingly, as a response to declining commercial performance. One element of this programme involves a thoroughgoing attention to the company's safety culture. Each avoids the generalisations, the broad prescription, and the empirical and analytical looseness which, for us, infects much of the current literature on 'culture'.

There is now a vast body of literature around organisational or business cultures, which deals with issues of measurement, categorisation and, in particular, cultural change[1]. Again reference to the context of occupational health and safety management illustrates this issue. Notions of cultural measurement and change are central to the current thinking of the Health and Safety Executive (HSE), the regulatory agency which is charged with overseeing the control of risks to employee and public health, safety and welfare in British workplaces; such notions are also inherent within a series of corporate initiatives around safety and health. For example, the HSE has repeatedly emphasised the importance of developing an effective safety culture as the precondition for successful safety management. Yet despite the great vogue of safety culture, there is hardly any agreement on what a safety culture is, how it can be measured, and - crucially - how it can be created (see, for example, Pidgeon, 1997; Reason, 1998). In 1993, the UK Advisory Committee on the Safety of Nuclear Installations, defined safety culture as

> "the product of individual and group values, attitudes, perceptions, competencies, and patterns of behaviour that determine the commitment to, and the style and proficiency of, an organisation's health and safety management" (ACSNI, 1993)

According to this broad definition, safety culture is a neutral term, which is not necessarily something that can be immediately considered as positive or negative.

Organisations are said to have effective or ineffective safety cultures, depending on those attitudes and competencies that prevail within them.

A component element of this broad definition of safety culture is the claim that "the creation of a positive culture" can only be secured by "involvement and participation at all levels" (HSE, 1991, 1997). This particular view of safety culture is informed by academic definitions of "culture" which typically embed safety culture into the disciplinary matrix of a specific academic field such as organisation studies and learning, or human factors analysis. Underlying such views is the assumption that culture can encapsulate the motivational template of an organisation and that the organisation can, in turn, take "ownership" of safety culture. This is only superficially plausible.

There are clearly some problems with this notion of culture. Chief amongst these is the assumption that a homogeneous, all-pervasive, stable set of beliefs can be generated within or imposed on the modern workplace. Workplaces, as we know them, are dynamic, complex, and often highly fragmented entities and have become even more so in recent years when organizational change has been both rapid and often unpredictable. In a world of subcontracting, out-sourcing, high turnover, short-term contracts, and 'flexible' periphery workforces, the notion of a stable organisational ethos is at best questionable. At its worst, such an assumption represents an inability to recognise the realities and insecurities of modern employment.

Apart from the question of heterogeneity of the workplace, an even more striking problem arises from the question as to who should generate the cultural beliefs adopted by the organisation. Most managers or management consultants involved in the creation of a "safety culture" have a clear idea as to who defines the term in practice, what it contains, and to what degree it is applied to the workforce. Less clear are the outcomes of these initiatives and how they should be measured. Thus 'culture' is itself seen as a deeply ideological term. Ideological, because built into the cultural mode of analysis is the assumption that within any society there are common goals and values which all members share. What is sometimes excluded in this view of culture as ideology is the fact that, far from being universally agreed upon, definitions of reality are often a matter of fierce controversy and conflict. In other words, within social arenas, such as the workplace, there are competing and sometimes conflicting views about the nature of existing problems and their potential solutions. Reality thus becomes an artificial construct, the dominant interpretation of which becomes a function of the influence of the powerful. In these contexts, notions of culture de-value and de-legitimise alternative views and ideas. More than this, because the notion of a workplace safety culture tends to presuppose a unified system of values and ideas, it misdirects attention from the context of power within which the respective culture is embedded.

Some writers see culture as being expressed in terms of the (deep) core beliefs, values and assumptions of senior (and influential) individuals within the organisation (see Pauchant and Mitroff, 1992; Sabatier, 1987). Turner (1976; 1978) saw the importance of a cultural readjustment in the wake of disasters. He argued that organisations needed to be willing to absorb the lessons of failure (expressed in terms of a disastrous event) that could arise from both within and outside of the organisation.

This "failure of hindsight" (Turner, 1978) invariably proved to be a major factor in allowing organisations to incubate the potential for failure within their systems, protocols and modes of decision making. Effective organisational learning – a term also plagued by problems of definition – has been seen as a central element in developing an organisational culture that is constantly seeking to adapt to the risk of failure (see Toft and Reynolds, 1994; Smith and Elliott, 2000). Effective learning can de-stabilise the command dynamics of an organisation by calling into question the nature and acceptability of knowledge, beliefs and assumptions. Indeed, such a paradigmatic shift is essential in ensuring that organisations develop both a culture that is open to question and decision-making procedures that recognise the legitimacy of knowledge in its various forms.

To link back to our previous discussion of quantification and audit, the more critical view of culture being highlighted here entails the argument that the complexities of the modern workplace require *multiple* sources of auditing and safety assessment, one of which must be "safety auditing from below" (see Whyte, this volume). This is the internally generated critique of existing procedures by legally and organisationally empowered workgroups. Yet increasingly the language of safety management systems, quantitative risk assessment, human factors analysis, and safety cases, is exclusionary rather than inclusive. Safety culture, in its current operational usage, may therefore present dangers to the creation of safe working environments, whilst at the same time redefining the parameters of "workplace safety" in manner that makes a true detection of safety problems impossible.

Crucially, then, many notions of workplace safety culture obscure the recognition of the real-life underlying realities and tensions associated with power imbalances between employers and employees and their respective influence in a changing and uncertain work environment. Culture, used in its current managerialist form, is often little more than a manipulative tool for the control of actions and even the beliefs of the workforce. This volume illustrates some of the more positive versions in which a focus on safety cultures can and have become manifest in relation to risk management (Hopfl, Smith and Elliot this volume) - but we need to be clear that such contributions do not exhaust, indeed are relatively rare within, the discourses of corporate culture in general and safety culture in particular. This is one issue upon which post hoc case studies, demonstrating the nature and effects of cultural change within organisations, seem to us to be crucial. There is an urgent need to move beyond unreflexive - if motivated - exhortation.

Management

This volume, as with most discussions of risk, centres partly on the issue of management and its processes for dealing with hazards. A particular focus of many of the papers in this volume centres on the limitations of those processes. Such a critique of 'management' is important but only if it is measured and balanced. However, the term management needs further exploration and discussion if this critique is itself to be effective. Perhaps the most fundamental question to be addressed in this context is how 'management' is to be defined, and more bluntly, who does, or should, 'manage' risk within that definition. This is more than a mere semantic issue as management processes (and, therefore, risk management) are predicated upon quite strict notions of control, effective information dissemination and (rational) decision making processes. Management has, however, proved to be a difficult term to define. This has been due principally to its abstract nature. For some management is closely associated with notions of organisation (see Fores, 1985), control (see Simons, 1995), learning (Senge, 1990) and the coordination of resources and there is little (if any) agreement on the absolute nature of the process. Kast and Rosenzweig (1985), for example, observe that,

> "Typical definitions suggest that management is a process of planning, organizing, and controlling activities. Some increase the number of sub-processes to include assembling resources and motivating; others reduce the scheme to include only planning and implementation. Still others cover the entire process with the concept of decision making, suggesting that decisions are the key output of managers" (p. 5).

Burgoyne (1985) outlines two distinct ways of looking at management:

> "On the one hand, managers organize resources external to themselves to get things done; whilst on the other hand the question of how, and how well they do it, is answered in terms of 'inner' psychological characteristics: knowledge, skills, attitudes, personality characteristics etc" (p. 47).

These personal characteristics are then brought to bear in the organising of resources and the people that utilise those resources It is these transactions, that exist at all levels of the organisation, which are an important aspect of the management which is seen as the creation of,

> "arrangements with these people, and maintaining these arrangements by continuous regeneration" (Burgoyne, 1985, p. 59).

This complex web of interactions creates difficulties for management due to those properties that can emerge out of them in ways unforeseen by those who designed the system. The process of management has been seen essentially as one of abstraction, it is the tasks of managers that give us insights into the process (Liliethal, 1967). Dörner (1989) points to the limitations of our knowledge and the false assumptions that we make within this process of emergence as being of central importance in the precipitation of failure.

In many cases, risk management seeks to emulate the broader actions of the management process. Management, in this case, can be considered to have four basic elements – an objective-driven approach, which is achieved through people, using established techniques and set within an organization and its boundaries (Kast and Rosenzweig, 1985). This approach, when applied to risk management, suffers from severe limitations, rendering the extent to which risk can be effectively managed highly problematic. Amongst these are: a faith in technocratic expertise, and concomitant assumptions concerning the irrationality of 'non-expert' groups; a penchant for avoiding interference, with the effect that risk is always a secondary issue, to be attended to for essentially negative reasons; assumptions, as with many claims around 'culture', that there exists some kind of unitary interest within business organisations which can be pulled unproblematically into one direction. In short, these assumptions, these elements, are unlikely to be conducive to the rooting out of 'pathogens', to the questioning of core assumptions, to the genuine recognition of the legitimacy of views contrary to those that dominate within certain cultures or ways of seeing the world. That is, within this approach to, or definition of, management, dominant organisational paradigms are likely to survive, relatively unchallenged, and thus the maintenance and reproduction of risk-producing organisations is ensured.

Lest it be thought that these attributes of management have been consigned to the dustbin of Taylorist or Fordist history, it is worth noting that some of the more recent trends in management thinking and practice share essentially similar assumptions; and, indeed, in some respects, they are more pernicious for their very claims of being 'softer' or more enlightened forms of management. Thus, for example, many contemporary approaches to risk management have linked effective risk management systems closely to the provision of Human Resource Management (HRM), and to quality or total quality management. These links are questionable for several reasons. HRM practices, on the whole, have introduced a tendency towards the more effective disciplining of, and control over, labour. Moreover, there is evidence that labour management practices characterised by the term HRM - or at least the "hard" and increasingly dominant versions of this - are symbiotically related to attempts at deregulation. Such deregulatory initiatives include, the removal of employment protection in general, as well as attacks on health and safety regulation in particular. Reduced worker autonomy, the intensification of work, and the increasing commodification of labour, cast within an image of consensual workplace relations and by-passing existent trades union structures, provides a troubled background for an effective health and safety system (on these points, and for a more general critical treatment of HRM, see Legge, 1995, Townley, 1994).

Notwithstanding these regressive developments in management theorising and practice, it remains possible - as Smallman, Hopfl and Whyte detail in their contributions to this volume - that alternative, more democratic, approaches to management in general, and risk management in particular, can be conceptualised. In brief, a more democratic form of management opens up the possibilities of wide scale participation and of a genuine dialogue within organizational cultures where communication and disagreement are encouraged and required. It seems to us that these are also amongst the key requirements or elements of effective organisationally based risk management. To give just one beneficial instance of such an approach to risk management, this is one means by which many of the problems of misinformation (see, for example, Turner, 1976; 1978; Smith, 1995; Wilson and Smith, Smallman, Whyte, and Beck and Woolfson, in this volume) might be eradicated, at least potentially. There are, of course some predictable objections to such a view of the management process, and it is perhaps worth noting some of these here.

One objection is that democratic organisations are likely to be slow and cumbersome. This is seen as a major criticism in an age in which flexibility and responsiveness to markets and technological changes is called for. Whilst such a concern has some validity, recent evidence from organisational theory suggests that this need not be the case (Clegg, 1991: 220-233). Nevertheless, in relation to risk management, we accept that an investigation into the nature of democratic forms of organisation might conclude that such forms are inappropriate, especially when these risks are actually realised, that is, in the context of some crisis event. This does not, however, present any overwhelming objection to the significance of more democratic structures in organizational attempts to manage the production and incubation of latent forms of risks. In such cases, it can be argued that the more democratic forms of organization might go some way towards preventing such incubation in the first place by challenging the core assumptions held by senior managers (see Tombs and Smith, 1995).

A second likely objection is that democratic forms of 'management' are likely to be inefficient. More specifically, through bad, yet democratic, decisions they may actually generate further risks and potential crises. Again, this criticism must be accepted as a possibility, although two points are worth making in this respect. First, as the voluminous literature on risk indicates, the existing (and currently predominant) non-democratic corporate forms manage perfectly well to make bad decisions (or non-decisions), and thus generate risk-laden contexts. Indeed, since a key element of democratic forms of risk management is the widest possible sharing of, and deliberation over, information, knowledge and expertise, then on this level the potential for bad decisions seems to us to be reduced. The surfacing of core beliefs, values and assumptions and their subsequent challenge by organizational members and affected stakeholders, might go some way towards dealing with the potential for risk incubation (see Turner, 1976 1978). Secondly, arguments about the 'bad' outcomes of democratic decision-making structures usually make reference to the problems of participation on the part of the uneducated - and this is, of course, a central element of paternalism. Yet, again, as almost all studies of post-disaster corporations indicate, the knowledge required to prevent the production or realisation of risks was present either in or around the organisation prior to that disaster event. The problem is often that such information is either not properly communicated or those in positions of power choose to discount that knowledge on the grounds that it is somehow invalid. In other words, it is less the case that those critics who are currently excluded from decision-making in corporations do not 'possess' the 'necessary' knowledge to participate; rather, what counts as knowledge has been defined in such a way as to exclude their potential contributions and legitimacy. 'Knowledge' is a social construction rather than a phenomenon conforming to certain essential pre-requisites (Feyerabend, 1978); and social constructions can, by definition, be altered.

Such brief considerations would seem to indicate that, far from being part of the solution to the control of risk, management may actually be a key source of the risk problem. This in turn indicates the need for some radical rethinking of what we mean by 'management', a task to which many of the papers in this volume make a contribution.

Risk Revisited

As this book demonstrates, the risks that present themselves to be 'managed' are incredibly diverse in nature. They range from industrial risks through the somewhat more esoteric science-technology risks associated with bio-technology and genetic engineering, to the truly global risks associated with climate change. Moreover, these risks may take forms that are either acute or chronic. Many of these risks are products of science-technology, others arise out of the organisation of social sub-systems, while perhaps most are a synergistic effect of socio-technical structures. Thus it is vital to bear in mind that most of these risks arise out of human decision making and managerial activities. Consequently, they are subject to intervention and thus prone to change by those very humans. Two questions emerge from this relationship which are of particular concern to our discussions here. One appears to be a quantitative question: are risks becoming more prevalent? The second poses itself largely as a qualitative question: are there emerging risks, which are different in form to those with which we have historically been faced?

Risk and the Quantitative Dimension

On the question of frequency, it is useful to turn to one area of industrial activity within which considerable and longstanding attention has been given to issues of risk, and within which one might expect the frequency and scale of incidents to be relatively well documented, namely the chemicals industries. Indeed, this is also an interesting context in which to consider the question of frequency, since there are some - notably Kharbanda and Stallworthy (1991) - who have made claims that the frequency of accidents is declining.

In contrast, however, Shrivastava (1992) has expressed concern over the increasing frequency of industrial accidents in the chemicals industries. In addition, an earlier study by Carson and Mumford (1979) documented an increasing incidence of major accidents in the UK during the period 1954-1979. While recognising the limited nature of their data set, the authors argue that a greater number of serious accidents (with multiple fatality potential) occurred in the period of study than in previous years (ibid.). The data presented by Carson and Mumford illustrates both a considerable rise in the number of such incidents as well as an increase in near fatal accidents.

The 1980s did little to alleviate public concern over major hazard risks. While perhaps the best known of such events, the accident at Bhopal was not the only catastrophic event of this decade. The accident at Mexico City in 1984, for example, resulted in over 500 deaths and it illustrates the considerable potential for such accidents that persists within industrialized societies (Chapman 1984; Pearce 1985). However, the accidents at Bhopal and Mexico City were not isolated incidents but represented the worst cases of what could be seen as an alarming trend. A survey commissioned by the US Environmental Protection Agency (EPA) revealed that between 1963-1988, there occurred seventeen potentially catastrophic releases of deadly chemicals in the US, in volumes and levels of toxicity, which exceeded that at Bhopal. While 'only' five people were killed in these incidents, this was seen on several occasions as being a result of 'sheer good luck' (New York Times, 30 April 1989). Moreover, all but two of these incidents occurred in the 1980s. In other words, they took place in the final third of this twenty five year period and at a time when one might have expected that safety standards had improved. A more recent survey by the US National Environmental Law Center found that almost 35,000 toxic chemical accidents occurred between 1988 and 1992 in the US. At least one in sixteen of these events caused immediate injuries, deaths or evacuations; furthermore, these accidents represented only a small proportion of near misses and they were concentrated in a relatively small number of densely populated US states (Chemistry & Industry, 3 October 1994, p. 796). King has recently examined accidental losses in the chemicals industries for the period 1958-1987, concluding both that not only is the magnitude of these losses increasing, but also that the recent record of the industry can be seen as "truly alarming" and one which "gives no room for complacency" (King 1990, p. 6).

Of course, none of this necessarily demonstrates that there is any greater risk associated with individual sites or plants. It may well be the case that such data obscures the fact that many people in the US and the UK, for example, are safer now than they were 100 years ago from chemical accidents. Whilst the frequency of initiating events may have increased quantitatively, and their nature altered qualitatively, improved regulation and systems defenses around those hazards has led to some reduction in the risk of catastrophic failure. We would certainly agree that in terms of major hazard regulations, such as those involving Notification of Installations Handling Hazardous Substances (NIHHS), the Control of Industrial Major Accident Hazard (CIMAH) and, more recently, COMAH, then the developments since 1974 have been progressive (if a long time coming). Nonetheless, we cannot assume that the chemical industries are now 'safe' and neither can we accept the usual corollary of such claims, namely that improving safety may have led to the possibility (and some might even argue the desirability) of a regulatory moratorium or the removal of particular 'regulatory burdens' (on this, see Smith and Tombs, 1995; Pearce and Tombs, 1998). There is still sufficient potential for harm inherent within the industry to justify continued vigilance and control. In addition, the processes of globalisation may have shifted the requirements of control and have led to the claim that hazard is being exported to developing economies where regulatory frameworks may be less developed that in the USA or Europe (see, for example, Jones, 1988; Smith and Blowers, 1992; Weir, 1987).

The chemical industry provides us with a useful focus for any discussion of globailisation, since it raises a further dimension of the scale of risk - for this is one industry in which production is truly global. Consequently, one might expect that the risks associated with such production would be global too. The increasing incidence of chemicals accidents and near misses is a case in point. This data can only be understood in the context of the spectacular expansion - both in terms of the quantities of production as well as spatially - of the international chemical industries in the post 1945 period (Aftalion 1991; Vilain 1989). The global nature of production and its associated risks might also lead to the exploitation of any weaknesses in the regulatory regimes. The result might be that poor operational practices or hazardous activities would become 'exported' to those areas where weak controls were existed. Bhopal itself illustrated the international dimensions of hazard generation by multinational companies, representing one instance of the 'export of hazard' (Castleman 1979; Ives 1985; Smith and Blowers 1992; Smith and Sipika, 1993). Indeed, while the international dimensions of the chemical industry means that the export of hazard has long been possible, trends towards a 'liberalisation' of the international economy are likely to have increased the opportunity for, and attractiveness of, such a strategy for chemicals companies.

Moving beyond the case of the chemicals industries, what is clear is that the hazards that we face within modern societies are not always the most obvious or visible. There is little doubt that an industrial installation, which stores significant quantities of chemicals, is hazardous. In contrast, sports stadia, for example, have not been considered to be major sources of risk (accepting the obvious caveat regarding hooliganism). Despite this belief, 152 people died in just two events at sports stadia in the UK alone (see Elliott and Smith, 1993). The portfolio of hazard generating activities that societies face may be seen by many as increasing, and yet this comes at a time when developed societies have better health and social provision than at any other time in history. Herein lies the obvious paradox. The risks that we face are becoming more complex and, often, less visible. They involve the truly global hazards such as the greenhouse effect and the "unseen" hazards such as virus transmission and food related problems such as BSE. In such a complex environment, notions of risk and its associated management strategies have to assume a greater level of sophistication. An expression of this sophistication has involved a growing recognition of the importance of the qualitative dynamics of risk generation and management. What this work suggests is that there is considerable scope, in many industries, for managerial or latent error in incubating the potential for harm (see Turner, 1976; 1978; Reason, 1990; 1997; 2000; Smith, 1995). What is of interest here is the manner in which management itself can contribute to the catastrophic failure of systems.

Risk and the Qualitative Dimension

Failure within organizations can be a cumulative process in which subtle changes can delude those who attempt to control the system. Dörner (1989) describes this process through his observation that that,

> "Failure does not strike like a bolt from the blue; it develops gradually according to its own logic" (p. 10).

Dörner makes the point, and echoes the seminal work by Turner (1976; 1978), that the incubation of failure assumes a certain inevitability:

> "As we watch individuals attempt to solve problems, we will see that complicated situations seem to elicit habits of thought that set failure in motion from the beginning. From that point, the continuing complexity of the task and the growing apprehension of failure encourage methods of decision making that make failure even more likely and then inevitable" (Dörner, 1989, p. 10).

As systems become ever more complex and as technology and science operate at the boundaries of our knowledge, then such a potential for failure takes on a dynamic perspective. This point is developed further by Smith and McCloskey (this volume) who point to the central role of technical experts in both shaping modern techno-scientific crises an incubating their potential. Such concerns lie at the heart of the post-modernist critiques surrounding risk.

As we indicated earlier in this chapter, the past ten years have witnessed the emergence of a body of literature, originally developed within sociology, and which can be traced back to the work of the German social scientist Ulrich Beck, but more recently popularised by Giddens. The consequent 'risk society' thesis has become a common, and perhaps even the dominant, reference point in conceptualisations of risk across social sciences. Within this literature, risk is no longer treated as a marginal issue - indeed, Beck had originally identified the emergence of what he termed the Risk Society. For those working within this emerging tradition, risk is treated not simply as one aspect of contemporary social life, but as a central or defining characteristic of a reflexively modern social order (see, for example, Beck, 1992, Beck et al., 1994, Giddens, 1998). Some of the contributions to this volume discuss this thesis directly (Pearce and Tombs, Smallman), while almost all others bear upon some of its central aspects. At this point, we simply wish to raise some general problems with the thesis that we live in a society marked by qualitatively different forms of risk, to the extent that this is a different form of society *per se*.

One of the most striking things about the contemporary work on risk is that it has focused on risks to consumers, the public, various communities and so on, and barely at all on risks to producers or workers, and the mundane risks that they face in the process of production. This is in many respects a necessary, yet at the same time a revealing, omission. It is necessary because the risk society literature is organised around a claim that the risks faced in contemporary society are qualitatively new - and this thesis depends upon an obscuring of long standing risks, such as those faced by workers. Clearly, workers who are dealing with new technologies are also exposed to these 'emergent' risks. Second, this is also necessary given the claim that risk is ubiquitous - and to focus on the workplace might privilege certain types of risks, and to recognise that for some groups of social actors, risks are both structured or organised into their experiences. Indeed such risks can be the object of struggle and may be organised out of those experiences - that is, on the basis of class politics. Here we find why this omission is revealing - for the risk society literature is one based upon an assumption that contemporary social orders are no longer organised around a fundamental cleavage in terms of class (see, for example, Giddens, 1998, and Beck's 1992 arguments for a new politics). Indeed, in the (necessary) myopia towards the class-based production, distribution and experience of risk which the risk society literature displays, some of the general problems with this area of work are exposed, and these are of importance in our general consideration of 'risk management and society'.

First, "risk" is used within such literature in an over abstracted sense. While it is important to theorise about risk, about the nature of a risk society and the shifts in the governance of risk (amongst others), these questions and considerations need to be rooted within, and indeed developed via, a consideration of specific risks in concrete circumstances. There is a point here of general importance regarding the appropriate levels of analysis and their integration, and it is one to which we return in the conclusion to this chapter. Second, as argued by Pearce and Tombs (this volume), this research contains a tendency towards an idealist understanding of scientific rationality. Missing from this work is any real analysis of power in general, or capital in particular (see Pearce and Tombs, this volume, for an extended consideration of this point). The third problem, and one that develops the argument made previously, the risk society literature fails to address the unequal distribution and experiences of risk. While there is an important truth to this observation regarding the ubiquitous nature of risk victimisation, two points of clarification must be made. There is an obvious sense to the claim that 'we' are all victims when we are speaking of environmental risk, for we are all exposed to the environment and we are all consumers, to greater or lesser extents. But it is perhaps less obvious why 'we' all experience victimisation in the case of other forms of risk. Indeed, beyond environmental risks, many risks seem to be highly discerning in terms of likely victims. Thus if financial risks have increased, as Smallman claims (this volume), then the distribution of these risks has had the effect of further impoverishing what he calls variously the 'dispossessed' or an 'underclass', while various elites have benefited enormously from financial instability (see also, and most famously, Hutton, 1995, for a development of such an argument). The second point of concern is that these considerations need further refinement. Thus while it is accurate, at one level, to point out that "we" are all, ultimately, victims of risk, it is crucial to be sensitive to the fact that speaking from societies driven by cleavages of class, gender, race and ethnicity, degrees of able-bodiedness, and age, then it is also non-sensical beyond a certain level of abstraction to speak of "we". This can be starkly illustrated with respect to the effects of environmental pollution, since environmental risks are frequently represented as the most ubiquitous of all. Thus, according to one currently dominant trend in social thought, we are now living in an era characterised as a risk society, where risks are ubiquitous and cannot be escaped by anyone (Beck, 1992: 22, 53, Beck et al., 1994). On one level, this is accurate as an indication of the qualitative shifts in the nature of those risks introduced by chemical, nuclear, and bio-technologies. On the other hand, there is clear evidence that environmental risks are borne to a disproportionate extent by those experiencing the harshest effects of other forms of social and economic inequality. As Welsh has put it,

> "the idea that there are global risks which are 'somehow universal and unspecific' recognising none of the social categories which have stratified societies .. is only true at the level of rational abstraction used in global risk assessments" (Welsh, 1996: 20).

Thus one must at least take seriously the scepticism of Fagan, when he writes that, "the dynamics and dimensions of [the] risk society look remarkably similar to the class society" which risk theorists claim has been transcended (Fagan, 1997: 16). Thus there is now a significant body of evidence which points consistently to the unequal distribution of environmental risks in the United States, and in which there is agreement concerning the association between the class, racial and ethnic composition of geographical areas and the extent of any exposures to environmental pollutants (see Bryant and Mohai, 1992; Clark et al., 1995; US General Accounting Office, 1983; Gould, 1986; Steretsky and Lynch, 1997; Bullard, 1990; Hofrichter, 1993). Whilst it is clear that environmental risk is unequally distributed on a global scale (see, for example, Hofrichter, ed., 1993, Williams, ed., 1996), there is at present little work carried out within "environmental justice studies" in the UK, nor indeed in Western Europe. However, there is no reason to expect that the environmental effects in Europe do not reflect these other factors.

Similar observations regarding the unequal distribution of victimisation and risk might be made beyond the realm of environmental risks. For example, research on victimisation by consumers to risks associated with the products of pharmaceutical industry, clearly points to the particular victimisation that has historically been experienced by women. This gendered victimisation to risk follows from the construction of women as reproducers (see Draper, 1991), so that women are differentially victimised by the products of the pharmaceutical industry (Szockyi and Fox, 1996) or are excluded from working within certain types of hazardous environment (Draper, 1991). However, as Draper observes, "Fertile women are often excluded on the basis of insufficient, inconclusive scientific information" (Draper, 1991, p. 68). She goes on to observe that,
> "Many of the chemicals from which women are excluded can harm future children through male workers by way of sperm damage or mutagenic effects" (Draper, 1991, p. 69).

Of course, such exclusion is not only gender specific but also logically age discriminatory as well because only those women of reproductive age should be excluded from that particular workplace. Draper also observes that such exclusion is often sectorally biased, with fertile women still being allowed to work in hazardous environments in such industries as health care, where they are the dominant employees in the workforce. Draper argues that,

> "Women are usually barred not from all jobs that entail toxic risks but only from the unskilled, relatively high-paying production jobs traditionally held by men" (Draper, 1991, p. 71).

But there are also class and ethnic dimensions that overlay these gendered aspects of victimisation. For example, Finlay notes that one of the reasons why DES victims were able to mobilise effectively was due to the "demographics" of the drug. DES was an expensive drug and one that was dispensed largely by private physicians, as opposed to those serving public hospitals or clinics:

> "Most of those who were exposed to DES are middle- or upper middle-class, white, well-educated women. The characteristics of the affected population, which came to be known as DES daughters, later contributed to their grassroots activism, pursuit of medical information, and inclination to file a large number of lawsuits. The injured women had the education to do research and become involved in their own medical treatment; and they are form a racial and economic group that tends to regard legislatures and courts not with alienation and distrust but with the expectation that they will produce justice" (Finlay, 1996: 67-8).

This work on the crimes of the pharmaceutical industry thus documents the extent to which women experience victimisation both as producers and consumers of unsafe "medical" or cosmetic products. In addition, they are also victimised in the labour market and within workplaces through a range of illegal exclusionary and discriminatory practices (Draper, 1991; Finlay, 1996). Indeed, while women work in sectors that are increasingly being recognised as particularly unsafe and unhealthy, representations of the hazard have traditionally been associated with male occupations. Despite the fact that research, largely conducted by men, has mostly ignored womens' occupational health and safety issues (Szockyi and Frank, 1996: 17), trends in data indicates that those areas in which women are over-represented, notably services, are those which exhibit both persistently high, and rising, rates of injuries and ill-health (see Craig, 1981, Labour Research Department, 1996).

More generally, where workers are exposed to risks against their health and safety, these are most likely to be those in poorest protected and most poorly paid occupations rather than those working in "inherently" dangerous occupations. This point has also been made by Carson and, more recently, by Woolfson and colleagues in relation to the UK offshore oil industry (which has drawn upon labour from the unemployment "blackspots" of Scotland and Northern England). That there are gender and ethnic, as well as class, dimensions to this unequal victimisation is also evidenced in the work of John

Wrench (Wrench, 1996, Lee and Wrench, 1980, Wrench and Lee, 1982; see also Boris and Pruegl, eds., 1996). Understanding the victimisation of groups around risk, and the differential responses made to, and public knowledge of, such victimisation, thus requires an understanding of various forms of class-race-gender articulations.

In short, therefore the risk literature within the Beck paradigm fails to provide the basis of any political economy of risk (Fagan, 1998: 7; Woolfson, this volume) and is one of a number of social-scientific trends which treat individuals in an abstracted sense, rather than in their real contexts, as women, workers, members of an ethnic grouping, and so on. Thus, as Fagan has noted, social scientific analysis

> "should at least illuminate and make comprehensible the details of individuals' lives in the context of changes that are taking place on a wider - even global - scale. The new risk discourse addresses this global aspect, but fails to relate it to the lives of individuals" (Fagan, 1998, p. 16).

By seeking to address risks at various levels (the individual, organisational, institutional and social), by exploring issues of risks within a variety of sectoral contexts, and by seeking to combine empirical and theoretical considerations, this volume seeks to avoid such abstraction, and in this way, following Fagan, to add greater substance to contemporary risk debates.

Conclusions

The papers collected in this volume speak to risk management from a variety of disciplines and from a range of political and theoretical perspectives. Taken together, they emphasise both the range of work contributing to risk management, and the significant areas of difference, even tensions, that exist within this body of literature. Indeed, if this volume achieves anything, it demonstrates that 'risk management' is not an easily identifiable, homogenous, nor closed area of academic and practitioner activity. There are at least two reasons for this diversity and openness. One is simply the range of problems and issues that are encapsulated within the umbrella term 'risk management' - thus the 'management' of these issues defies any reducibility to a standard recipe, or protocol for success. Second, as a sub-disciplinary area of intellectual endeavour, risk management is dependent, and some would even say parasitic upon, a series of other disciplines. Thus chapters in this volume draw upon business and management studies, criminology, industrial relations, insurance economics, political science, psychology, and sociology. This broad disciplinary base is a potential strength of the work, not least given the fact that genuine multi-disciplinarity is a rare, but powerful, aspect of academic endeavor. At the same time, however, such a multi-disciplinary approach is also a source of potential weakness, since at best it creates the danger of epistemological, theoretical and even political eclecticism. At worst, it can generate an internal paradigm incommensurability, removing the possibility of different contributors to the field from being able to engage in any meaningful dialogue, and thus progress. It is our view, however, that while there are inevitably problems generated by difference and openness, and that these must be recognised, this diversity should be treated as fruitful, and thus welcome. Moreover, in practical terms, if the issues and problems captured by the increasingly familiar term 'risk' are to be addressed successfully than a plethora of approaches, forcefully and rigorously debated, is required.

The contributions to this volume do at least provide, when taken together, a set of parameters within which a working definition of risk management can be gathered - though there remains much about the details of this definition to be debated and, potentially, resolved. For our the purposes of our current discussion, the papers in this collection have generally followed the definition developed by Nedved which defined risk management as

> "the set of ongoing management and engineering activities of a business that ensures that risks are effectively identified, understood, and minimised to a reasonable achievable and tolerable level. The activities include feedback mechanisms and continuing performance monitoring" (Nedved, 1998, p. 1).

Such a definition requires, of course, that it is developed through adding greater specificity to particular terms. Thus, we need to know what constitutes the process of management in general and especially risk management (it could be argued that all management is concerned with the management of risk). One might ask here what role is there or should there be for those outside those 'businesses' who bear the consequences of those risks? A further question concerns the role of the various techniques that are available for risk management. Of particular interest here is the question of how robust these techniques are and upon what assumptions are they based? Ultimately, there is an important question to be asked concerning the relationships that do and should exist between the organisational and the technical. What is the role of the technical expert as a mediator between risk generators and victims and how valid is that expertise under conditions of emergence? These issues provide further clarification of the definition of risk management that is set out in Nedved's definition and the remainder of this volume will seek to explore them in more detail.

NOTES

[1] Parts of this section draw upon work conducted by Charles Woolfson (see, for example, Woolfson and Beck, 1999, Woolfson et al., 1996), and one joint project on which Steve Tombs collaborated with him (see James and Walters, eds., 1999, chapter 2).

REFERENCES

ACSNI Human Factors Study Safety Group (1993) Third Report: *Organising for Safety*, Sudbury: HSE Books.

Adams, E.K. and Young, T.L. (1999) Costs of smoking: A focus on maternal, childhood, and other short-run costs, *Medical Care Research and Review*, 56(1), pp. 3-9.

Aftalion, F. (1991) *A History of the International Chemical Industry*, Philadelphia: University of Pennsylvania Press.

Aronowitz, R.A. (1998) Making sense of illness. *Science, Society, and Disease*. Cambridge: Cambridge University Press.

Beck, U. (1992) *Risk Society: Towards a New Modernity*. London: SAGE.

Beck, U., Giddens, A. and Lash, S. (1994) Reflexive Modernisation. Politics, Tradition and Aesthetics in the *Modern Social Order*, Cambridge: Polity.

Bennett, P. and Calman, K.C. (Eds) (1999) *Risk communication and public health*. Oxford: Oxford University Press

Bernstein, P. (1996) *Against the Gods. The remarkable story of risk*. New York:Wiley.

Boris, E. and Pruegl, E. eds. (1996) *Homeworkers in Global Perspective. Invisible no more*, London: Routledge.

Bryant, B. and Mohai, P., eds. (1992) *Race and the Incidence of Environmental Hazards*, Boulder, Co: Westview Press.

Bullard, R. (1990) *Dumping in Dixie. Race, Class and Environmental Quality*, Boulder: Westview Press.

Blowers, A. (1984) *Something in the Air*. London: Paul Chapman Publishing.

Boehmer-Christiansen, S. (1994) The precautionary principle in Germany - enabling government, in O'Riordan, T. and Cameron J. (1994) (Eds) *Interpreting the Precautionary Principle*. London: Earthscan. pp. 31-60.

Boggan, S. (1999) BNFL Hit By Claims that it faked Tests on Atom Fuel for Japan Power Station, *The Independent*, 30 September.

Bourriau, J. (Ed.) (1992) *Understanding Catastrophe*. Cambridge: Cambridge University Press.

Burgoyne, J. (1985) 'Self-Management', Elliott, K. and Lawrence, P. (Eds) *Introducing Management*. Harmondsworth: Penguin. pp. 46-59.

Calman, K.C. (1996) Cancer: Science and Society and the Communication of Risk; *British Medical Journal* 1996; 313 :pp. 799-802

Carson, R. (1962) *Silent Spring*. London: Penguin.

Castleman, B. (1979) The Export of Hazard to Developing Countries, *International Journal of Health Services*, 9, 4.

Centre for Disease Control and Prevention (1998a) *Isolation of Avian Influenza A(H5N1) Viruses from Humans* - Hong Kong, May-December 1997. p. 5. **www.cdc.gov/ncidod/diseases/flu/h5mmwr.htm** Updated Thursday September 03 1998 18:44:42. Accessed 23/11/99

Centres for Disease Control and Prevention (1998b) *Isolation of Avian Influenza A(H5N1) Viruses from Humans* - Hong Kong, May-December 1997. p. 4 **www.cdc.gov/ncidod/diseases/flu/h5mmwr2.htm** Updated Thursday September 03 1998 18:44:43. Accessed 23/11/99

Centres for Disease Control and Prevention (1999) *Influenza A(H9N2) Infections in Hong Kong*. National Center for Infectious Diseases, Atlanta. **www.cdc.gov/ncidod/diseases/flu/H9N2Info.htm** Accessed 23/11/99.

Carson, P.A. and Mumford, C. J. (1979) An Analysis of Incidents involving Major Hazards in the Chemical Industry, *Journal of Hazardous Materials*, 3.

Chapman, P. (1984) Mexico's catalogue of gas disasters, *New Scientist*, 1432, 29th November.

Clark, RD, Lab, S. and Stoddard, L. (1995) Environmental Equity: a critique of the literature, *Social Pathology*, 3, (1).

Clegg, SR (1991) *Modern Organisations. Organisation studies in the postmodern world*, London: Sage.

Craig, M., New Edition by Phillips, E. (1991) *Office Workers' Survival Handbook*, London: The Women's Press.

Cohl, H.A. (1997) *Are we scaring ourselves to death? How pessimism, paranoia, and a misguided media are leading us towards disaster*. New York: St Martin's Griffin.

Collingridge, D. and Reeve, C. (1986) *Science Speaks to Power*. London: Francis Pinter

Committee on Risk Perception and Communication (1989) *Improving Risk Communication*. Washington DC: National Academy Press.

Davies P. (1999) *Catching Cold. 1918's forgotten tragedy and the scientific hunt for the virus that caused it*. London: Michael Joseph.

Deville, A. and Harding, R. (1997) *Applying the precautionary principle*. Sydney: The Federation Press.

Dörner, D. (1989) *Die Logik des Misslingens* translated into English (1996) *The Logic of Failure. Recognizing and Avoiding Error in Complex Situations*. Reading, Mass.: Perseus Books.

Douglas, M. (1980) Environments at risk, in Dowie, J. and Lefrere, P. (Eds) (1980) *Risk and Chance*. Milton Keynes: Open University Press.

Douglas, M. (1985) *Risk Acceptability According to the Social Sciences*. London: Routledge and Kegan Paul.

Douglas, M. and Wildavsky, A. (1984) *Risk and Culture*. Berkley: University of California Press.

Draper, E. (1991) *Risky Business. Genetic testing and exclusionary practices in the hazardous workplace*. Cambridge: Cambridge University Press.

Elliott, D. and Smith, D. (1993) Football stadia disasters in the United Kingdom: Learning from tragedy, *Industrial and Environmental Crisis Quarterly*, 7(3) pp.205-229.

Elliott, D. and Smith, D. (1997) 'Waiting for the next one' in, Frosdick, S. and Walley, L. (Eds) (1997) *Sport and Safety Management*. Oxford: Butterworth-Heinmann. pp. 85-107.

Epstein, S. (1996) *Impure science. Aids, activism, and the politics of knowledge*. Berkley: University of California Press.

Erikson, K. (1994) *A New Species of Trouble: Explorations in Disaster, Trauma and Community*. New York: Norton.

Fagan, T. (1997) *Risk and Social Science - why risk? Why now?*, Open D-bate,19, May, 13-16.

Feyerabend, P. (1978) *Science in a Free Society*, London: New Left Books.

Finlay, L.M. (1996) The Pharmaceutical Industry and Women's Reproductive Health' in Szockyi, E. and Fox, J.G., eds., *Corporate Victimisation of Women*, Boston: Northeastern University Press, 59-110.

Fischer, F. (1980) *Politics, Values, and Public Policy: The Problem of Methodology*. Boulder: Westview Press.

Fischer, F. (1990) *Technology and the politics of expertise*. Newbury Park: Sage.

Fores, M. (1985) 'Management: Science or Activity?', in Elliott, K. and Lawrence, P. (Eds) *Introducing Management*. Harmondsworth: Penguin. pp. 17-33.

Fortune, J. and Peters, G. (1995) *Learning from Failure: The Systems Approach*. Chichester: Wiley.

Furedi, F.(1997) *Culture of Fear. Risk-taking and the morality of low expectation*. London: Cassell.

Giddens, A. (1990) *The Consequences of Modernity*. Cambridge: Polity Press.

Giddens, A. (1998) *The Third Way*, Cambridge: Polity.

Health and Safety Executive (1991, 1997) *Successful Health and Safety Management*, London: HMSO.

Hofrichter, R., ed. (1993) *Toxic Struggles. The Theory and Practice of Environmental Justice*, Philadelphia, PA: New Society Publishers.

Hutton, W. (1995) *The State We're In*, London: Jonathan Cape.

Irwin, A. (1995) *Citizen Science*. London: Routledge.

Irwin, A., Dale, A. and Smith, D. (1996) Science and Hell's Kitchen - The local understanding of hazard issues, in. Irwin, A. and Wynne, B. (Eds.) (1996) *Misunderstanding Science? The public reconstruction of science and technology*. Cambridge: Cambridge University Press. pp. 47-64.

Irwin, A. and Wynne, B. (Eds.) (1996) *Misunderstanding Science? The public reconstruction of science and technology*. Cambridge: Cambridge University Press.

Ives, J., ed. (1985) *The Export of Hazard. Transnational corporations and environmental control issues*, Boston: Routledge and Kegan Paul.

James, P. and Walters, D., eds. (1999) *Regulating Health and Safety at Work: the way forward*, London: Institute of Employment Rights.

Jones, T. (1988) *Corporate Killing: Bhopals Will Happen*, London: Free Association Books.

Kast, F.E. and Rosenzweig, J.E. (1985) *Organization and Management*. 4th Edition. New York: McGraw-Hill.

Kharbanda, O.P. and Stallworthy, E.A. (1991) *Industrial disasters – Will self regulation work?*, Long Range Planning, 24(3),

King, R. (1990) *Safety in the Process Industries*, London: Butterworth-Heinemann.

Labour Research Department (1996) *Women's Health and Safety*, London: Labour Research Department.

Lasch, C. (1995) *The revolt of the elites and the betrayal of democracy*. New York: W.W. Norton.

Leape, L. L. and Berwick, D. (2000) Safe health care: are we up to it? *British Medical Journal*, 320, pp. 725-726.

Lee, J. and Wrench, J. (1980) "Accident-Prone Immigrants: an assumption challenged", *Sociology*, 14, (4), 551-566.

Legge, K. (1995) *Human Resource Management. Rhetorics and Realities*, London: Macmillan.

Lilienthal, D.E. (1967) *Management: A Humanist Art*. New York: Columbia University Press.

Lupton, D. (1999) *Risk*. London: Routledge.

Nedved, M. (1998) *System Safety as a Tool in Risk Management*, paper presented at Hazards and Sustainability: contemporary issues in risk management, Durham University Business School, 26-27 May.

Neisser, U. (1980) On 'social knowing *Personality and Social Psychology Bulletin*, 6, pp. 601-605.

Nolan, T.W. (2000) System changes to improve patient safety *British Medical Journal*, 320, pp. 771-773.

Morgan, B. (1999) Regulating the regulators: Meta-regulation as a strategy for reinventing government in Australia, *Public Management: An international journal of research and theory*, 1(1), pp. 49-65.

Miké, V. (1991) Understanding uncertainties in medical evidence: Professional and public responsibilities, in Mayo, D.G. and Hollander, R.D. (Eds) (1991) *Acceptable Evidence. Science and Values in Risk Management*. New York: Oxford University Press. pp. 115-136.

O'Riordan, T. and Cameron J. (1994) The history and contemporary significance of the precautionary principle in O'Riordan, T. and Cameron J. (1994) (Eds) *Interpreting the Precautionary Principle*. London: Earthscan. pp. 12-30

Pauchant, T. and Mitroff, I.I. (1992) *The crisis-prone organization*. San Francisco, CA: Jossey-Bass Publishers.

Pearce, F. and Tombs, S. (1998) *Toxic Capitalism: corporate crime and the chemical industry*, Aldershot: Ashgate.

Pearce, Fred (1985) *After Bhopal, who remembered Ixhuatepec?*, New Scientist, 1465, 18th July.

Pidgeon, N. (1997) 'The limits to safety? Culture, politics, learning and man-made disasters, *Journal of Contingencies and Crisis Management*, 5(1), pp. 1-14.

Porter, T. M. (1995) *Trust in numbers. The pursuit of objectivity in science and public life*. Princeton, NJ: Princeton University Press.

Reason, J. (1990) *Human error*. Cambridge: Cambridge University Press.

Reason, J. (1997) *Managing the risks of organizational accidents*. Aldershot: Ashgate.

Reason, J. (1998) 'Achieving a safe culture: theory and practice', *Work and Stress*, 12(3), pp. 293-306.

Reason, J. (2000) Human error: models and management *British Medical Journal*, 320, pp. 768-770.

Reddy, S.G. (1996) Claims to expert knowledge and the subversion of democracy: the triumph of risk over uncertainty, *Economy and Society*, 25(2), pp. 222-254.

Rose, N. (1996) The death of the social? Re-figuring the territory of government, *Economy and Society*, 25(3), pp. 327-356.

Rowe, W.D. (1977) *The Anatomy of Risk*. New York: Wiley.

Ryan, F. (1996) *Virus X. Understanding the real threat of the new pandemic plagues*. London: Harper Collins.

Sabatier, P. (1987) *Knowledge, policy-oriented learning, and Policy change, Knowledge: Creation, Diffusion, Utilization*. 8(4), pp. 649-692.

Schwartz, M. and Thompson, M. (1990) *Divided we stand: redefining politics, technology and social choice*. Hemel Hempstead: Harvester Wheatsheaf.

Senge, P. (1990) *The Fifth Discipline. The arts and practices of the learning organization*. New York: Currency Doubleday

Sethi, S.P. (1975) Dimensions of corporate social performance: An analytical framework, *California Management Review* 17(3) pp. 58-64.

Sethi, S.P. (1983) A strategic framework for dealing with schism between business and academe, *Public Affairs Review*, 1983, pp. 44-59.

Sheldon, T.A. and Smith, D. (1992) Assessing the Health Effects of Waste Disposal Sites: Issues in Risk Analysis and some Bayesian Conclusions in, Clark, M., Smith, D. and Blowers, A. (Eds) (1992) *Waste Location: spatial Aspects of Waste Management, Hazards and Disposal*. London: Routledge. pp 158-186

Shrader-Frechette, K.S. (1991) *Risk and Rationality*. Los Angeles: University of California Press.

Shrader-Frechette, K.S. (1993) *Burying Uncertainty: Risk and the case against geological disposal of nuclear waste*. Los Angeles: University of California Press.

Shrivastava, P. (1992) *Bhopal: Anatomy of a Crisis*, 2nd Edition. London: Paul Chapman Publishing.

Simons, R. (1995) *Levers of control. How managers use innovative control systems to drive strategy renewal*. Harvard: Harvard University Press.

Smith, D. (1990) Corporate Power and the Politics of Uncertainty: Risk Management at the Canvey Island Complex. *Industrial Crisis Quarterly*, 4 (1) pp.1-26

Smith, D. (1991)'The Kraken wakes - the political dynamics of the hazardous waste issue *Industrial Crisis Quarterly* 5(3), pp. 189-207.

Smith, D. (1995) 'The Dark Side of Excellence: Managing Strategic Failures, in Thompson, J. (Ed) (1995) *Handbook of Strategic Management*. London: Butterworth-Heinemann. pp. 161-191.

Smith, D. (2000) Living on Factory Row. Issues in risk, public health and the precautionary principle. *Mimeo*. Centre for Risk and Crisis Management Occasional Papers Number 1. University of Sheffield. (www.cracm.com/papers/20.1)

Smith, D. and Blowers, A. (1992) Here Today, There Tomorrow: the politics of transboundary hazardous waste transfers, in Clark, M., Smith, D. and Blowers, A., eds., *Waste Location. Spatial aspects of waste management, hazards and disposal*, London: Routledge. pp. 208-226.

Smith, D. and McCloskey, J. (1998) Risk Communication and the Social Amplification of Public Sector Risk. *Public Money and Management*, 18 (4) pp. 41-50

Smith, D. and Sipika, C. (1993) Back from the brink - post-crisis management. *Long Range Planning*, 26(1) pp.28-38.

Smith, D. and Toft, B. (1998) Issues in Public Sector Risk Management. *Public Money and Management*, 18 (4) pp. 7-10

Smith, D. and Tombs, S. (1995) Self regulation as a control strategy for Major Hazards *Journal of Management Studies*, 32(5), pp. 619-636.

Smith, R. (1998) Regulation of doctors and the Bristol inquiry. Both need to be credible to both the public and doctors, *British Medical Journal*, 317, pp. 1539-1540.

Steingraber, S. (1998) *Living Downstream: an Ecologist looks at Cancer and the Environment*. London: Virage

Stellman, J. and Henifin, M.S. (1983) *Office Work Can Be Dangerous to Your Health. A handbook of office health and safety hazards and what you can do about them*, New York: Pantheon.

Stretesky, P. and Lynch, M. (1997) *Class Structure and Predictions of Distance to Accidental Chemical Releases: spatial geography, urban justice and chaotic strange attractors*, paper presented at the Annual Meeting of the American Society of Criminology, San Diego, November, 19-22.

Szockyi, E. and Fox, J.G., eds. (1996) *Corporate Victimisation of Women*, Boston, Mass.: Northeastern University Press.

Thagard, P. (1999) *How scientists explain disease*. Princeton: Princeton University Press.

Tombs, S. and Smith, D. (1995) Corporate responsibility and crisis management: some insights from political and social theory. *Journal of Contingencies and Crisis Management*, 3(3), pp. 135-148

Townley, B. (1994) *Reframing Human Resource management: power, ethics and the subject at work*, London: Sage.

Treasure, T. (1998) Lessons from the Bristol case. More openness – on risks and on individual surgeons, *British Medical Journal*, 316, pp. 1685-1686.

Turner, B.A. (1976) The organizational and interorganizational development of disasters, *Administrative Science Quarterly*, 21, pp. 378-397.

Turner, B.A. (1978) *Man-Made Disasters*. London: Wykeham.

Vilain, J. (1989) The Nature of Chemical Hazards, their Accident Potential and Consequences, in Bourdeau, P. and Green, G., eds. *Methods for Assessing and Reducing Injury from Chemical Accidents*, Chichester: John Wiley.

Walshe, K. and Sheldon, T. (1998) Dealing with Clinical Risk: Implications of the Rise of Evidence-Based Health Care *Public Money and Management* 19 (4) pp. 15-20

Weick, K.E. (1988) Enacted sensemaking in crisis situations *Journal of Management Studies*, 25, pp. 305-317

Weick, K.E. (1993) The collapse of sensemaking in organizations: The Mann Gulch Disaster, *Administrative Science Quarterly*, 38, pp. 628-652.

Weick, K. (1995) *Sensemaking in organizations*. Thousand Oaks: Sage Publications.

Weick, K.E. and Roberts, K. H. (1993) Collective minds in organizations: Heedful interrelating on flight decks, *Administrative Science Quarterly*, 38, pp. 357-381.

Weinberg, A.M. (1972) Science and Trans-science. *Minerva*, 10, pp. 209-222.

Weir, D. (1986) *The Bhopal Syndrome: Pesticide Manufacturing and the Third World*, Penang: International Organization of Consumers Unions.

Welsh, I. (1996) *Risk, Race and Global Environmental Regulation*, paper presented at the British Sociological Association Annual Conference, University of Reading, 1-4 April.

Whyte, A. and Burton, I. (1980) *Environmental Risk Assessment*. Chichester: Wiley

Whyte, D., Tombs, S. and Smith, D. (1995) *Offshore safety management in the "New Era": Perceptions and experiences of workers*, in Institution of Chemical Engineers (1995) Major Hazards Offshore and Offshore II. Rugby: IChemE. Symposium Series 139. pp. 35-53.

Williams, C., ed. (1996) Social Justice. Special Issue: *Environmental Victims*, 23, (4), Winter.

Woolfson. C. and Beck, M. (1999) 'Safety Culture – a concept too many?', *The Safety and Health Practitioner*, January.

Woolfson, C., Foster, J. and Beck, M. (1996) *Paying for the Piper? Capital and labour in the offshore oil industry*, Aldershot: Mansell.

Wrench, J. (1996) *Hazardous Work: ethnicity, gender and resistance*, paper presented at the British Sociological Association Annual Conference, University of Reading, 1-4 April.

Wrench, J. and Lee, J. (1982) Piecework and Industrial Accidents: two contemporary case studies, *Sociology*, 16, (4), 512-525.

A CASE STUDY IN RISK MANAGEMENT: THE UK PUMPED STORAGE BUSINESS

A.IAN GLENDON
School of Applied Psychology,
Griffith University,
Gold Coast Campus, PMB50 Gold Coast Mail Centre, Queensland 9726,
Australia

BRYAN O'LOUGHLIN AND RICHARD T.BOOTH
Health and Safety Unit,
Department of Mechanical and Electrical Engineering,
Aston University, Aston Triangle, Birmingham B4 7ET, UK

Case studies of safety and risk often focus on learning from disasters. However, learning can also result from studying risk management practices within organisations in which disasters have not occurred. This organisational case of the Pumped Storage Business (PSB), at the time it was part of the UK National Grid Company (NGC), demonstrates integrated management of safety and financial risks in a successful business. This chapter describes the main aspects of risk management that were central to the organisation at the time of the study, before considering these within the context of a risk management model.

Reasons for studying NGC's Pumped Storage Business include:
- it has very evident hazards, which require continuous application of risk management principles and has had no major accidents, suggesting that safety and risks have been successfully managed;
- it is a highly successful business, indicating appropriate application of financial risk management principles - one PSB manager said, ' ... this ... is a risk management business';
- although part of a larger organisation, it can be studied as a relatively autonomous business;
- it is unique within the UK and unusual in respect of its production process even on a world-wide scale, providing future opportunities for international comparative studies;
- the UK electricity supply industry is the first of a number of such privatisations around the world - and continues to experience significant changes, making it an interesting example of this genre.

E. Coles, D. Smith and S. Tombs (eds.), Risk Management and Society, 31–52.

The UK Electricity Supply Industry and NGC

As part of the UK government's privatisation program, most of the electricity supply industry was privatised in 1990. In England and Wales, the 12 regional electricity companies (RECs) were sold, as was the Central Electricity Generating Board's (CEGB) non-nuclear generating capacity, forming National Power and PowerGen - and the nationwide transmission system, which became The National Grid Company Plc (NGC). While the main generators entered the private sector as quoted companies, NGC's shareholders were the RECs and, through this mechanism, NGC became the world's first privatised national grid system. NGC's shareholders floated the company on the stockmarket in 1995, thereby opening the shareholder base to general investors.

The UK electricity industry has sustained high profitability, and there have been mergers within the sector, for example between RECs and other utility companies. A number of companies, including NGC, have growth ambitions beyond the UK. There is also scope for selling technical and managerial expertise to companies in countries in which utilities are being privatised.

NGC operates the high voltage transmission network throughout England and Wales, taking electrical energy from the generating companies and transmitting it to the RECs via their bulk supply points. NGC also operates the spot market for electricity - the 'Pool'. Generating companies sell electricity into the Pool, which is bought by RECs to sell to their customers. Through the Pool, NGC facilitates fair competition between generators and also ensures that existing and potential users have fair access to its transmission system.

NGC's Pumped Storage Business

When the CEGB was privatised, its only two pumped storage power stations were allocated to NGC. Table 1 outlines generating details for the two stations. These stations, together with a commercial office, formed the Pumped Storage Business (PSB) within NGC. PSB became the fourth largest electricity generator in capacity terms in England and Wales. All three PSB sites are in North Wales. Following the flotation of NGC, PSB was demerged from NGC in November 1995 as First Hydro Plc. First Hydro was subsequently sold to Mission Energy, a subsidiary of South Californian Edison Corporation, in December 1995. This paper relates to the period when PSB was part of the National Grid Group.

PSB is a competitive generator within the electricity market in England and Wales. Because NGC operates the Pool and is obliged not to discriminate between generators, confidentiality between PSB and the rest of NGC is needed - described as an 'arms length' relationship by one PSB manager. This necessary distance reinforced the separate identity of PSB within NGC, ensuring that there was little interplay between the cultures of the two parts of the organisation. PSB's power stations are physically distant from NGC's other operations and have a separate community identity - for around two thirds of their 180 or so staff, Welsh is their first language. Just under 200 people work for PSB, the balance being made up of commercial office staff.

Table 1: PSB power stations - capacity information

Power Station	Dinorwig	Ffestiniog
Generating since	1984	1963
Number of units	6	4
Capacity per unit (MW)	288	90
Total capacity (MW)	1730	360
Response time to full capacity (seconds)	<15	60
Hours of generation at full load	5½	4

PSB has three sources of income. As a generator, the company sells energy and capacity to the Pool by bidding in a daily price and availability for each of its generating units. A second source of income is provision of ancillary services. Because of the rapid response characteristics of the plant, PSB has a contract with NGC to provide reserve capacity and fast response to assist in maintaining both system stability and economic operation.

The third source of income is 'contracts for differences'. RECs are obliged to publish fixed tariffs and thus too much volatility in Pool prices causes them concern. They insure against the possibility of very high Pool prices by taking out contracts for differences with generating companies. These financial transactions ensure that, in return for a premium payment, generators undertake to pay RECs the amount by which the Pool price exceeds an agreed figure. Generators hedge against paying the RECs by bidding to the Pool a price which should ensure that they are called upon to generate - and therefore receive payment from the Pool - at such times. Contracts for differences thereby serve to reduce market price volatility effects. PSB's main costs are: electricity purchase (for pumping), use of system charges (for transmission), rates, salaries and maintenance.

PSB's mission statement reads: 'PSB aims to be the most dynamic generator of electricity, totally responsive to customer needs'. Maximising shareholder value, consistent with meeting its other obligations, is a prime focus for the business. However, license conditions imposed on NGC restrict it to pumped storage generation and only within England and Wales. Therefore, to increase profitability, PSB must improve efficiency and productivity within its existing business and seek opportunities overseas - acquisitions and new developments. As part of its financial risk management strategy, PSB spread risk attendant upon its operations by acquiring one or more hydro stations within EU countries as well as pursuing business development ventures in Asia. This represents a conventional business risk management strategy - protecting an existing business focus, extending this through sufficiently profitable acquisitions to spread short-term risks, and looking further ahead to new developments where profitability is less certain but which represent potential longer-term revenue and profit. PSB reduces its financial risk exposure to new development ventures by forming partnerships with major construction companies to share such risks. PSB launched International Hydro in May 1994 to pursue business developments overseas.

There are limited opportunities for significant cost savings through reducing headcount in a business with less than 200 employees and for which staff costs are relatively low. Despite this, PSB staff are concerned about the long-term future of their jobs. Although formal benchmarking exercises have not been carried out, comparison with a German company with five pumped storage power stations showed a higher staffing level than PSB - and PSB has a commercial function not present in the German company. A plant which is already technologically advanced might appear to have little scope for improving operational efficiency. However, the engineering case described below indicates that this assumption can be challenged.

A Brief History of Dinorwig

Because many safety problems in a wide range of industrial plant and equipment arise as a result of design shortcomings, the design and construction of Dinorwig as PSB's main asset is briefly considered. Unusual problems associated with Dinorwig's construction make this power station an interesting case example, one that is the subject of a book (Williams 1989).

Six generating and pumping units were installed - all with identical components so that parts were interchangeable. Ninety-five percent availability and 99% reliability criteria were imposed as plant operational parameters. Two units were initially built and commissioned and were operational while excavations for the other four units were completed.

A design feature incorporated was that a single system fault should not take out more than two of the six units. Incompatible with this principle was the single high pressure shaft - which was not duplicated mainly due to the construction cost involved, considering also that it would require a catastrophic event to affect this part of the system. The high pressure shaft at Dinorwig can be inspected using remote operated vehicles - obviating the need to dewater (drain) the station. Technical parameters attendant upon the construction of critical system components, notably the high pressure system, were of a very high standard. For example, the steel used in the high pressure shafts was monitored and tested from its point of manufacture so that its complete history was known. The high standard required means that monitoring these structures is required very infrequently.

A design feature which would be of consequence in the event of a worst case disaster (rapid catastrophic flooding of the station), is that tunnels excavated to monitor integrity of the rock in advance of the main excavations, like the rest of the station complex, are below the level of the lower reservoir. Thus, if the station did flood catastrophically the only escape routes would be through the main entrance (plant access) tunnel and the construction access tunnel. The tunnels above the main caverns were not in the original station design as they were dug to monitor the main excavation operations and were never designed as escape routes.

Pumped Storage Operation

The operating principle of a pumped storage system is that the power station uses water from an upper reservoir to generate electricity during the day when demand (and Pool price) is relatively high and uses electricity to pump water back from a lower reservoir to the upper reservoir when demand (and price) is low - usually at night. Typically, PSB is a very significant user of 'off peak' electricity for its pumping operations - perhaps taking up to one tenth of the entire night-time market during periods of lower summer demand. This makes PSB a very influential customer at night and means that it can have a material effect on the setting of the night-time Pool selling price.

The principal advantage of a pumped storage system is its very rapid response time, meaning that it is a valuable reserve capacity for coping with sudden increases in demand on the network or sudden shortfalls caused by plant failures elsewhere on the system. Dinorwig is the largest pumped storage station in Europe, one of the largest in the world, and is regarded as being at the leading edge of technology in terms of response speed. While Ffestiniog is a traditional site adjacent to its lower reservoir, the Dinorwig station complex is built deep inside a mountain in a complex of large caverns and 16 km of tunnels. These excavations house the generating plant, transformers, switchgear, control room, workshops and stores. Both stations operate a five-shift system around the clock.

From the upper reservoir at the Dinorwig system, water enters the 1695m primary low pressure tunnel via the headgates before arriving at a vertical shaft at the intersection of the high pressure shaft and the surge shaft. The high pressure shaft is 412m deep and 10m diameter. From the bottom of the high pressure shaft, the water enters the high pressure tunnel for 446m down a 1:10 gradient before entering the high pressure manifold system which feeds six steel-lined high pressure pipes (penstocks), each 170m in length, which end at the six main inlet valves (MIVs). Each MIV supplies a turbine from which the water passes to the lower reservoir via the draft tubes, draft tube valves, tail race tunnels and tail works.

For normal operations and maintenance the MIVs are the point of isolation for the water supply into the power station. It is also possible to isolate the system at the top reservoir by closing the headgates, although these take about 30 minutes to close. If it was ever necessary to dewater the power station, the headgates would have to be closed and the high pressure shaft dewatered slowly over several days to avoid damage to the shaft by water pressure imploding the concrete lining. As a single tunnel supplies water from the upper reservoir, the station is then out of action. Thus, dewatering would be a drastic step involving losing several millions of pounds revenue and would only be considered if no other option were available.

Formal Structures within PSB

PSB's General Manager (GM) reports to an NGC Executive Director. The GM's executive team comprises:

- Plant Manager - responsible for all aspects of day to day plant management;
- Engineering Development Manager - projects beyond short term and routine maintenance;
- Commercial Manager - marketing, sales and trading activities;
- Business Strategy Manager - strategy, planning, business development and finance.

These five, together with other senior managers, meet monthly as the PSB Management Executive (ME). The PSB Board meets quarterly and includes the five PSB senior managers, the NGC Executive Director for Engineering and the NGC Group General Manager for Finance. The PSB Management Executive is the key body as far as managing risk, along with all other aspects of the business, is concerned. The prevailing team culture ensures collective responsibility for business issues and all senior managers can represent the business in any of its main aspects. Other regular meetings include PSB's local Council, which meets bi-monthly and is the forum for consultation between PSB management and staff representatives, and separate Health and Safety Advisory Committees (HESACs) at Dinorwig and Ffestiniog.

Since privatisation, the Management Executive has been constantly aware of the requirement to address shareholder interests. PSB is in the relatively unusual situation of being owned by its customers (via NGC) - two stakeholders effectively being one party. Employees are the third main stakeholder, while interests of local communities and other groups (e.g. environmentalists or hillwalkers desiring access to PSB-owned premises) are also considered. Another important party is the electricity market regulator (OFFER). As far as relations with OFFER are concerned, PSB has opted for a consistent and explicable trading strategy in the electricity market so as not to incur adverse regulatory scrutiny.

The long-term planning process operates on a ten-year timescale - which takes a prospective view of the business's future. The ten-year plan is revisited annually to determine whether the core business philosophy remains constant. Five-year and two-year plans are used to guide medium-term management decisions. Scenario planning identifies the most likely range of trading conditions and profit expectations. While risk or risk management are not permanent ME agenda items, most ME decisions involve risk, so that plant and personnel safety is a continuing agenda item. Managers report on their particular function and discussion focuses upon business costs and ventures as well as opportunities and balances between them.

A Substantive Issue Illustrating Risk Management Decision Making

Major engineering work at Dinorwig, involving a technique developed within PSB in conjunction with the equipment supplier, was commenced in September 1994. If successful, the technique would provide huge commercial savings to PSB, but was technically extremely challenging and involved a meticulous approach to managing considerable risks. The technique had never previously been used anywhere in the world and therefore had commercial potential beyond immediate benefits to PSB.

After ten years of operation, more frequent maintenance than originally planned had been carried out on a crucial part of the system - the main inlet valves (MIVs). It was considered that replacement of key components (trunnion bearings) was probably necessary. The original design life of the trunnion bearings was around half the life of the station and no prior provision had been made for changing them during normal maintenance. Standard procedure for replacing the bearings would require dewatering the station (the 'dry' method) and would take over six weeks, with consequent considerable revenue loss. Because of the costs of a complete station shut down, an alternative proposal was explored which would allow trunnion bearing replacement with the station remaining operational (the 'wet' method).

As well as short-term revenue loss which would result from a station shut down, PSB management was also very aware of a longer-term commercial risk. If Dinorwig were out of action for six weeks, NGC and its customers - essentially the RECs - would need to adapt to the changed situation. A concern of PSB management was that its main customers would have to adapt to operating without Dinorwig's rapid response capability and that it could be difficult winning back those customers in an increasingly competitive market.

PSB management was therefore highly motivated to seek a method for this major maintenance task that did not involve dewatering the station. In 1988, a working party was established comprising PSB staff and designers of the MIV trunnion bearings, to investigate an alternative to the dry method. The working party came up with a proposal that would allow the work to be done without dewatering. In 1990, a design study was commissioned to determine the feasibility of the proposed technique. The PSB Head of Mechanical Engineering - reporting to the Engineering Development Manager - led the team that developed the wet method. The team included the Principal Design Engineer from the manufacturers, experienced fitters and a Shift Production Engineer, who was responsible for safety and human resource management.

The technique was tested on site in 1993 and the first full trunnion bearing removal and replacement was successfully completed on the first MIV unit in September 1994. During the 1993 trial, and again during the 1994 replacement, the team worked long hours dedicated to this single task. From a series of design studies and associated testing and monitoring of plant, a method statement was prepared. This was used during the 1993 trial, modified in the light of that experience and adopted for the 1994 work. The trunnion bearing replacement work on the other five Dinorwig units was then scheduled to be completed over the following 3-4 years. The second MIV trunnion bearing change was successfully completed in 1995 using the wet method and incorporating the experience gained on the first occasion. By May 1997, only two units required the work to be completed.

In assessing both physical and financial risks of the wet method, the company produced a report identifying the various stages of risk assessment and considered three major aspects of risk:

- water release;
- injury to personnel;
- unit out of action indefinitely or delayed return.

In each case, the risk was first analysed in respect of the various ways in which it could be manifested before setting out measures for controlling the risk. The risk assessment was incorporated into the method statement. As part of the risk assessment process, PSB submitted the proposed plan to the National Nuclear Corporation (NNC), a body with considerable experience of power station design, for intensive scrutiny. The NNC's external assessment approved the work proposed and confirmed that the procedure was satisfactory.

The two main hazards to personnel associated with carrying out the work by the wet method were the presence of high pressure water in the near vicinity and handling heavy equipment in a restricted space. Potential consequences of the high pressure water hazard range from the worst case scenario of an uncontrollable flood to a relatively minor leakage, which could still represent a risk to personnel. Risks associated with handling heavy equipment are comparable with those encountered in normal maintenance operations. An additional risk with this novel procedure is that a stage may be reached at which it is not possible either to continue with the work or to restore the unit to its operational state. In this case, the unit would have to remain out of service until the problem could be rectified. In the worst case, this could mean no generating or pumping capability from that unit until it was repaired dry – i.e. by dewatering the entire station.

Calculations showed that the estimated cost to the business of such a prolonged absence of capacity from one unit was £17.5m. The probability of such an outage was assessed at less than 1% - the product of these two figures representing the (business) risk, notionally £175,000. When making a decision on whether to use the wet or the dry method, the ME had estimated that the dry method would cost £6m more per unit than the wet method – i.e. £36m for the six units. After completing the risk analyses, the decision was made to use the wet method on the ground of considerable cost savings – set against the one percent probability that a unit would be out of operation for a period and that associated costs would be spread over a period of time.

The project to undertake the MIV trunnion bearing replacement without dewatering the station demonstrated one aspect of cultural change within PSB following privatisation. It was an example of PSB management seeking opportunities to maintain a commercial edge over their competitors. It is interesting to speculate on whether, if the problem had arisen prior to privatisation, the development work necessary to achieve a successful outcome of the wet method would have been undertaken – i.e. would the work have been undertaken using the dry method without a complete risk analysis?

Once PSB was competing with other generators, non-availability of Dinorwig for the period necessary to adopt the dry method would have had considerable consequences, both in the direct revenue loss and in the anticipated perception of NGC management that the pumped storage facility was less' essential' than previously thought. The high value of savings inherent in the wet method as well as commercial potential for application of the procedure elsewhere in the world, meant that this was a high profile project.

From a technical and safety aspect, the work took place in the most sensitive part of the plant in respect of potentially catastrophic consequences if something went wrong. Therefore, the project had to be managed with great care. The approach was extremely cautious and systematic, with meticulous attention to detail. The procedure was developed over an extended period and incorporated skills and knowledge from designers, operations and craft workers, using a combination of consultancy contracts and team work. Considerable attention was paid throughout to integrating design and development experience with operating and craft knowledge. Safety and practical considerations advanced in parallel with technical developments. Due weight was given to training as well as keeping not only team members informed but also other groups within the business who had a need to know. Considerable thought and effort also went into engineering the new components required, not only from a functional aspect, but also from a materials handling perspective. Simulations using real size mock ups were used. At all stages the approach was to identify what could go wrong and to devise a means of controlling the situation if it did. Worst case scenarios and conservative assumptions were made on many aspects of the procedure. The method statement was subject to continuous development prior to and after the 1993 trial.

Further Business Aspects

Arthur Andersen (1994) notes that management of newly privatised companies requires new tools for new times so that they can focus on market signals and continually re-evaluate their strategies for creating value. These tools, required to shift the company from managing book financial performance to managing shareholder value, include discounted cash flow (DCF) and risk assessment models. The report notes that, 'The shift in financial management from budget control to strategic resource allocation and asset management requires new methods of risk assessment, new ways of integrating strategy into financial planning ... and streamlined procedures to provide flexibility in resource deployment'.

As players in an immature market, PSB managers had to develop techniques for assessing market risks by learning from experience. This process was complicated by frequent electricity market changes following privatisation. The medium-term (3-5 years) future remained uncertain as the RECs developed increasingly acquisitive ambitions or were themselves takeover targets. Other relevant aspects of the business environment included greater regulatory interest in the major generators, plant divestment and Pool price movements. Predicting exactly how changes in their main customers' (the RECs) behaviour and structure would affect PSB operations was not easy. A key marketing activity for PSB was maintaining close personal contact with its customers and seeking to match their demands with PSB products. PSB targeted the 'quality' end of the electricity market, ensuring that premium price and quality were perceived as consistent by their customers. With a small customer base, personal selling and networking were prime promotional tools. PSB's differential advantage was its ability to meet demand peaks - for which the market is often at a premium level. Thus, some competitors could also be customers, the electricity market being essentially divided between base load and peak requirements. PSB's marketing function had confidence in plant reliability and availability and therefore considered itself to be at the 'cutting edge' in developing products for its customers.

PSB positioned itself at the quality end of the market, i.e. not being a volume player but rather maximising profit consistent with plant safety and integrity in both the short and long term. Short-term gain might conflict with long-term interests. For example, it would not be good (risk) management to work the plant hard for short-term gain if the business was degraded in the longer term. PSB management thereby regarded long-term asset management as a key objective.

Privatisation provided both NGC and PSB with opportunities to acquire new businesses overseas. As a result of privatisation, NGC - described by one manager as an 'inherently cautious' and 'risk averse' organisation, moved increasingly towards a clear focus on maximising shareholder value. Contrasting PSB with NGC operations in general, one manager said, 'We're at the other end of the spectrum from NGC in terms of risk; we've got a high risk, potentially high profit business'. Another PSB manager said, 'We make our profit by accepting risk and by taking on other people's risk and as long as we can get a good premium for that we're keen to do business'. Within PSB, DCF analysis was a key method used for driving investment decisions, for example in respect of plant refurbishment. In making decisions, for example that on the MIV trunnion bearing work, a set of parameters was employed, the first of which was the safety of people at plant level. Neither safety nor plant integrity was traded against cost. Commercially, decisions were concerned with balancing physical risk to the plant and financial benefits of marketing PSB's products.

In assessing risks associated with ventures in other countries, political stability, cash availability (ability to pay) and currency stability were key factors in initial decision making. Other factors included availability of spares, cultural factors affecting expected staffing levels and wage costs. While an IRR of around 12% might be acceptable for an acquisition in good condition in a stable market (e.g. Europe), an IRR of at least 20% would be sought for a shared new venture in a developing country. A prime objective in managing the risk of such a venture would be apportioning risk between parties involved.

An acquisition might take between six and nine months to complete the process of:
- initial 'due diligence' surveys;
- present case to ME and revise as necessary (iterative process);
- present case to management board and revise as necessary;
- follow up 'due diligence' surveys;
- final offer.

Criteria would include:
- does it fit PSB's portfolio requirements?
- is the price right?
- is it worth having (in respect of rate of return)?

Changes within the electricity market - for example in respect of regulatory aspects of Pool operation - such as the introduction of an agreement on bid prices between the Regulator and the major generators in April 1994, triggered a review of the way in which market risks were analysed. Like most aspects of risk management, the process involved within the ME was an iterative one. Other components of the risk management process were considered on an occasional basis; for example, biennial assessment of insurance risks as a prelude to premium review sought to assess the risks faced by PSB in the event of a major catastrophic plant failure.

Safety Management Systems

When Dinorwig was under construction, a 40-page Station Safety Report (CEGB 1983) was produced. The report details all likely sources of hazard and its purpose was to show that special risks associated with the plant had been properly considered and that suitable measures had been taken to ensure acceptable standards of safety during operations. Engineering safety of the system is secured by avoidance of common mode failures, interlock systems and tripping facilities with in-built redundancy.

The station safety report provided a standard against which continuing safety provision for engineering aspects of Dinorwig could be assessed and as a point of reference when developing new work. The report gives systematic consideration to flood risk. Other risks reviewed in this report are fire and smoke, toxic fumes and civil engineering works - including dams and reservoirs. The report also considers means of escape, emergency power supply and operating and maintenance manuals. However, it does not address managing these risks on a continuing basis.

One way of ensuring that risks and control measures identified in the station safety report are adequately considered, is to subject these to rigorous safety management audit process - ideally one which involved both station staff and independent auditing staff. NGC auditors used a standard version of a proprietary safety audit system to audit PSB. However, while PSB managers appreciated the audit assessments, they were not much involved in their production. One criticism of safety auditing is that it can be misused - for example to obtain 'stars' rather than as a way of improving site safety (Eisner and Leger 1988). A standard proprietary audit is not designed to take account of the unique mix of hazards that exist at a site and its use therefore imposes fixed parameters for all types of workplace, irrespective of the nature of their risks. Glendon (1994) summarises advantages and shortcomings of safety auditing.

Although independent audit reports provided considerable detailed information on the state of safety in the two power stations as well as guidance and recommendations for improvement, the results were problematic in respect of overall risk management. Independently of the management safety audit, PSB management had considered a number of 'worst case scenarios' in assessing the most obvious major hazards, such as a transformer fire or an oil spillage - which could produce a small amount of local environmental damage. Risks from other major hazards were minimised through 'belt and braces' design features, such as both containment and insulation of high voltage switchgear, which made human injury through maloperation virtually impossible.

A key aspect of managing risk is to ensure that it is accompanied by active learning. To an extent, there is comfort in safety reports which indicate that all systems have been adequately catered for, audit returns suggesting that a reasonable level of safety is being achieved and a safety track record, which while it has some interesting incidents, indicates that accident rates are low. The management effort of managing risk by exception is probably less in the short term than seeking to manage risk proactively - meaning that management is never satisfied with the present state of affairs but always seeking to advance to higher levels of excellence. When problems require engineering and human skills to solve them, these also provide windows of opportunity to ask a wider range of 'what if' type questions to see what may be learned from the experience.

When disasters occur and inquiries are held into the circumstances surrounding them, it is frequently found that one or more similar incidents have preceded them, usually with relatively minor consequences (see for example, Glendon and McKenna 1995, Toft and Reynolds 1997). An organisation's ability to extrapolate from such incidents - to imagine what could have happened if circumstances had been less favourable, is an indication of their adroitness in learning from minor misfortunes which do not have serious consequences, but which could have done.

One way of increasing opportunities to learn from such occasions is to implement a suitable system for recording 'near misses' (van der Schaaf *et al* 1991). PSB has a safety suggestion scheme for reporting near misses and a Confidential Hazard or Incident Reporting Procedure (CHIRP). Successful and effective use of such a system requires user training and imagination by those using it. The CHIRP report form has entry spaces for recording: location of hazard or incident, description of the activity or situation, suggestions for removing or rectifying the hazard, date and time of the observation. Each completed form is given a serial number and a documented tracking record identifies proposed actions, date completed, HESAC comments, need for further action and manager's final endorsement.

Where there are very obvious hazards in a workplace, such as water under high pressure, toxic substances and high voltages, there may be reduced awareness of more 'mundane' or less spectacular but higher probability risks, such as are produced by vehicle movements. An example of an incident from a PSB site occurred when a new lorry was aquired for one of the power stations and on which reverse gear was in the same position as first gear on the previous lorry. A driver had inadvertently put the vehicle into reverse gear instead of forward gear and had reversed into a wall. The driver injured his back in the accident and it was therefore entered into the accidentbook. Investigation revealed that another driver had made the same error some weeks earlier but had not hit anything and had neither reported the event nor mentioned it to anyone. Individuals tend to blame themselves for such 'errors' rather than identifying system shortcomings which promote them and thus there cam be an understandable reluctance to report. Nevertheless, reversing incidents can result in fatal accidents and it remains unknown as to whether even the second incident would have been reported if it had not resulted in injury but merely in damage to a wall or to the vehicle. The solution adopted was to label the gear positions clearly.

Monitoring Risks and Plant Integrity

A key aspect of continuing risk assessment is the operating life of the stations and the plant and equipment they contain. The risk here is of commercial loss in the event of equipment failure where, for example, the life of a component has been incorrectly assessed. All items of plant were subjected to the assets assessment process. The combined engineering and financial assessment, involving the Engineering Manager, the Management Accountant and the General Manager, was used to determine the predicted life of a component, replacement time and cost involved. From this could be calculated the replacement program and spares requirements. These assessments were linked with the financial success of the company and assumptions were made in respect of projected future values, discounted to obtain net present values – for example for carrying out maintenance or replacement work.

Because of the nature and scale of the structure enclosing Dinorwig, monitoring is required to detect suspected or potential problems. A six-monthly survey report on rock movements is compiled from readings at several hundred monitoring points. Future scheduled maintenance is to be supplanted by maintenance as required, based on data from increasingly sophisticated condition monitoring equipment. Measures of vibration, temperature, pressure and other plant variables during normal operation, can identify developing faults. Costs are reduced as maintenance is only undertaken when required, rather than on a regular schedule.

While civil engineering works at Dinorwig are scheduled for a life of 80 years, many operational plant items have a life of 40 years or less. Major projected maintenance has to be built into forward business plans. Because of the plant's uniqueness, it can never be certain that substantial refurbishment will not be necessary within the normal 40-year life of major plant items. Managing this risk crucially depends upon monitoring plant performance and availability of control measures, including spares and maintenance capability.

A specific example of monitoring in part of the high pressure system, is a reinforced bung that was inserted into one of the tunnels to plug a hole originally used to excavate rubble from the high pressure shaft, which had accumulated during construction. The substantial reinforcement surrounding the plug ensured that it could not fail catastrophically, although slow leakage is possible. In a system involving so great a volume of water under pressure, leaks are almost inevitable and another construction feature that has left a 'leaking legacy' is the use of and failure to remove pipes used to drain surplus water during welding operations on the steel penstock linings. The first unit to be installed had a leak from the pipe that was concreted in during construction, although none of the other five units was affected, even though the same technique was used on all units. Possibly the technique for sealing in the pipes was better developed for subsequent units or it could be that the first unit experienced more strain and vibration than the others did.

While catastrophic flooding could not result from the penstock leak, repair would be costly and the monitoring strategy was to measure the flow on a weekly basis. It was considered that the leak was likely to remain within manageable bounds and parameters were set - based upon the station's normal drain pumping capacity - for when action would need to be taken. The leak performed a valuable function by reminding PSB management of one of the risks of running plant at fast response full capacity - when vibrations are likely and component wear-out rate increased. Such risks have to be set against business advantages of operating plant exclusively at the behest of consumer demand. The balance that has to be maintained is also affected by what major competing generators can provide.

If the penstock leak led to a situation where dewatering the station in order to effect repairs became a real possibility, because of the costs involved it would have required a civil engineering report to the ME. Financial ramifications would be explored and consultants engaged to advise on specialist areas. A leak repair was eventually carried out in 1996 using a high pressure grouting technique, taking advantage of a 24-hour period when the station was shut down and the head of water lowered to allow inspection of the low pressure tunnel.

All work involving non-recurring revenue requires a risk assessment by the ME. This body, in which engineering and plant operational interests are balanced by commercial and business requirements, coordinated by the GM, showed itself to be an effective structure for integrating assessment of engineering and business risks. Conflicts and disagreements could be accommodated through participants' mutual professional respect, and decisions which had been vigorously debated in this forum were more likely to be robust than those which had not been subjected to this process.

Other Risks and their Management

1.OPEN DAYS

In Dinorwig's daily program, groups of visitors are collected from a field centre in the nearby village and taken by bus into the station complex. Guides take the groups round a planned route and there is a limit of 50 visitors inside the station at any one time. Up to 80,000 visitors every year see Dinorwig in this way. Dinorwig has also had two open days. In the first, in 1987, many thousands of visitors were allowed to walk from the main entrance and to wander around inside the power station. PSB staff were stationed at places where visitors might diverge from the planned route.

The second open day was in 1994, and while most members of the Management Executive had changed in the intervening years and there was no formal report of the earlier open day to use as a direct learning experience, the second open day was organised more methodically. Managing the risks associated with a large influx of visitors was assigned to the Plant Manager and Site Warden. Using their existing experience of allowing members of the public underground, this group identified the hazards and assessed what could go wrong before determining a method to minimise the risks of visitors gaining access to unauthorised areas and to allow for evacuation if necessary.

The method adopted was to use a shuttle bus service to bring groups of 50 people into the station with a maximum of 250 allowed in at any one time and no more than two buses in transit at any one time. The final agreed route emerged through an iterative process that took place over a two-month period. Visitors were permitted to walk around certain parts of the station. A duty manager kept in continual radio contact with other PSB staff at strategic locations.

In terms of health hazards to which visitors might be exposed, the only one identified by management was noise. Noise levels are monitored every two years. Although the station has conspicuous notices informing staff and visitors that hearing protection must be worn, ear plug dispensers are sometimes empty and the 'rule' is not enforced. Noise surveys indicated that even if a person was adjacent to a unit on full load for forty minutes, their exposure would not approach statutory limits. Therefore, noise exposure risk for visitors was judged to be acceptable.

2. UNUSUAL AND PERIPHERAL RISKS

Many organisations are likely to have risks associated with their activities which cannot reasonably be foreseen and for which little provision is made, particularly when they are rarely realised. Such risks might be contrasted with that of water leaks in a hydro powered system - which are an obvious constant threat to operational integrity, unless actively assessed, monitored and, where necessary, controlled.

While no fatalities arose at any PSB site as a result of operational activities there were tragic deaths in the vicinity of these sites. Because the PSB sites are located within an area of natural beauty, used by many people keen to explore the outdoors, restriction of access is not an option. In 1993, a member of a group exploring an underground passage that was part of an old slate mine fell to his death in the Ffestiniog locality. The fatality occurred very close to land owned by PSB, and the incident - the first of its type known in thirty years in the area - prompted a review of all access points to old mine workings on PSB land around the power station sites. The accident was followed by a review undertaken by the Site Warden, in conjunction with an engineering consultancy firm. Natural and other hazards at both power station sites were identified from an exhaustive survey carried out by this group and detailed in a report to the Plant Manager.

The report, with its assessment of the risks, was evaluated by senior managers and by the ME and legal issues were addressed. It was determined that different requirements pertained for adults and children – for the former, warning signs might be enough, while for children secure barriers were needed. Control measures were modified to meet concerns of environmental groups, who argued that bats might live in the shafts and that access had to be maintained for these animals. While there was no evidence for the presence of bats in the passages, nevertheless the environmentalists' concerns were accommodated by barriers consisting of steel bars on concrete sills across mine shaft entrances. To ensure continued security the Plant Manager and Site Warden undertake long-term monitoring of the barriers.

An example of an unusual event – which can be useful in respect of encouraging reappraisal of safety management systems – occurred when contractors were carrying out refurbishment at the headworks that involved scaffold construction. In the course of this work a contractor dropped a scaffold pole into the hydraulic system, which because it fell end on, breached the strong netting required in the contract safety specification. On being informed of this event, an engineering manager performed a rapid risk assessment (up to his waist in water!) to establish the water flow rate that would levitate an aluminium scaffold pole and carry it into high pressure system – and thence to one of the turbines. Station output was reduced so that the low pressure system water flow was maintained at no more than one-third of the estimated critical flow rate. Efforts to locate the scaffold pole using a diver were unsuccessful.

It had been determined that because the 2m pole was made of aluminium, it could probably be ingested by one of the turbines without serious damage resulting and, indeed, three pieces of the pole were subsequently found the next time one of the tail races was opened for inspection. However, this incident led to speculation about the likely outcome if the pole had been made of steel rather than aluminium. While it would have been less likely to have been levitated by the water in the low pressure system, particularly if uncapped, had it been ingested by a turbine the outcome would probably have been the destruction of that unit - cost approximately £10m. As a result of this incident, more stringent requirements were demanded for such work - including scaffold screens that would not allow through even a steel scaffold pole dropped end on.

Another obvious source of hard foreign bodies entering the hydraulic systems is from the upper reservoirs of the two stations. While coarse screens (with 12cm x 60cm apertures) at the intakes would prevent large solid objects entering the hydraulic system, smaller debris could pass through. Thus, when an amateur helicopter pilot ditched into the upper Dinorwig reservoir, there was concern that debris from the crash could enter the station's hydraulic system and cause considerable damage.

The helicopter incident was a very low likelihood event - yet it happened, although without consequential damage to plant or personnel. When assessing risks relating to nuclear power stations, risk assessors are known to consider the 'worst case scenario' of a full passenger jumbo jet landing on such an installation and the reckoning is that such a scenario is too unlikely to be considered a credible risk. While PSB management did discuss with their insurers the possibility of banning aircraft from overflying reservoirs, this was deemed to be impracticable.

In the history of the two power stations, there have been three staff suicides. While there can be no suggestion of any adverse reflection upon the organisation or its management, each loss represents a personal tragedy as well as a temporary blow to staff morale and a waste of human resources. It has been suggested that the rather 'gloomy' cavernous aspects of both the station sites may predispose workers to depression, although the locality has a high suicide rate and such deaths among workers at the power stations might therefore to be 'expected' from a statistical point of view. It is a moot point as to how far risk management extends into the non-work lives of employees - even accepting that work and non-work lives are separate. Some organisations have established employee assistance programs which offer counselling services and other support facilities for employees, which are deemed to be cost-effective.

3 HUMAN RESOURCE ISSUES

Compared with its parent company and other parts of the UK electricity supply industry, PSB had a remarkably stable workforce. However, other pumped storage power stations operate without staff for some of their operation cycles - being operated from remote locations. For example, Wivenhoe in SE Queensland was designed for remote operation and is only staffed during daylight hours. While there is a difference between plants that are designed to be operated remotely and those, like Dinorwig, that are not, it would be technically feasible to convert Dinorwig so that it could be operated remotely. Even if this is achieved, there remains a difference between operating a station remotely and having no staff on site.

At present a team of six people is left in charge of Dinorwig on the night shift. In the event of a worst case emergency, a minimum of five people would be required to handle the situation, at least some of whom would need to be trained BA users. There is thus effectively no scope for reducing numbers of people on shift and the alternative would be a remotely operated station which would not put the lives of personnel at risk in the event of an emergency. However, set against this is possible risk to local communities of a disaster at one of the stations - albeit a highly unlikely event. However, Dinorwig staff have had to deal with a few emergencies. For example, in August 1994, an intruder slipped under the perimeter security gate, entered the station, made off with a vehicle and drove around for a time inside the station complex. Police eventually caught the intruder.

Even if a remotely operated station were possible, whether cost savings from reducing staff numbers set against risk to plant if a situation should arise that required human intervention, could justify destaffing would remain problematic until a full risk assessment had been conducted.

Conclusions

Most management decision making involves risk of one sort or another and PSB is a good example of a company with integrated risk management practices. There is no separate risk management function, the Management Executive operating as a continuous monitoring function with respect to known risks associated with operating the plant. Where 'one off' risk assessments are required, specialist groups are formed to undertake these.

Specific risk management principles revealed by this study include:

- use multiple methods - e.g. audits, ad hoc studies, management reviews;
- continuous monitoring and review processes;
- risk management processes are never complete because new risks are always emerging, from human behaviour in response to technology as much as by technological changes.

Because the main risks are so evident, a proactive risk management program is a standard part of managing this company's business. PSB operates according to standard risk management principles in both engineering (safety) and commercial aspects of its business. Both pure and speculative risks are identified and assessed as appropriate in respect of their damage potential or commercial benefit. Appropriate control measures are implemented and monitored. Where action is demanded as a result of monitoring, this is discussed and implemented. Reactive risk management is practiced as required when low likelihood events reveal gaps in the safety management system.

Figure 1: Waring and Glendon (1998) Risk Management Model

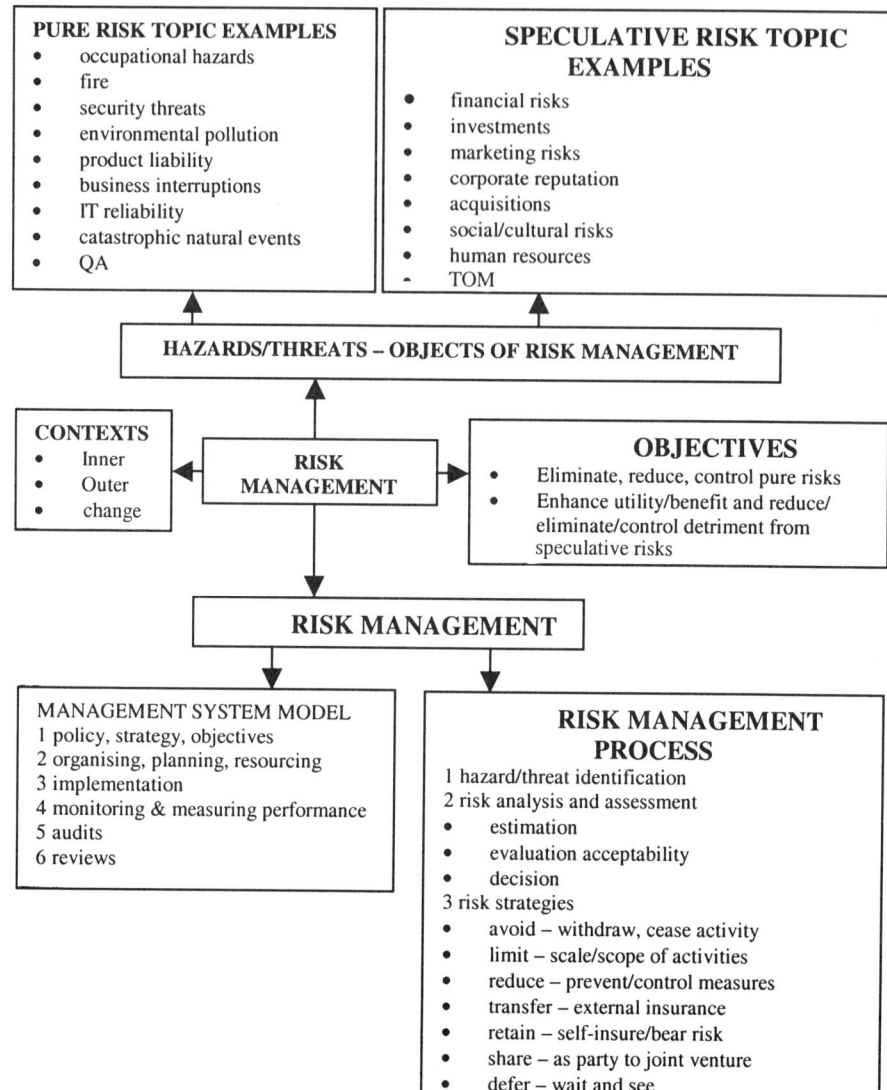

PURE RISK TOPIC EXAMPLES
- occupational hazards
- fire
- security threats
- environmental pollution
- product liability
- business interruptions
- IT reliability
- catastrophic natural events
- QA

SPECULATIVE RISK TOPIC EXAMPLES
- financial risks
- investments
- marketing risks
- corporate reputation
- acquisitions
- social/cultural risks
- human resources
- TOM

HAZARDS/THREATS – OBJECTS OF RISK MANAGEMENT

CONTEXTS
- Inner
- Outer
- change

RISK MANAGEMENT

OBJECTIVES
- Eliminate, reduce, control pure risks
- Enhance utility/benefit and reduce/eliminate/control detriment from speculative risks

RISK MANAGEMENT

MANAGEMENT SYSTEM MODEL
1 policy, strategy, objectives
2 organising, planning, resourcing
3 implementation
4 monitoring & measuring performance
5 audits
6 reviews

RISK MANAGEMENT PROCESS
1 hazard/threat identification
2 risk analysis and assessment
- estimation
- evaluation acceptability
- decision
3 risk strategies
- avoid – withdraw, cease activity
- limit – scale/scope of activities
- reduce – prevent/control measures
- transfer – external insurance
- retain – self-insure/bear risk
- share – as party to joint venture
- defer – wait and see

More generally, this organisation's approach to risk management reflects many aspects of the model of risk management proposed by Waring and Glendon (1998), and shown in figure 1. The model has four main components:

- threats - the objects of risk management, both pure and speculative;
- objectives of risk management;
- contexts - inner, outer and change;
- methods - system and process.

Most of the pure risk examples in the Waring and Glendon model, as well as a majority of those shown as speculative risks, are encompassed within decision making processes revealed by this case study. The case study also demonstrates strategic objectives for pure and speculative risks respectively, which are shown in the model. Examples of control of pure risks and enhancing the utility while reducing the detriment from speculative risks are also found within the PSB case study.

The inner context reflects a rational approach to managing risk, based within an engineering culture that characterises PSB. This rational process incorporates all the elements of the management system model as well as the main aspects of the risk management process, which together comprise the methods component of the model. The outer and change contexts are represented by a variety of environmental influences upon the organisation. Of particular importance is the changing market environment within which PSB operates - for example the industry privatisation that occurred prior to the study period. Thus, for both business risks and major engineering risks confronting the business since privatisation, PSB's approach to managing these risks can be described as being proactive and following a rational model.

However, management of peripheral and unusual risks over an extended period - i.e. including those from before privatisation, tended to be ad hoc and reactive rather than following a rational model.

There are a number of ways in which the case study described in this paper could be complemented in order to develop insights into risk management in commercial and other organisations. Two of these extend the unit of analysis adopted in this case study - the organisation, first to a more macro level, and second at a more micro level.

By undertaking further case studies both within the same industry and in other industry sectors and countries, more generalised models of risk assessment and risk management could be developed. For example, organisations in private, corporatised and public sectors can be compared in respect of their risk management practices. By undertaking further studies of this type, political, social, cultural and economic influences - increasingly recognised as critical to risk management (e.g. Royal Society 1992) - can be evaluated for their impact upon risk management practices in organisations.

A second major way in which research may be extended is to consider in greater detail risk perceptions of decision makers, and factors and dimensions that affect their decisions on risk. The literature in this area (e.g. Pérusse 1980, Royal Society 1983, HSC 1993) suggests a number of dimensions of risk perception as a basis for decision

Dimensions that might be suggested from this research include those shown in Table 2.

Table 2: Dimensions of risk perception affecting decision making in this case study

Risk type	Dimension	
	To plant	To people
	Known	Unknown
	Likely to be realised	Unlikely to be realised
	Large	Small
Physical	Obvious	Not obvious
	Long term	Short term
	Easy to control	Difficult to control
	Results from design	Results from contemporary activity
	Arise from natural events	Created by human activity
	Can be reduced/eliminated remotely/automatically	Only resolved by human intervention
	Known	Unknown
Financial	High	Low
	Controllable	Not controllable
	Long term	Short term

Acknowledgements

This paper is based on research undertaken between 1994 and 1996, using interviewing, informal discussions, analysis of documentary material, site visits and observations. The research was sponsored by The National Grid Company Plc and the Health & Safety Unit at Aston University. We wish to express our gratitude for the help that we have received from PSB management and staff in the course of the research.

Glossary

BA	Breathing apparatus
CEGB	Central Electricity Generating Board (prior to 1990)
CHIRP	Confidential Hazard or Incident Reporting Procedure
DCF	Discounted cash flow (type of financial analysis)
EU	European Union
GM	General Manager (of PSB)
HESAC	Health and Safety Advisory Committee
HSC	Health and Safety Commission
IRR	Internal rate of return (for assessing investment risk and opportunity)
km	Kilometres
m	Metres (as appropriate)
m	Million (as appropriate)
ME	Management Executive (of PSB)
MIV	Main inlet valve (of pumped storage power station)
MW	Megawatt
NGC	The National Grid Company Plc (UK)
NNC	National Nuclear Corporation (UK)
OFFER	Office of the Electricity Regulator (UK)

PSB	Pumped Storage Business
REC	Regional Electricity Company
SE	South east
UK	United Kingdom

References

Arthur Andersen Consulting (1994). *Predictable patterns: navigating the continuum from protected monopoly to market competition*. London: Arthur Andersen.

Central Electricity Generating Board (1983). *Dinorwig Power Station: station safety report*, revised March 1983. Barnwood: CEGB.

Eisner, H. S. and Leger, J. P. (1988). The International Safety Rating System in South African mining. *Journal of Occupational Accidents*, 10: 141-60.

Glendon A. I. (1994). *Safety auditing*. Module RA6, Occupational Health and Safety Training Unit, University of Portsmouth.

Glendon, A. I. and McKenna E. F. (1995). *Human safety and risk management*. London: Chapman & Hall.

Health and Safety Commission (1993). Advisory Committee on the Safety of Nuclear Installations, ACSNI Study Group on Human Factors, *Third Report: organising for safety*. London: HMSO.

Pérusse, M. (1980). *Dimensions of perceptions and recognition of danger*. PhD thesis, Birmingham: Aston University.

Royal Society (1983). *Risk assessment: report of a Royal Society study group*. London: The Royal Society.

Royal Society (1992). *Risk: analysis, perception and management - report of a Royal Society study group*. London: The Royal Society.

Toft, B. and Reynolds, S. (1997). *Learning from disasters: a management approach*. (2nd edn). Leicester:

van der Schaaf, T. W., Lucas, D. A. and Hale, A. R. (eds) (1991). *Near miss reporting as a safety tool*. Oxford: Butterworth-Heinemann.

Waring, A. E. and Glendon, A. I. (1998). *Managing risk*. London: Thomson.

Williams, E. (1989). *Dinorwig: the electric mountain*. London: The National Grid Company.

Royal Society (1983). *Risk assessment: report of a Royal Society study group*. London: The Royal Society.

Royal Society (1992). *Risk: analysis, perception and management - report of a Royal Society study group*. London: The Royal Society.

Toft, B. and Reynolds, S. (1997). *Learning from disasters: a management approach*. (2nd edn). Leicester:

van der Schaaf, T. W., Lucas, D. A. and Hale, A. R. (eds) (1991). *Near miss reporting as a safety tool*. Oxford: Butterworth-Heinemann.

Waring, A. E. and Glendon, A. I. (1998). *Managing risk*. London: Thomson.

Williams, E. (1989). *Dinorwig: the electric mountain*. London: The National Grid Company.

CHALLENGING THE ORTHODOXY IN RISK MANAGEMENT:

The Need for a paradigm shift?

CLIVE SMALLMAN
Judge Insitute of Management Studies
University of Cambridge
Trumpington Street
Cambridge, CB2 1AG, UK

Rising instability in society

The structure and nature of society in the developed and nearly-developed countries has been radically transformed since the beginning of this century (Drucker, 1995) and no moreso than in the last 30 years (Thurow, 1996: p. 21), although the seeds of change date back much further (Giddens, 1990). The industrial economy has overwhelmed an agrarian way of life, an evolved form of which is still required for the production of food, but is no longer deemed central to humankind's existence. Society in the developed free-market countries now has a different configuration from that which it had in 1900. It works through different processes and suffers from a steadily widening range of problems. There has been a qualitative and quantitative shift in people's professional and personal lives.

Polity (the organisation of society) has seen the transformation of social institutions and new modes of life. This wide-scale change has effectively swept aside and lost (perhaps permanently) traditional social structures (Drucker, 1995; Giddens, 1990). This is evidenced in changing patterns of employment and work (Drucker, 1995). At the beginning of the century farmers and live-in domestic staff were a major part of the employment force. Domestic staff now are rarely seen, except in the houses of the super-rich; indeed if this group were another specie they would be on the endangered list. The traditional farmer too is nearly "extinct".

In 1900 the blue collar worker represented an exploited proletariat. By the 1950s they had risen close to the nadir of their political power. Strong trade unions and weak management allowed the development of an industrial and political hegemony that lasted well into the 1970s. The 1980s and 1990s have seen a retreat, fuelled by the breaking of trade unions (with a subsequent decline in workers' rights (Hutton, 1995:

53

E. Coles, D. Smith and S. Tombs (eds.), Risk Management and Society, 53–79.

pp. 11-15)), as the post-Fordist era develops. It is perhaps too early to say, as does Drucker (1995), that the early 21st century will see the final demise of manual work in the developed countries. However, the comparative advantages of cheap and plentiful labour that developing countries hold now, will enlarge and accelerate the decline and fall from power of the skilled manual worker in the developed world. The modern labour force is increasingly heterogeneous, and under the pressure of economic change is likely to become increasingly fragmented as established patterns of work are fundamentally and radically altered.

Whatever else work represents, one of its main characteristics is that it is no longer a source of certainty in the lives of many people. In the UK, in the late 1990s only 40 percent of those in employment have secured tenure as employees or as self-employed. A further 30 per cent are in insecure self-employment, in involuntary part-time work or in causal employment. The bottom 30 per cent are marginalised, either idle or working for poverty wages, and often caught in a "benefit trap".

New society? New societal forms?

THE DISPOSSESSED

The decline in established patterns of society and work falter has apparently seen the further development of Marx's "lumpen proletariat" (Thurow, 1996: p. 29) into Murray's (1990, 1995a, 1995b) *"under class"*, equating to the UK's "bottom" 30 per cent (Hutton, 1995: pp. 11-15). A general growth in inequality means this socially disadvantaged group suffer living standards and conditions that are at best comparable with some of the worst in developing countries (Thurow, 1996: pp. 20-22 and 29-30).

In the late 1990s unemployment (in the UK approximately one quarter of males of working age have no job (Hutton, 1995: p. 1)), crime (marked by a rising prison population (Hutton, 1995: p. 2)), drug abuse (be it alcoholic or narcotic), squalid housing and poor education (marked by rising illiteracy (Hutton, 1995: p. 2)), all effectively supported by a shrinking welfare state, typify what is an increasingly miserable existence, and one which lies too close to the heart of many of the developed world's largest cities. Adding to this group, in the UK and America, are the new working poor; people who have suffered and continue to suffer from declining living standards resulting from the increasing insecurity and instability that goes with the fashion for "down sizing", cost cutting and casualisation of labour that is the result of the ongoing drive for productivity growth that typifies modern business practices (Marr, 1995: pp. 317-318; Hutton, 1995: p. 2; Thurow, 1996: pp. 26-29).

According to Murray (1990, 1995a, 1995b) and other commentators (Hutton, 1995: pp. 1-2; Thurow, 1996: pp. 29-30), there are signs that this dispossessed, disinherited and largely disowned group will increase in size, whilst suffering continued deterioration of what remains of their family and cultural life.

The Knowledge Society

Outside the underclass the societal signs are better. Several authorities (Drucker, 1980 and 1995; Thurow, 1996: pp. 65-87; Woodall, 1996: p. 43; OECD, 1996) point to the emergence of what is termed the *knowledge society*. The thesis is that, given the decline of skilled manual work, *knowledge* (that subjective combination of information and perceived meaning) will become the new currency (Brundtland, 1994: p. 65). Thurow (1996: pp. 65-76) goes further, citing the rise of the knowledge society as symptomatic of the disappearance of classical comparative advantage, whereby knowledge and skills become *the* resource base for industry.

Claiming this currency as their own are the *knowledge workers*, a highly educated and successful "cognitive elite" (Murray, 1995a and 1995b). Their jobs, reflected in the rise of the so-called professions, rely on an individual's grasp of knowledge and upon continuous learning as a tool of industry and technology. It is claimed that these people are forming a new *"over class"* (not to be confused with the existing upper class). Rather, the over class is larger, independent of social background and inherited wealth (with incomes in the top five to ten per cent of earnings), and is fundamentally dependent on intellectual talent (Murray, 1995a and 1995b). This is not solely about formal education (although this is a critical issue), but rather about the ability to apply theoretical and practical knowledge in the analysis of problems (Drucker, 1995). Over and above this, the emerging over class, by the very nature of their occupations as managers and professionals, have taken and will strengthen a profound grip and influence upon the great social institutions of commerce and government (Murray, 1995a and 1995b).

The new middle class, whilst not sharing the same power or riches as the elite, nevertheless are now demanding greater access to good education, a higher standard of living, and interesting, rewarding work. They too have a place in the knowledge society.

The knowledge society requires a different approach to work and requires a different mind set, driven by major changes in attitudes, values and beliefs, indeed in the human condition (Drucker, 1995). The knowledge society is a response to the problem of raising productivity to meet increasingly intense competition on the world stage, and so is at the heart of the process of globalisation.

All that said, neither the elite nor the middle class can properly regard themselves as secure. No one is immune to the economic and political dynamics that drive both the developed and developing nations. People of all groups are prey to a chaotic economic cycle of success and failure that is singularly difficult to follow or forecast (Marr, 1995: p. 317; Sampson, 1995; Sasakura, 1995; Stacey, 1989).

Equality Under Risk?

Ulrich Beck's (1992) polemic view is that humankind is developing a societal structure that overlays both the knowledge society and the under class. Rather than post-modernity, Beck (1992) sees society in a state of "advanced modernity". This, more closely defined by Giddens (1990) as "high modernity", is a condition defined by a complex interplay of institutions (see figure 1) which lies at the heart of radicalised change in the world's political economy, driven by the exponential growth of production forces (Beck, 1992).

Figure 1: Institutional dimensions of high modernity

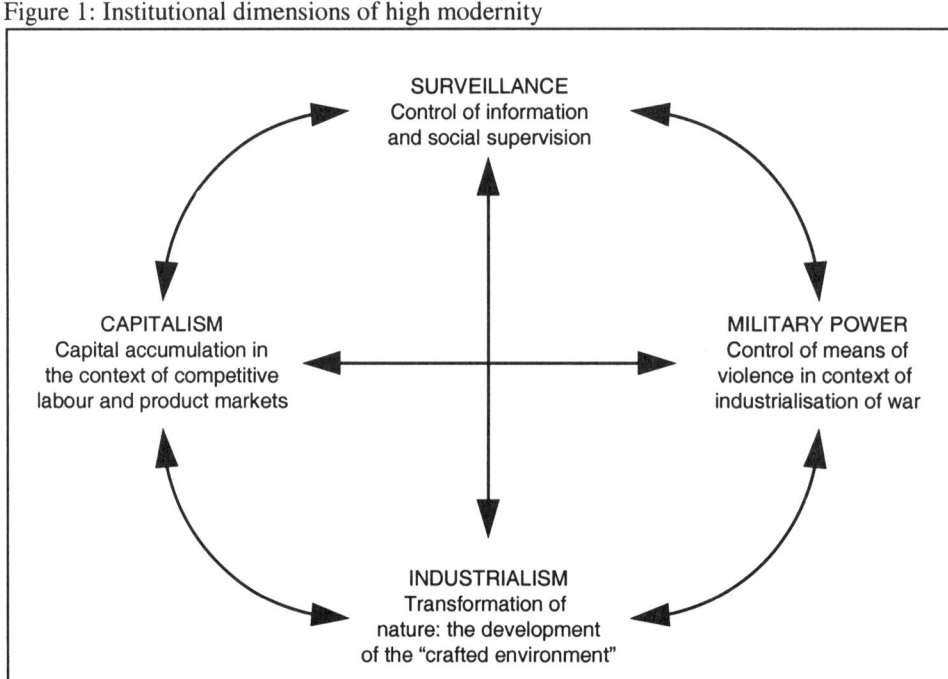

After Giddens (1990: p. 59).

This high modernity is typified neither by solely the rising power of an elite, supported by a swelling middle class, nor just by the continued expansion of the underclass, but also, by more sombre issues such as rising insecurity, declining trust in institutions and in "experts", and markedly increased risk (Giddens, 1990). Such phenomena it seems are inherent in a world increasingly fraught with danger (Giddens, 1990), hazards and potential threats to a previously unknown extent (Beck, 1990).

According to Beck (1992), advanced modernity equates to the emergence of the paradigm of the *risk society*. Herein the tolerable limits of the ecological, medical, psychological and social spheres are close to being breached, if that has not already occurred. The modernisation process, it would seem, has systematically produced *latent side effects* (Reason, 1990: p. 173), which humankind has only a limited ability to distribute away.

Beck's (1992) argument lays over the knowledge society and underclass in terms of risk distribution and management. The advantage of comparative riches which fall to the over and middle classes lies not in terms of risk distribution, but riches do offer an advantage in terms of people's abilities to deal with risk. Inequalities in occupational and educational strata that relate almost directly to income distribution, it seems, directly influence our skills in understanding and acting upon information. *And*, in the risk society moreso than in the knowledge society, *information is power*.

The effect of comparative riches in the risk society is to amplify and filter information on risk. Such filtering further disadvantages the under class and the lower reaches of the middle class in coping with risk, just as much as it causes crises and errors of judgement in modern business (Bella, 1987). Hence, the *nouveau pauvre*, an underclass already dispossessed, are further threatened by the production forces and societal changes that have already imposed upon them a poor quality of life (Beck, 1992). Further immiseration threatens those least able to deal with it. In short the members of humankind share a common anxiety of risks in the modern world. They do not share a common ability to deal with such risks.

This social inequality takes place in the context of a rapily evolving world economy. Fuelled by the failure of attempted socialist economies (Barratt Brown, 1995: pp. 181-286), the activities of newly (and established) independent countries freed from colonial shackles and the continued development of established market economies, there exists a rising trend of marketisation across the globe. The whole world is now moving towards a broadly homogeneous economic system where most nations will play to the "same rules"; for the time being those of global capitalism. However, the game is not yet being played on a "level field". There is a very real need for an economic system that allows first for social transition (Barratt Brown, 1995: pp. 334-356) and subsequently for a new social order (Barratt Brown, 1995: pp. 357-399; Thurow, 1996: p. 326), broadly in line with calls for sustainable development and planetary equity (Brundtland, 1994; Meadows et al, 1972; Independent Commission on International Development Issues, 1980; Independent Bureau on International Development Issues, 1983; World Commission on Environment and Development, 1987).

In the short to medium term this does not bode well for economic stability. Rather a period of increasingly dynamic and fractious change is slowly descending upon the world of trade and finance. This period of "punctuated equilibrium" with everything in flux, disequilibrium as the norm and uncertainty as the dominant characteristic (Thurow, 1996: p. 8), has had and will continue to have inevitable economic, political and social consequences.

In this chaotic context, it is pertinent to ask, how well equipped are our poicy makers and business leaders in approaching decision making in such an uncertain period? At the heart of this question lies a need to examine how the management of risk is currently approached and how well suited this is to the current context and future scenarios. For example, the Economist Intelligence Unit & Arthur Anderson (1995) questioned 3,000 executives world-wide. They found that *"business risk assessment and control practices are in desperate need of attention"*. Only 5.4 per cent of respondents to the survey expressed "absolute confidence" that they are properly managing all significant risks.   Even adding those organisations who have some confidence in their risk controls means that less than half of the organisations surveyed have some confidence in their firms' risk management. Not surprisingly a majority of respondents also noted that plans for change are in place. It seems that *"substantial revisions are under way"*. Findings such as this reinforce the need to examine the legitimacy of the current orthodoxy in risk management.

Homeostatic risk management: an inappropriate paradigm?

Risk homeostasis theory was originally proposed by Wilde (1982) as an explanation of driver behaviour relating to vehicle accidents. Reinterpreted at a more general level in the context of organisational behaviour (Hood et al, 1992; Hood and Jones, 1996: pp. 205-207), the *homeostatic* approach relies on institutions to set pre-determined risk tolerances and to convert these goals into quantified decision rules, following a feed forward loop (see figure 2). These guidelines can be applied by experts to given cases and can be incorporated into organisational design and operations.   This requires anticipation, quantification and the specification of outputs.

This is what Hood (1996: p. 209) labels the SPRAT (social pre-commitment to rational acceptability thresholds) approach to risk management. Hood (1996: p. 212) suggests that its great strengths are that it allows for the establishment of stable agreed targets, allows the "danger" thresholds to be operationalised in advance, and so enables correcting mechanisms to come into play as soon as the system moves outside the control limits. However, it is limited by:

- the breadth of rationality and cultural variety through which targets are set,
- the danger threshold is an artificial creation and one that does not allow for noise, opportunism in setting more appropriate targets or for flexibility in operation, and
- systems change is very slow (the system does not change until crisis looms).

In addition, the system may be deemed acceptable by those in "control", but those at the receiving end frequently view the "official" settings as just plain wrong (Hood, 1996: p. 213).

This is very clearly mirrored in current insurance practice whereby premiums are set on the basis of quantified actuarial models of risk and follow a set pattern and process. Changes to the models generally only occur in the wake of negative events (for example, poor institutional performance or an increases in socio-technical failures). It is also narrow, considering only those risks which fall inside the orbit of the model or models that the company perceives as being as an immediate threat.

Recent research (Smallman, 1997: pp. 322-331) into managers' perceptions of organisational risk indicates that this is the dominant model, and that managers fundamentally react to risks, following a strongly homeostatic doctrine (Glendon et al, 1996; Hood et al, 1992) wherein:

- Chronic risk is accepted as commonplace and thus the approach seems to be one of accepting failure, but expecting the management of risk to change following catastrophe, crisis or fraud.

- Responsibility for risk is almost always attributed to someone else, and managers claim in general to be able to do very little about risk (and have done most of what they can anyway), this suggests that *blame over risk is attributable*, but always somewhere else.

- Safety and similar risks are not cited nearly as frequently as commercial risks suggesting that *there is an implicit trade off* in managing different types of risk.

There is support in and amongst the same research that this is a weak approach (Smallman, 1997: pp. 322-331):

- A general lack of efficacy (the ability to do something) and control over risk that is felt by managers.

- Managers may attribute responsibility for risks upwards, but they also frequently cite doubts about their seniors' knowledge and ability (Hood, 1996). This suggests a problem for those who set the standards and rules for risk management and assessment.

- Homeostasis relies on reaction to agreed stable targets. However, in almost all cases managers find that risks are increasing and maintaining the stability is difficult (not least since stability is a word not easily associated with the current socio-economic context).

- The current orthodoxy makes the assumption of complete knowledge of risk, yet subjects frequently cite problems relating to the supply and quality of information concerned with business risks.

- The orthodoxy consistently refuses to acknowledge the role of politics in risk assessment and management (Hood and Jones, 1996: pp. xi-xiii). The attribution of responsibility, coupled to subjects' perceptions with regard to their own and others efficacy in dealing with risks directly contradicts this stance.

The context of the risk society, the evidence cited above and the structure of the homeostatic approach, suggests that the current orthodoxy in risk management is not well-suited to the current chaotic environment, and is out of step with increasingly dominant patterns of socio-economic change. But how should risk be managed?

Matching the context: collibrational risk management

Against the orthodoxy stands the collibrational paradigm (Hood et al, 1992; Smallman, 1996). This recognises that forecasting is limited by scientific uncertainty, and that the picture is further disturbed by cultural dynamics (Hood et al, 1992; Schwarz and Thompson, 1990). Hence, it is extremely difficult to build models and so set decision-making rules. Adams (1995: p. 163), provides a lucid summary of the issue wherein people attempt to predict futures based on a series of accepted and "proven" models; the choice of model is determined by the assumptions made about a process the nature of which is not fully understood. Hence, the assumptions are made beyond the range of available data, and on the basis of evidence that is contentious and not at all firm. This makes it difficult to transfer risk and dangerous to accept it without good management; it must therefore be avoided, prevented and reduced, especially where organisations have retained major risks.

This approach falls very much in line with emergence of *"holistic risk management"*:

"A systematic, statistically based, and holistic process that builds a formal risk assessment and management, and addresses the set four sources of failure within a hierarchical multi-objective framework:
(1) hardware failure
(2) software failure
(3) organisational failure
(4) human failure."   Haimes (1991)

Collibration works through forces that act in different directions in the same system at the same time (feedback and feedforward – see figure 3). Hence, a system governed by collibration is in a state that is the product of multiple forces held in tension (Dunsire, 1978: 181, 207-8; Dunsire, 1986; Dunsire 1990). The essence of collibration is that all organisational risks and their inter-relationships are actively considered, driven by potential hazards and learning from events (Shrivastava et al, 1988; Toft and Reynolds, 1994; Kleiner and Roth, 1997 lay out clear methods for the development of organisational learning points from past events and projects).

This approach, its antecedents and derivatives have been and continue to be derided by a range of critics, drawn from the engineering, statistical and natural science communities. Their principal objection is that the "human factor" is difficult to enumerate and ground in their chosen approaches (Hood and Jones, 1996: xi-xiii). In short they have little or no desire to accommodate politics within the dominant homeostatic model. This in turn is matched by the committed call of more radical thinkers (Hood and Jones, 1996) to encompass risk rhetoric and the perceptions of "non-experts" within the risk management process, in an attempt to more fully understand and so better handle risk in what is an increasingly hostile business and societal context (Smallman, 1996). There

Figure 2: Risk homeostasis: feed-forward control

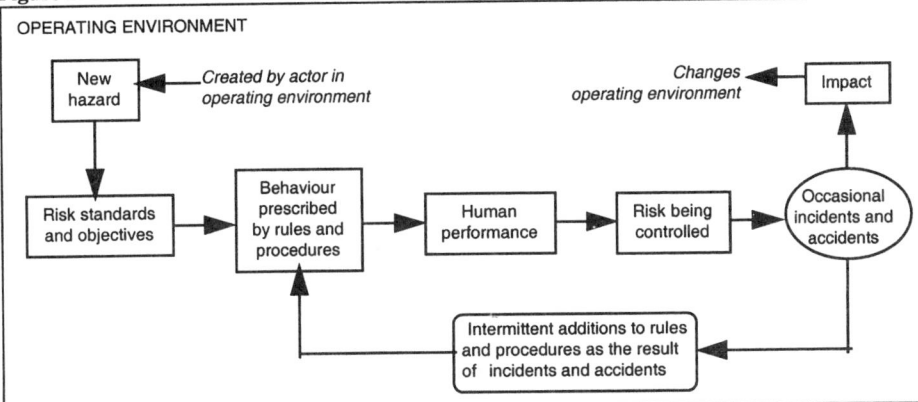

Adapted from Reason et al (1995).

Figure 3: Collibrational risk management

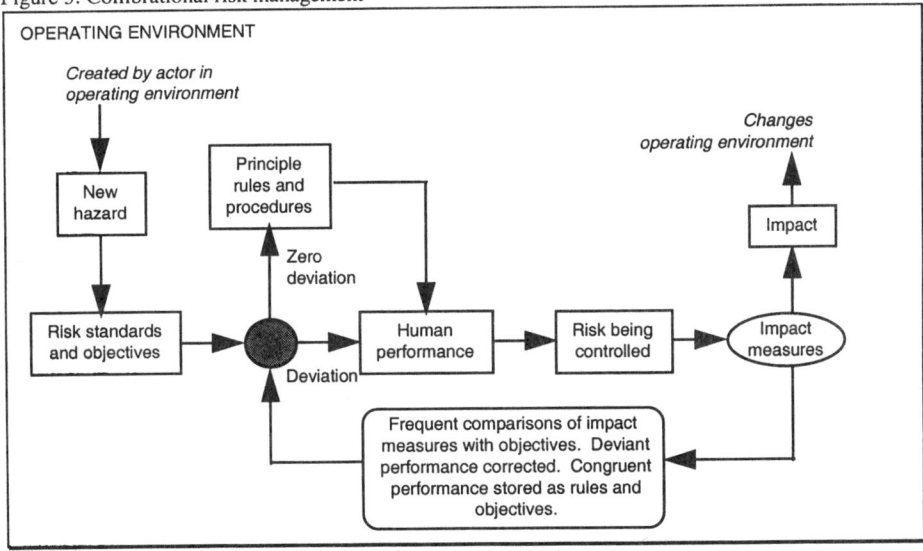

Adapted from Reason et al (1995).

should be little doubt that the most critical constituency of "non-experts" is those managers who run organisations and so deal with risk issues on a daily basis.

The risk management dialectic

In justifying the need to adopt the collibrational model it is important to examine key doctrines and debates that exist within the discipline of risk management and their relationship to collibrational risk management (Hood et al, 1992; Hood and Jones, 1996). Hood et al (1992) highlight a number of current and important debates in modern risk management. These are summarised in table 1.

Table 1: Seven doctrinal contests in risk management policy summarised

Doctrine	Justification	Counter-Doctrine	Justification
Anticipationism	Apply causal knowledge of system failure to ex ante actions or better risk management	Resilienism	Complex system failures not predictable in advance and anticipationism makes things worse
Absolutionism	A "no-fault" approach to blame avoids distortion of information and helps learning	Blamism	Targeted blame gives strong incentives for taking care on the part of key decision-makers
Quantificationism	Quantification promotes understanding and rationality, exposes special pleading	Qualitativism	Proper weight needs to be given to inherently unquantifiable factors in risk management
Designism	Apply the accumulated knowledge available for institutional design	Design agnosticism	There is no secure knowledge base for or real market for institutional design
Complemetarism	Safety and other goals go hand in hand under good management	Trade-offism	Safety must be explicitly traded off against other goals
Narrow participationism	Discussion is most effective when confined to expert participants	Broad participationism	Broader discussion better tests assumptions and avoids errors
Outcome specificationism	The regulatory process should concentrate on specifying structures or products	Process specificationism	The regulatory process should concentrate on specifying institutional processes

Source: Hood et al (1992).

These doctrinal contests may be aligned with the collibrational and holistic paradigms as represented in table 2.

Anticipation

Anticipation is defined in the *Collins English Dictonary* as "confident expectation". Anticipationists focus upon the need to detect potential threats and so prevent latent failures from building up, calling for risk audits of organisations and locations.

Table 2: Doctrines aligned with risk management paradigms

Collibrational risk management	Homeostatic risk management
Anticipationism	Resilienism
Absolutionism	Blamism
Qualitativism/Intelligent quantificationism	Blind quantificationism
Designism	Design agnosticism
Complementarism	Trade-offism
Broad participationism	Narrow participationism
Process specificationism	Outcome specificationism

Derived from Hood et al (1992).

The case for taking such an active approach has been argued in different ways (Hood et al, 1992). Tait and Levidov (1992) argue for the use of the "precautionary principle" as advocated by the German Government and other authorities throughout the European Union. In the context of environmental risks the principle calls for an extension of precautions for highly complex and uncertain risks. Operating in this framework authorities do not wait for scientific evidence of damage from a particular risk nor do they weigh costs against benefit in regulating the risk for business and the public. This rests on the principle that "prevention is better than a cure".

From systems failure research comes a slightly different argument in favour of anticipation, relating to the need for the development of knowledge about networks of causation (Shrivastava et al, 1988). Baldissera (1987), Pidgeon (1988), Toft and Reynolds (1994), Turner (1976, 1978, 1989, 1991), Turner et al (1989), and Turner and Pidgeon (1997) each discuss related aspects of the impact of latent failures (Reason, 1990). The research indicates that many disasters, as a result of inherent organisational cultural and structural factors, are simply "waiting to happen" (Richardson, 1993a). Worse still failure to act upon hindsight and the experiences of other organisations means that organisations seldom learn from previous mistakes. Horlick-Jones et al (1991) locate this "failure of hindsight" (Toft and Reynolds, 1994) in the context of the way in which organisations act within their environment. Poor organisational response to the environment is a frequent source of latent failures. The solution lies in frequent audits designed to identify failure networks and nodes through an organisational learning system (Toft and Reynolds, 1994).

From a somewhat different tangent (that of natural hazards research) Jones (1996: p. 30) makes well the case for anticipation:

> *"... natural scientists continue to be confident that anticipatory measures can considerably reduce losses and limit 'surprises' ... Hazards are inevitable: disasters are not."*

Yet, whilst "natural" systems may well be predictable, the problem of forecasting anthropocentric systems are legendary. Social scientists are slowly gaining ground in the development of analogous anticipatory measures, but few managers seem to want to hear about risks or bad news; it is simply not a strategic priority. For example, Peter Hayes, retired Deputy Chief Constable of South Yorkshire Police and an authority on crisis and contingency management cites this as a major issue that he faces in his consulting activities with large public and private sector organisations. His observation is simply that executive management seldom seem to take the issue of risk seriously until it is too late (*pers. comm.*).

Against the "anticipationists" stand those in favour of taking a more fatalistic approach. Typified by the work of Douglas and Wildavsky (1982a and 1982b), this school finds that latent failures are easy to identify only in hindsight. Their counter-argument rests on the notion that it is impossible to predict the behaviour of highly complex systems and so identify potential problems. This they argue is because of the complexity and stochastic behaviour of the systems. Recent developments in chaos theory (Gleick, 1987) seem to run counter to this argument: that in fact seemingly random behaviour does actually follow complex, but never the less predictable patterns. Rather than "waste time" forecasting and searching for faults, the resilienists argue for preparatory work the better to cope with disasters.

However, this is not as negative as perhaps it first seems. Rather, as Collinridge (1996) argues, it is an argument for building in flexibility and diversity, so promoting resilience to damage creating processes. This is about anticipation of events, since the choices made in organisational or process design must reflect beliefs and forecasts about potential hazards. As with most elements of the risk management dialectic, it is a question of balance, and in practice elements of both arguments are followed. However, recent legislation has tended to argue the benefits of preparing for the worst (Cullen, 1990).

Absolution: liability and blame

In the case of every major accident and the majority of minor ones, it is common policy for organisations, victims or their relatives to seek scapegoats (Smith and Sipika, 1993), be they innocent or guilty. The frequent use of the phrase "human error" in transport disasters is indicative of the perceived need for somebody to be at fault: somebody that relatives and investors can blame. In line with their chaotic approach to risk regulation, governments traditionally take up the public cry for retribution and offer up the "sacrificial" individual or group. This is what Jenkins (cited in Horlick-Jones, 1996: p. 61) terms the process of "ritual damnation" - a conjunction of the socio-psychological constructs of guilt and vindication, and the need for retribution and deterrence (Johnston, 1996: pp. 76-77).

Yet, rarely it seems is it an organisation that is faulted by ministers; more usually it is an individual (though seldom a member of the government). For, despite the upward spiral of vengeance that seems to epitomise modern Britain (Wells, 1996: p. 50), and the credence given to this perceived need by politicians of all parties (Bagehot, 1996), the successful prosecution of seemingly "obvious" cases of corporate manslaughter are rare indeed. As Wells (1996: p.54) notes the unsuccessful prosecution of P&O for manslaughter as a result of the capsizing of the *Herald of Free Enterprise* in 1987, was only the third such case brought in English legal history. More recently, a successful prosecution of "outward bounds" firm OLL Ltd. on the same charge was made; this following the deaths in the English Channel of four school children in a poorly organised canoeing trip in 1994.

This shows the key legal issue in blame: the definition of liability and its relation to corporate manslaughter is wide indeed and very much open to interpretation. Hence, it seems that the legal process is equally as unpredictable as the reaction of the public. As Wells (1996: p. 60) advises, this is a difficult set of contingent issues for managers to consider.

Despite these legal and societal issues, blame is very much a part of the traditional approach to risk management (Hood et al, 1992). Herein organisations commonly place strict legal and financial liability for a variety of risks upon those individuals and groups who are directly responsible for taking actions to prevent or minimise risks. Such organisations may argue that a "socially inefficient" system, where blame is not effectively pre-determined, may result in the occurrence of "avoidable" failures. This the "blamists" argue is because there is too little incentive for decision makers to be worried about hazards. Effective "criminalisation" of poor management practice and laying down sanctions against key decision makers is the way to prevent accidents. Such an approach may be implemented in terms of instant dismissal where an employee is found "guilty" of safety violations. Alternatively, blame may be placed on system designers rather than operators, since they are effectively responsible for faulty design.

Such approaches are grounded in the principles of "scientific management", the school of managerial thought embodied in the work of Frederick Winslow Taylor

(1947). Blaming and the criminalisation of individuals in a culture of strict liability resembles the cold rationality and bureaucratic approach embodied in "Taylorism". It conceives of individuals as a part of a perfect cybernetic system within which their performance can be improved through logical system design and simplistic economic incentives. As has been proven time and again this may work in very specific instances, but in general it promotes an atmosphere of fear, uncertainty and doubt in people: the very antithesis of much of current management thinking, which relies on empowerment and knowledge sharing.

The basic problem is that the blamist approach ignores one of the basic principles of human behaviour: that people do not do the same thing every time they perform the "same" task whatever the period between repeats. Hence, they cannot be regarded as a perfect part of a cybernetic system and consequently any socio-technical system will always operate in degraded mode (Weir, 1993; Weir, 1996: pp. 116-117). Also, more recent theories of motivation seem to indicate that financial reward or loss, whilst important to individuals, is not the only consideration people make in performing their duties. Horlick-Jones (1996: pp. 64-65) cites the "politics of blame" as further evidence of the dysfunctional nature that pervades organisations where scapegoating is the norm, since blame transference and avoidance seem to play major roles in advancement. Indeed in such organisations, failure is regarded as either overtly damaging or at the very least subversive (Weir, 1996: p. 117). Blame too acts as a major barrier to organisational learning, since protecting managers and designers is more likely to exacerbate problems than it is to identify true cause.

Those adherents to the "no-blame" view are extremely sceptical of the efficacy of attaching strict liability. Largely on the basis of criticisms similar to those levelled at Taylorism, the critics of the "blamists" argue that rather than produce an efficient system, strict liability is ineffective and counter-productive. Indeed by limiting their liability organisations effectively inhibit real changes in behaviour which might better prevent risks developing further. The blamist approach does produce a tendency for managers to "go by the book" rather than adopt the most appropriate actions in the circumstances. Quarantelli (1988) points towards the inadequacy of such an approach in dealing with disasters, whilst Richardson (1993b) points towards the need for creative skills and flexibility in those assigned to deal with crises. Such policies do not necessarily result in the reduction of overall risk. Rather than accept liability managers will move the activity or problem to an area or individual where criminalisation of risk is either minimal or non-existent.

Absolutionists have moved away from scapegoating preferring instead to promote an atmosphere of "openness" and learning in a move to manage complex socio-technical systems through more creative means (Horlick-Jones, 1996: p. 66). Such an environment, it is claimed, helps to reveal the full facts with little or no corruption of their original meaning. Where strict liability operates facts may be distorted or obscured by both actors and the accompanying quasi-legal or legal and consequently adversarial process. Clear information, the absolutionists claim, provides a strong basis for improving risk management and risk exposures. Attaching blame to or sanctioning individuals it seems is a recipe for instilling silence. Feedback is arrested (both qualitatively and quantitatively), silence ensues and learning falters (Johnston, 1996: p. 83). Hence, improvement of risk management techniques and exposures cannot occur. Good risk management relies on an active flow of information. "Beating" employees will not activate that flow, least of all at a time of crisis when cultural norms become distorted (Smallman and Weir, 1996; Mars et al, 1997).

The absolutionists call is for "calm stock taking" rather than media generated hysteria, the latter of which prevents pooling of vital information. Apportioning blame through hindsight reviews, heavily penalising individuals and simply blaming people will not deliver the results. In summary, the blamists look to limit risks through punitive sanctions on individuals, whilst the absolutionists expect to learn from mistakes. Yet there is the need for pragmatic balance. Blame-free cultures are constrained by contingencies in the business environment, and must acknowledge the very real socio-political pressures that demand a scapegoat. Hence, rather like the "zero defects" of the quality movement, "blame-free" remains an illusory, if laudable, aspiration.

Qualitative methods

> *" 'What is truth?' asked Pontius Pilate ... ... 'it ain't statistics', said a voice in the crowd."* Darrell Huff (1974: p. 110).

> *"If a guy tells me the probability of failure is 1 in 10^5, I know he's full of crap."* Richard Feynmann (1988)

Risk management is almost totally dominated by the "quantificationists". Their argument runs that systematic quantification of risk is the only method by which risk may be rationally analysed and measured against pre-determined objectives. Such supposed rationality, coupled to technically sophisticated risk assessment methods which parallel cost benefit analysis sits well with bureaucrats and legislators. The key point in favour of the application of numerical methods in risk assessment is that it provides a convenient tool for resource allocation. For example, much of Government safety policy and regulation is based around the use of quantitative risk assessment (QRA). The very technical sophistication of QRA also seems to offer a feeling of security, particularly where the assessment is automated and the final verdict delivered via a computer programme. Hence in socio-technical risk assessment and in the evaluation of natural hazards, QRA has no equal: it effectively dictates public and corporate policy and resultant resource allocation. With such vested interests its proponents strongly argue that there is no alternative (Hood and Jones, 1996: pp. 84-85).

The danger is, as always with numbers, that "analysis leads to paralysis", whereby the use of numbers is used as a means of abdicating management responsibility.

In other words the risk with QRA is that decision makers may "use it to do more than it can" (Foster, 1996: p.156). Furthermore, when probabilities are assigned to risks it is difficult for risk managers to be sure of properly communicating the meaning of something such as "10^{-9} failures per hour is an acceptable risk".

Grose (1990) highlights the shortcoming of using quantitative risk assessment:

- Complexity in human behaviour cannot be reduced to numbers.
- Numbers for the intricacies of real life will always be over-simplifying.
- Lack of applicable data will often force dishonest guessing.
- Decision makers may inadvertently substitute numbers for reasoned judgement.
- Risk can be too simplified and unrealistically traded off against benefit by relying on numbers.

There are those within this dominant camp who admit the need for careful consideration of other elements within the risk envelope, to go beyond the limits of the traditional approaches (Cohen, 1996: p. 87). Furthermore, as expressed by Cohen (1996: p. 91), there is the view that, as reliable and objective as QRAs are, they are only as good as the decision making process of which they are a part. They can be subject to misinterpretation and the judgement of their value belongs to the decision maker to whom they are offered up. The problem with this view is that the majority of decision makers who are confronted with expert opinion such as this, are often poorly qualified in the interpretation of the proffered assessment (nor are many managers properly trained in rational decision making (Bazerman, 1994: p.1)). Because of this they will turn to the experts for their professional advice. Given the professional background of such experts (Toft, 1996: pp. 99-100) it is hardly surprising that a narrow "technocentric" view dominates.

Underpinning all of the belief in probabilistic risk assessment as "right and proper", there lies the assumption that "... normal standards of organisation, management and inspection prevail" (Cohen, 1996: p. 91). However, the evidence presented in chapters one and two suggests that the prevailing environment predicates against any form of *normality*, and the variation in standards of management and organisation within this uncertain atmosphere conspire to create business and associated risks in the first instance.

At the other side of the debate are those who view QRA with unease, running against a massive weight of "professional" and governmental interests (typified by the work of Toft, 1996). Their main criticism relates to the "value laden and implicit" assumptions made in QRA (Hood et al, 1992). Also, building upon Grose's (1990) arguments against numerical risk assessment, they are genuinely sceptical of claims to be able to quantify risks with high degrees of accuracy.

Lave and Malès (1989) argue that risk regulation is simply too complex and overly prone to variation in too many values to be managed effectively within one framework. Cost- and risk-benefit analysis are primarily economic tools aimed at maintaining efficiency, that is keeping socio-technical system life cycle costs at an acceptable minimum. In other words the focus of QRA is on economic efficiency, a primary determinant of organisational and governmental policy. However, these techniques pay little deference to social equity. Neither do they account for public opinion nor do they actively reduce risk: they simply allow for resource allocation to limit the (largely) financial cost of a potential hazard. There is little allowance for human cost and it seems that life is too frequently regarded as an economic commodity (Foster, 1996: pp. 158-160).

In the *qualitative versus quantitative* debate, for holistic risk management, the weighting must fall towards qualitativism, since the very essence of holistic is that *all* things must be considered. Yet there must be room for the *intelligent* use of quantitative techniques (Cohen, 1996: p. 97). The use of numbers in predicting risks is well advanced, but the problem with many risks is that all too frequently a human element is involved. Time and again modelling human behaviour with numbers has been proven to be fraught with problems. Finally, modelling risks is often based on historical data and whilst such modelling is a valid and proven technique, it is too frequently used as a substitute for good management judgement. However, a change of approach is unlikely since there is a major cultural barrier to the development of a broader approach; as Toft (1996: p. 107) notes quantitative analysis is probably too deeply ensconced in the natural scientists, engineers and actuarial mind sets to be changed even a little. Radical and potentially more robust approaches are thus unlikely to capture the orthodoxy's attention.

Institutional design

Frequently, in the aftermath of large scale disasters and subsequent public enquiries, recommendations are made for the change in social and administrative systems in organisations. The question is whether or not it is feasible to design risk management into complex organisation structures, specifically social and socio-technical systems.

The debate is polarised between those who believe there is sufficient evidence to support the conjecture that organisation design directly affects risk attitudes and so organisational "crisis proneness" (Richardson, 1993a), and those who are sceptical of "design for risk management" (Hood and Jones, 1996: pp. 111-113).

Much of the case "for" is based on the notion of "safety cultures" (Toft, 1994) whereby, rather like quality, safety is "built-in" through the members of organisations, rather than "inspected-in". Such cultures are generally grounded in a safety review and audit philosophy (Hood and Jones, 1996: pp. 111-112). Those who have difficulty accepting these idea point to the wide variation in the implementation of the "safety culture" concept by multi-nationals. The detractors also dispute that such cultures are based on a systematic approach. The case seems to centre on the issue of communications and here there is a relationship to the issue of blame. It seems that the majority of both socio-technical and business failures can be attributed to some form of breakdown in information flow (Mars et al, 1997; Smallman and Weir, 1995; Weir,

1996: pp. 118-119 and 123-125). The best known examples include the losses of *Challenger* and the *Herald of Free Enterprise* and the failures of *Barlow Clowes*, *Lloyd's of London*, *Polly Peck* and *Barings*, amongst many others.

Hence, the argument against organisation design for risk is based largely on perceived limitations in human management and perception of risk (largely relating to the unpredictability of human behaviour), and in the theory base of "organisational design". The detractors seem largely to be drawn from the engineering community where design knowledge is much more certain. However, this poses somewhat of a dilemma for engineers as they are required to design systems that are increasingly complex and so rely upon sophisticated human machine interfaces. Furthermore, designing sufficiently robust institutions is not without difficulty. Penning-Rowsell (1996: p. 127) lays out the problems associated with designing those public institutions that are responsible for dealing with natural disasters. The competing pressures of offsetting public cries for blame and responsibility when organisations fail must be reconciled with the requirements of "value for money" from the holders of the public purse. The net result is a short-term orientation to hazard alleviation and the design of related institutions. Hence, the overall approach may be fundamentally flawed. Analogous flaws are clearly visible in the design of those institutions designed to afford protection to the public and commercial organisations from the effects of business failures. For example, the regulation of the UK financial services sector was demonstrated to be flawed on several notable occasions in the 1980s (Weir, 1996: p. 123 and pp. 130-131). At least a part of the issue seems to relate to the artificially created linguistic barriers that exist between the professions (engineers, accountants, natural scientists and actuaries) and decision makers. As has been frequently asserted, the professionals and their "masters" seem not to speak the same language (Penning-Rowsell, 1996: p 128; Weir, 1996: p. 119).

Whilst communication issues seem to dominate the debate on design, Penning-Rowsell (1996: pp. 128-132) also notes a number of serious barriers to developing reliable and robust organisations:

- There is the problem of scale (and of responsibility). When do the regulators intervene? When should local authorities cry for help in a natural disaster? How much power can regulators safely be given?
- What constitutes good policy in mitigation of business failure or socio-technical disasters? The balkanisation of risk management is a critical issue in the future of this broad discipline. Too many groups claim responsibility for the protection of others, and express the view that they are best qualified for this role. As a result what seems to happen is often that the least appropriate authority holds too much power and that less than perfect policy results.
- Developing the previous point further, there is the issue of performance targets. As Penning-Rowsell (1996: p. 129) notes "a further difficulty with institutional design is 'design for what'?". With so many interest groups the difficulty facing the regulator

(ignoring for one moment the issue of appropriate choice of authority) is where to set protective goals?

Complementarism: the cost of risk reduction

The traditional view of risk reduction is centred on the opportunity cost model of economics (Hood and Jones, 1996: pp. 141-143). The traditionalists, in seeking lower life cycle costs for systems holding significant risks, argue that safety improvements must be at the cost of other objectives (for example, profit, competitiveness, environmental degradation, declining service), and risk management is a cost to be borne. Hood and Jones (1996: p. 141) note this problem in the BATNEEC (best available technique not entailing excessive cost) principle of environmental "protection"; the trade off is recognised, but no indication is given of definitions of either "best available" or "excessive".

This is in essence the short-termist approach that currently dominates public policy. Foster (1996) justifies this approach in a wide ranging discussion of safety costs, noting the necessity for the use of economic analyses as both an adjunct to [quantitative] risk assessment and to the regulatory process. His argument is that rational decision making rests on balancing differential benefits and costs, but with clear reference to objective assessment of uncertainty and organisational attitudes to risk, along with risk itself. Critically, whilst he briefly acknowledges the issue of subjectivity in decision making (Foster, 1996: p. 157), Foster fails to acknowledge the impact of the "human factor". This is a fault commonly associated with economic analysis.

Less traditional is the view that good risk management practice can complement other goals. Here investment in proactive risk management is taken as a sign of good management practice and leads not only to an improved risk profile, but also to improved effectiveness in other areas. The argument is that improved risk strategy (through good practice based on thorough and well-planned training) is self-financing, since prevention is cheaper than redressing the impact of a crisis event. This unorthodox view is a long-term approach that views risk management, like training, as an investment rather than a cost.

Horlick-Jones (1996) aligns the issue of safety and risk with the principles of total quality management (TQM). This is based on the principle that safety failures or disasters are essentially quality defects. This is a position that has been adopted by the environmental movement in the form of *total quality environmental management* (Welford and Gouldson, 1993). The primary guiding principle of what might be termed "total risk management" is that it is based upon a "safety culture" approach to risk management (Horlick-Jones, 1996: pp. 150-152). Safety cultures are in essence the embodiment of an holistic approach to risk in which all employees of an organisation are immersed, and one where there is a requirement for continuous vigilance and improvement in safety performance (the multiple forces required in collibrational risk management). This is potentially a very useful and effective way forward. However, as Horlick-Jones (1996: p. 151) notes there are two main pitfalls in adopting this approach:

- The focus of total risk management is on reliability and performance improvements. However, this may detract from consideration of Reason's (1990) resident pathogens, the ultimate conjunction of which can lead to the initiation of latent failures and disasters.

- Total risk management requires up front investment. Annual return on this will be small and elements of the return will be intangible. Hence, investment in this approach is very much long-term, and its appeal may be limited to those organisations that are large enough to afford such programmes. Consequently, as has been the case with quality and with environmental management, small and medium sized enterprise may pass on the opportunity to invest in the development of safety cultures. Furthermore this is not solely an issue of resources; there is also the question of availability of expertise.

Participation

Policy makers and professional groups traditionally associated with risk management (for example, insurance, finance, engineers) see no need to involve other people or professions, for that matter, in their deliberations (Hood and Jones, 1996: p. 112). In point of fact they are frequently hostile to the inclusion of outsiders, countering calls for broader participation with claims that such breadth might dilute the effectiveness of risk strategy. The narrow participationists claim that risk management decisions are best made by a few well-informed decision makers acting on information given to them by scientifically trained professionals, capable of reaching consensus through the balance of experience. They point out that the interests of those affected by their decisions will be protected by regulators, legislators and the courts.

As Pidgeon (1996: pp. 165-167) highlights this represents a significant inversion of norms in the organisation of the greater society, since it effectively implies that democracy is, in the domain of risk assessment at least, subordinate to the technocracy. The ascendancy of the technocrats in the assessment of risk is compounded by their desire to reduce risks to purely a matter of "science". Hence, they choose to ignore the body politic, so disenfranchising those who must take the consequences of any miscalculation, despite the contra-indication that they should truly be in control. The supposition throughout is that the technocrats take a moral and sensitive view of risk, but one that is rigorously objective, secure from overt political influences. This supposition in turn rests on the assumption that because the decision makers are "experts" they can be trusted. However, trust in experts seems to be a declining commodity, despite its key role as a foundation of societal attitudes to risk (Giddens, 1990).

Counter to this is the belief that through extending the community involved in risk decisions, a more open decision making and greater scrutiny will lead to more well-informed debate and a lower error rate in decision making. There is also the question of increasing the accountability of the professionals and the policy makers (Hood and Jones, 1996: p. 161).

At least a part of the argument for broader participation rests on the perceived need for legitimacy of due process in the consideration of risks. However, as Pidgeon (1996: pp. 168-169) notes there are problems with the quest for greater legitimacy. First with breadth comes the dissolution of responsibility. Given the debate over blame and absolution considered earlier, this is somewhat of a difficult issue from both the legal and societal standpoints. In addition there is the issue of multiple "plural rationalities" in risk. As discussed in the following chapter, the psychology of risk perception is complex indeed. Hence perhaps the key difficulties in broadening participation are how to effectively combine "competing" rationalities and who should make the decision on appropriate combinations? In such decisions the implacable opponents seldom see the opportunity for a "meeting of minds", and, on the evidence of recent UK public enquiries in road developments, there seems to be little room for meaningful dialogue between experts and the public.

At the root of all of this lies what Pidgeon (1996: p. 169) terms the "substantive argument": that concealed errors and assumptions may be uncovered by the involvement of greater numbers of stakeholders. Certainly theoretical and hearsay evidence seems to suggest that this is the case, and that a breadth of knowledge, experience and skills critically applied will improve the quality of decisions and implementation. Furthermore, active learning seems to be a direct consequence of the involvement of broader groups. Since risk is a social construct, this is somewhat of a mute point, but one that is well made none the less.

However, as is the case in all information based endeavours, the greater the number of signals entering the system, the greater the potential for noise (and exponentially so). As the picture of the risk phenomenon under discussion fills out, so it becomes increasingly complex and decreasingly clear.

At a more fundamental level, Funtowicz and Ravetz (1996) argue that broader involvement in risk decision making is a direct product of scientific advance. Implicitly aligning themselves with the work of Beck (1992) they see the "new science" as something that is outwith the control of its progenitors: a phenomenon that can only be controlled by harnessing the power of the educated masses. In short they see the expansion of the risk management community as a natural part of the process of democratisation. As such their argument rests on the basis that the historic roots of regulation, as the means of protecting a relatively uneducated society, have now been superseded by the "need to know" and the desire for involvement. This it seems is a refection of the rise to ascendancy of Drucker's (1995) and Murray's (1995b) "cognitive elite". However, developing this argument further in the context of societal change, in broadening the risk assessment community in this manner, there is a danger of the focus shifting away from the greater society towards an emphasis on the perceived threats to the elite and at the expense of the "underclass". Indeed the argument for broader participation may represent a regression towards more feudal times, when the relatively rich elements of society dictated public policy at the expense of the poor.

As Funtowicz and Ravetz (1996: p. 181) conclude, there is the need for control and for balance, and the democratisation of risk management is required in order to meet the extension of science. However, all-encompassing risk enquiries will prove to be expensive, time-consuming and largely fruitless, since many of the plural rationalities involved may ultimately be irreconcilable.

Specification and regulation

During the 1980s and early to mid-1990s there existed a broad trend in UK government policy that placed the onus for regulation on industries (Kenney, 1991). Legislation in health and safety, the environment and also in financial services adopted the approach of attempting to establish higher level objectives or performance standards (the specification of institutional processes) as opposed to detailing a series of compliance measures or methods (the specification of physical products or structures).

The key features of these "objective" regulations are that they provide greater flexibility for compliance and are less prone to obsolescence in technology or (in the case of service industries particularly) business philosophy. However, the chief problem for regulated groups relates to the difficulties they may face in demonstrating to regulators that their solution complies with both the letter *and* intent of high level objectives. In other words the burden of proof for demonstrating a "safe plant" or an "environmentally acceptable practice" rests entirely on the shoulders of the regulated industry.

In the past there has been heavy reliance upon prescriptive legislation which places requirements upon operators that they are duty bound to meet to prevent prosecution. Its value is that it provides a set of fundamental principles and standards that the engineering community must follow and which society can expect. However, the strength of prescriptive regulation is also its main weakness; it imposes artificial barriers which effectively delineate the "safe" boundaries at a given point in time. This means that organisations can claim to have done enough to meet legal requirements and that they can "stop" at that point. In other words they impose no moral duty or business stimulus to go beyond what the law requires.

Risk management: time for a paradigm shift?

The environment that both business and wider society face today is markedly different from the past and there is little doubt that it will continue to change rapidly in the future. Such change not only induces problems for organisations by its very rate and scale, but also in terms of the uncertainty that it brings to bear. Modern managers then are confronted with an increasing variety of political, economic, social, technological and environmental factors, coupled to increasing competitive pressures, that they must appreciate and adapt to in order to survive and thrive.

The homeostatic risk management paradigm is demonstrably not suited to this context and there is an established case for accelerated evolution to the collibrational approach.

REFERENCES

Adams, J. (1995) *Risk.* London: UCL Press.

Bagehot (1996) Jack Straw's balancing act. *The Economist,* May 24, p. 37.

Baldissera, A. (1987) Some organisational determinants of technological accidents. *Quaderni di Sciologica,* **33,** (8), 49-73.

Barratt Brown, M. (1995) *Models in Political Economy. A Guide to the Arguments.* Second Edition. Harmondsworth: Penguin Books.

Bazerman, M.H. (1994) *Judgment in Managerial Decision Making.* Third edition. New York: John Wiley and Sons.

Beck, U. (1992) *Risk Society. Towards a New Modernity.* London: Sage.

Bella, D.A. (1987) Organisations and systematic distortion of information. *Journal of Professional Issues in Engineering,* **113,** (4), 360-370.

Brundtland, G.H. (1994) What is world prosperity? *Business Strategy Review,* **5,** (2), Summer, 57-69.

Cohen, A.V. (1996) Quantitative risk assessment and decisions about risk. An essential input into the decision process. In C. Hood and D.K.C. Jones (eds.) (1996) *Accident and Design. Contemporary Debates in Risk Management.* London: UCL Press. pp. 87-98.

Collinridge, D. (1996) Resilience, flexibility and diversity in managing the risks of technologies In C. Hood and D.K.C. Jones (eds.) (1996) *Accident and Design. Contemporary Debates in Risk Management.* London: UCL Press. pp. 40-45.

Cullen, The Hon. Lord. (1990) *The Public Enquiry into the Piper Alpha Disaster.* London: HMSO.

Douglas, M. and Wildavsky, A. (1982a) *Risk and Culture. An Essay on the Selection of Technological and Environmental Dangers.* London: University of California Press.

Douglas, M. and Wildavsky, A. (1982b) How can we know the risks we face? Why risk selection is a social process. *Risk Analysis,* **2,** (2), 49-51.

Drucker, P.F. (1980) *Managing in Turbulent Times.* Oxford: Butterworth-Heinemann.

Drucker, P.F. (1995) *Managing in a Time of Great Change.* Oxford: Butterworth-Heinemann.

Dunsire, A. 1978. *Control in a Bureaucracy: the Execution Process.* Volume 2. Oxford: Martin Robinson.

Dunsire, A. 1986. A cybernetic view of guidance, control and evaluation in the public sector? In F-X. Kaufman, G. Majone, V. Ostrom (Eds.) *Guidance, Control and Evaluation in the Public Sector:* 327-346. Berlin: de Gruyter.

Dunsire, A. 1990. Holistic Governance. *Public Policy and Administration,* **5,** (1): 4-19.

Economist Intelligence Unit and Arthur Andersen (1995) *Managing Business Risks: An Integrated Approach.* London: The Economist Intelligence Unit.

Feynmann, R. (1988) *What Do You Care What Other People Think?* London: Unwin Hyman.

Foster, C. (1996) Risk management: an economists approach. In C. Hood and D.K.C. Jones (eds.) (1996) *Accident and Design. Contemporary Debates in Risk Management.* London: UCL Press. pp. 155-160.

Funtowicz, S.O. and Ravetz, J.R. (1996) Risk management, post-normal science and extended peer communities. In C. Hood and D.K.C. Jones (eds.) (1996) *Accident and Design. Contemporary Debates in Risk Management.* London: UCL Press. pp. 172-181.

Giddens, A. (1990) *The Consequences of Modernity.* Cambridge: Polity Press.

Gleick, J. (1987) *Chaos. Making a New Science.* London: Abacus.

Glendon, I., Hoyes, T.W., Haigney, D.E. and Taylor, R.G. (1996) A review of risk homeostasis theory in simulated environments. *Safety Science,* **22,** (1-3), pp. 15-26.

Grose, V.L. (1990) *Managing Risk: Systematic Loss Prevention for Executives.* Second edition. Englewood Cliffs, N.J.: Prentice-Hall Inc.

Haimes, Y. (1991) Total risk management. *Risk Analysis,* June.

Hood, C.C., Jones, D.K.C., Pidgeon, N.F., Turner, B.A., Gibson, R., Bevan-Davies, C., Funtowicz, S.O., Horlick-Jones, T. McDermaid, J.A., Penning-Rowsell, E.C. Ravetz, J.R., Sime, J.D. and Wells, C. (1992) Risk Management. In Royal Society. (1992) *Risk: Analysis, Perception and Management. Report of a Royal Society Study Group* London: The Royal Society. Chapter 6, pp. 135-192.

Hood C.C. (1996) Where extremes meet: "SPRAT" versus "SHARK" in public risk management. In C. Hood and D.K.C. Jones (eds.) (1996) *Accident and Design. Contemporary Debates in Risk Management.* London: UCL Press. pp. 208-228.

Hood, C.C. and Jones, D.K.C. (eds.) (1996) *Accident and Design. Contemporary Debates in Risk Management.* London: UCL Press.

Horlick-Jones, T. (1996) The problem of blame In C. Hood and D.K.C. Jones (eds.) (1996) *Accident and Design. Contemporary Debates in Risk Management.* London: UCL Press. pp. 61-71.

Horlick-Jones, T., Fortune, J. and Peters, G. (1991) Measuring disaster trends part two: statistics and underlying processes. *Disaster Management,* **4,** (1), 41-48.

Huff, D. (1974) *How to Lie with Statistics.* Harmondsworth: Penguin Books.

Hutton, W. (1995) *The State We're In.* London: Random House.

Independent Bureau on International Development Issues (1983) *Common Crisis. North-South: Co-operation for World Recovery.* London: Pan Books.

Independent Commission on International Development Issues (1980) *North-South: A Programme for Survival.* London: Pan Books.

Johnston, A. N. (1996) Blame punishment and risk management In C. Hood and D.K.C. Jones (eds.) (1996) *Accident and Design. Contemporary Debates in Risk Management.* London: UCL Press. pp.72-83.

Jones, D.K.C. (1996) Anticipating the risks posed by natural perils In C. Hood and D.K.C. Jones (eds.) (1996) *Accident and Design. Contemporary Debates in Risk Management.* London: UCL Press. pp. 14-30.

Kenney, G.D. (1991) The double-edged sword of objective regulations. *Proceedings of the Institution of Mechanical Engineers, Safety developments in the Offshore Oil and Gas Industry,* Mechanical Engineering Publications Limited, 125-128.

Kleiner, A. and Roth, G. (1997) How to make experience your company's best teacher. *Harvard Business Review,* September-October, 172-177.

Lave, L.B. and Malès, E.H. (1989) At risk: the framework for regulating toxic substances. *Environmental Science and Technology,* **23,** 386-391.

Marr, A. (1995) *Ruling Brittania. The Failure and Future of British Democracy.* London: Michael Joseph.

Mars, G., Smallman, C. & Weir, D.T.H. (1997) A cultural theory explanation of organisational failure in crises. Presented at *BAM97 - The British Academy of Management Conference,* Queen Elizabeth II Conference Centre, London, September.

Meadows, D.H., Meadows, D.L., Randers, J. and Behrens, W.W. (1972) *The Limits to Growth. A Report for the Club of Rome's Project on the Predicament of Mankind.* London: Pan Books.

Murray, C. (1990) *The Emerging British Underclass.* Choice in Welfare No. 2. London: Institute of Economic Affairs.

Murray, C. (1995a) *Underclass: The Crisis Deepens.* Choice in Welfare No. 20. London: Institute of Economic Affairs.

Murray, C. (1995b) The partial restoration of traditional society. *The Public Interest,* No. 121, Fall, 122-134.

OECD (1996) *The Knowledge-Based Economy.* Paris: Organisation for Economic Co-operation and Development.

Penning-Rowsell, E. (1996) Criteria for the design of hazard mitgatiion institutions. In C. Hood and D.K.C. Jones (eds.) (1996) *Accident and Design. Contemporary Debates in Risk Management.* London: UCL Press. pp. 127-140.

Pidgeon, N.F. (1988) Risk assessment and accident analysis. *Acta Psychologica,* **68,** 355-368.

Pidgeon, N.F. (1996) Technocracy, democracy, secrecy and error. In C. Hood and D.K.C. Jones (eds.) (1996) *Accident and Design. Contemporary Debates in Risk Management.* London: UCL Press. pp. 161-171.

Quarantelli, E.R. (1988) Disaster crisis management: A summary of research findings. *Journal of Management Studies,* **25,** (4), July, 373-385.

Reason, J. (1990) *Human Error.* Cambridge: Cambridge University Press.

Reason, J., Parker, D., Lawton, R. and Pollock, C. (1995) Organisational controls and the varieties of rule related behaviour. ESRC Conference on *Risk in Organisational Settings,* London, 16-17 May.

Richardson, B. (1993a) Why we probably will not save mankind: A "natural" configuration of crisis-proneness. *Disaster Prevention and Management*, **2**, (4), 32-59..

Richardson, B. (1993b) Why we need to teach crisis management and to use case studies to do it. *Management Education and Development*, **24**, (2), 138-148..

Sampson, A. (1995) *Company Man. The Rise and Fall of Coporate Life.* London: Harper Collins.

Sasakura, K. (1995) Political economic chaos? *Journal of Economic Behaviour and Organisation*, **27**, 213-221.

Schwarz, M and Thompson, M. (1990) *Divided We Stand: Redefining Politics, Technology and Social Choice.* Hemel Hempstead: Harvester Wheatsheaf.

Shrivastava, P., Mitroff, I.I., Miller, D. and Miglani, A. (1988) Understanding industrial crises. *Journal of Management Studies*, **25**, 4, 283-303.

Smallman, C. 1996. Challenging the orthodoxy in risk management. *Safety Science*, **22**, (1-3): 245-262.

Smallman, C. (1997) The strategic implications of uncertainty. Unpublished PhD thesis, University of Bradford Management Centre.

Smallman, C. and Weir, D.T.H. (1995) Culture and communications: countering conspiracies in organisational risk management. In L. Barton (ed.) *New Avenues in Risk and Crisis Management.* Volume IV. Las Vegas, Nvada: Board of Regents, University of Nevada, Las Vegas.

Smallman, C. and Weir, D.T.H. (1996) Risk communication and cultural distortion during crises. In G. Mars (ed.) Industrial Relations and Risk, London: Dartmouth. In preparation

Smith, D. and Sipika, C. (1993) Back from the brink - Post-crisis management. *Long Range Planning*, Vol. 26, No. 1, 28-38.

Stacey, R.D. (1989) Emerging strategies for a chaotic environment. *Long Range Planning*, **29**, (2).

Tait, E.J. and Levidov, L. (1992) Proactive and reactive approaches to risk regulation: the case of biotechnology. *Futures*, April, 219-231.

Taylor, F.W. (1947) *Scientific Management.* London: Harper and Row, 39-73. Reprinted in D.S. Pugh (ed.) (1990) *Organisation Theory. Selected Readings.* Third edition. Harmondsworth: Penguin Books, Chapter 11, 203-222.

Thurow, L.C. (1996) *The Future of Capitalism. How Today's Economic Forces Shape Tomorrow's World.* London: Nicholas Brealey Publishing.

Toft, B. (1996) Limits to the mathematical modelling of disasters. In C. Hood and D.K.C. Jones (eds.) (1996) *Accident and Design. Contemporary Debates in Risk Management.* London: UCL Press. pp. 99-110.

Toft, B. and Reynolds, S. (1994) *Learning from Disasters. A Management Approach.* Oxford: Butterworth Heinemann.

Turner, B.A. (1976) The organisational and interorganisational development of disasters. *Administrative Science Quarterly*, **21**, 378-397.

Turner, B.A. (1978) *Man-Made Disasters.* London: Wykeham Publications.

Turner, B.A. (1989) How can we design a safe organisation? *Second International Conference on Industrial and Organisational Crisis Management*, New York.

Turner, B.A. (1991) The development of a safety culture. *Chemistry and Industry*, April, 241-243.

Turner, B.A., Pidgeon, N.F., Blockley, D.J. and Toft, B. (1989) Safety culture: its importance in future risk management. *The Second World Bank Workshop on Safety Control and Risk Management*, Karlstad, Sweden.

Turner, B.A. and Pidgeon, N. (1997) *Man-Made Disasters.* Second edition. London: Butterworth Heinneman.

Weir, D.T.H. (1993) Communication factors in system failure or why big planes crash and big businesses fail. *Disaster Prevention and Management*, **2**, (2), 41-50.

Weir, D.T.H. (1996) Risk and disaster: the role of communications breakdown in plane crashes and business failures. In C. Hood and D.K.C. Jones (eds.) (1996) *Accident and Design. Contemporary Debates in Risk Management.* London: UCL Press. pp. 114-126.

Welford, R.J. and Gouldson, A. (1993) *Environmental Management and Business Strategy.* London: Pitman.

Wells, C. (1996) Criminal law, blame and risk: corporate manslaughter In C. Hood and D.K.C. Jones (eds.) (1996) *Accident and Design. Contemporary Debates in Risk Management*. London: UCL Press. pp. 50-60.

Wilde, G.J.S. (1982) The theory of risk homestasis: implications for safety and health. *Risk Analysis*, **2**, pp. 209-225.

Woodall, (1996) A survey of the world economy. *The Economist*, 28 September.

World Commission on Environment and Development (1987) *Our Common Future*. Oxford: Oxford University Press.

INCENTIVES FOR LOSS PREVENTION INSTEAD OF DISASTER MANAGEMENT BY THE STATE IN THE CASE OF CATASTROPHIC RISKS

WALTER R. STAHEL
The Geneva Association
18 chemin Rieu
CH-1208, Geneva
Switzerland

1.The changing objectives of risk management over the centuries

It has been proven by François Ewald that we have gradually developed into an "insurance" society over the centuries (Ewald, 1989):

- The 19th century was greatly influenced by the ideas of social security in the work place;
- The 20th century is characterised by the development of risk-management, which, being based on the theories of risk-minimisation and loss prevention, has been strongly influenced by areas of technology, followed by investment and finally the appraisal of technological consequences;
- Ewald established the theory that the 21st century would be greatly influenced by the use of the "precautionary principle"

2. The "Risk" market

There are a number of methodologies, which can be used to structure the "risk" market. This paper is, however, based on a division common in the insurance field (Frey, 1992), which determines three main areas, namely risk assessment, risk control and risk financing.

- Risk Assessment primarily differentiates between risk perception, risk identification and risk evaluation.
- In the more specific field of Risk Control the options of ignorance, avoidance and minimisation are taken in to account.
- Risk Financing allows for the options of bearing, sharing or transferring the risk along with insurance.

One should take into consideration that risk assessment is not determined by a single "risk" action alone, but at least as much so by anticipated profit. This profit is in some cases only recognised with hindsight, e.g. Newton and the apple; or it may be consciously adopted, e.g. big game hunters or casinos. Under some circumstances in

E. Coles, D. Smith and S. Tombs (eds.), Risk Management and Society, 81–100.

the second case, large "risks" can be consciously undertaken with a certain degree of control on behalf of the person or body carrying the risks.

Due to increasing integration between society and technology, as well as possible interaction at an individual, group and national level, both society and individuals themselves are taking more and more risks, resulting in the beneficiary no longer being the major risk-bearer, and thus eliminating any personal advantages from risk-profit-assessment. (see Figures 1&2)

Figure 1:(Private) Automobile Drivers Responsible for Accidents according to Age and Length of Time in Possession of a Driving Licence.

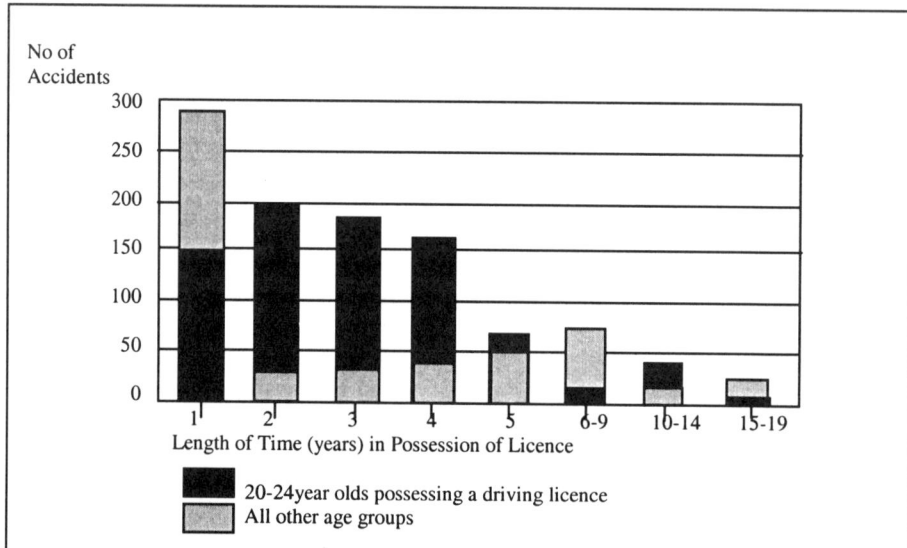

The column for 20-24 year olds possessing a driving licence shows the most significant results. The columns to the right of this one become gradually smaller until indicating a modest number of accidents as the length of time in possession of a licence increases.

Source: After Paul Gross Schlechte Verkehrsdisziplin ais'Zeiterscheinung'? NZZ vom 9.3.1989, S.103

Figure 2: Shift in Weighting of War Victims from the Armed Forces to Civilians (1st World War Onwards).

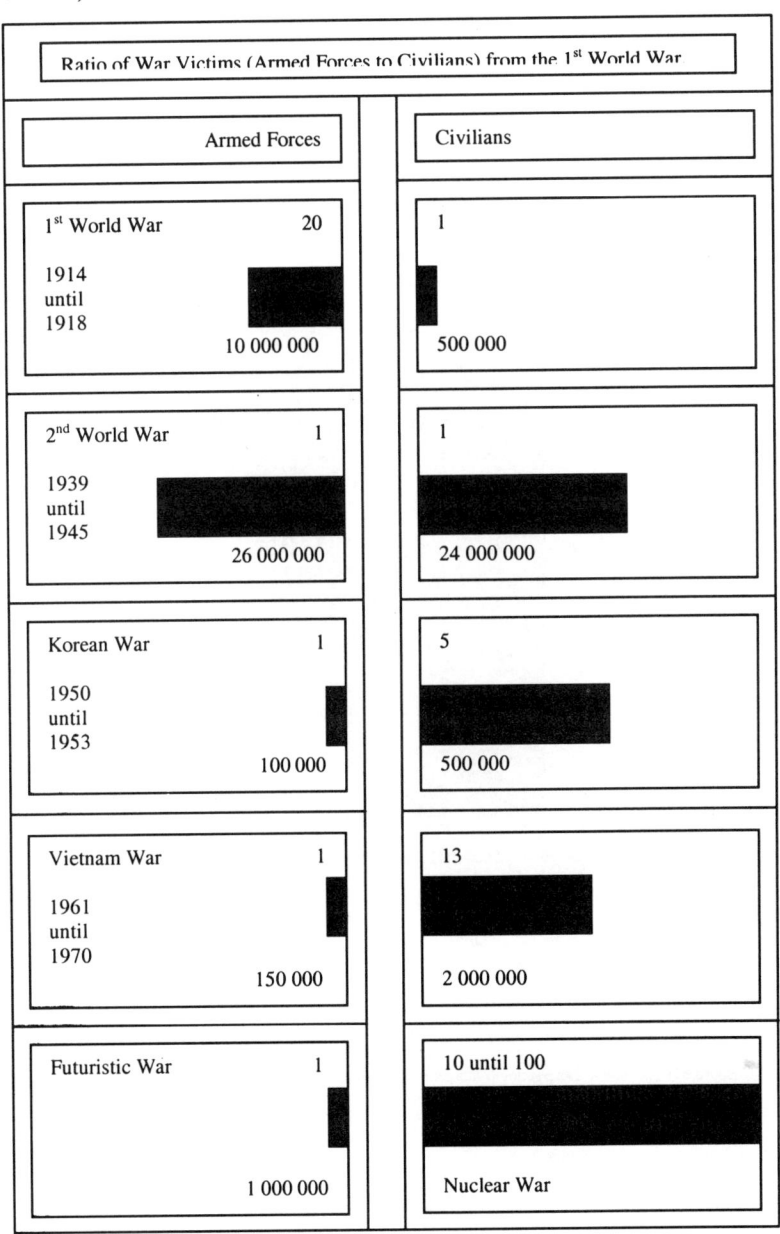

Source: After Schweizerische Ärztezeitung

An analysis of risk-assessment within the "Emergency Management Agencies" in the USA has proved that those with positions entrusted with preventing catastrophes have a view on risks, not directly apparent from any statistics, but based on the experience of involved parties: (see Figure 3)

Figure 3

The 10 Hazards Perceived as Major Concerns by Local Emergency Management Organisations 1985

(3107 jurisdictions were included in the survey sample)

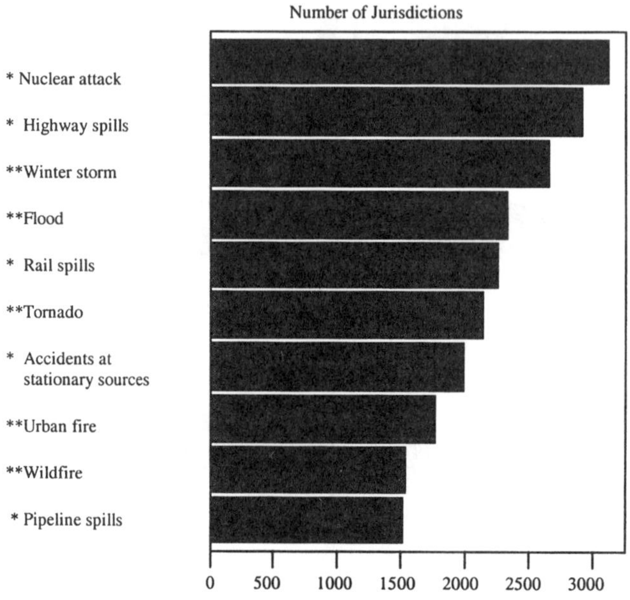

Number of Jurisdictions

* Nuclear attack

* Highway spills

**Winter storm

**Flood

* Rail spills

**Tornado

* Accidents at
   stationary sources

**Urban fire

**Wildfire

* Pipeline spills

0   500   1000 1500 2000 2500 3000

*   Concerns related to Hazardous materials
**   Other types of concerns

Source: After Federal Emergency Management Agency, Washington DC 1993

Here the transport of hazardous materials by road and rail are estimated to be as great a threat as winter storms, floods and tornadoes, ranking second as a major concern are accidents due to hazardous substances in stationary installations (factory and treatment plants) and urban fires. Last on the ten major hazard concerns are spills from the pipeline-transportation of hazardous substances and forest fires. (For a relative risk assessment of storms between Europe and America see Figure.5.)

3. Natural catastrophes

The position of villages, avalanches and churches can be read from the position of the "triangular avalanche protection forests" shown on pictures from the Swiss Alps. It is apparent from these pictures that not every avalanche is actually a *natural* catastrophe but only becomes so after a clash with mankind and its technology. Thus risk control regulations always apply to a particular policy objective.

Figure 4

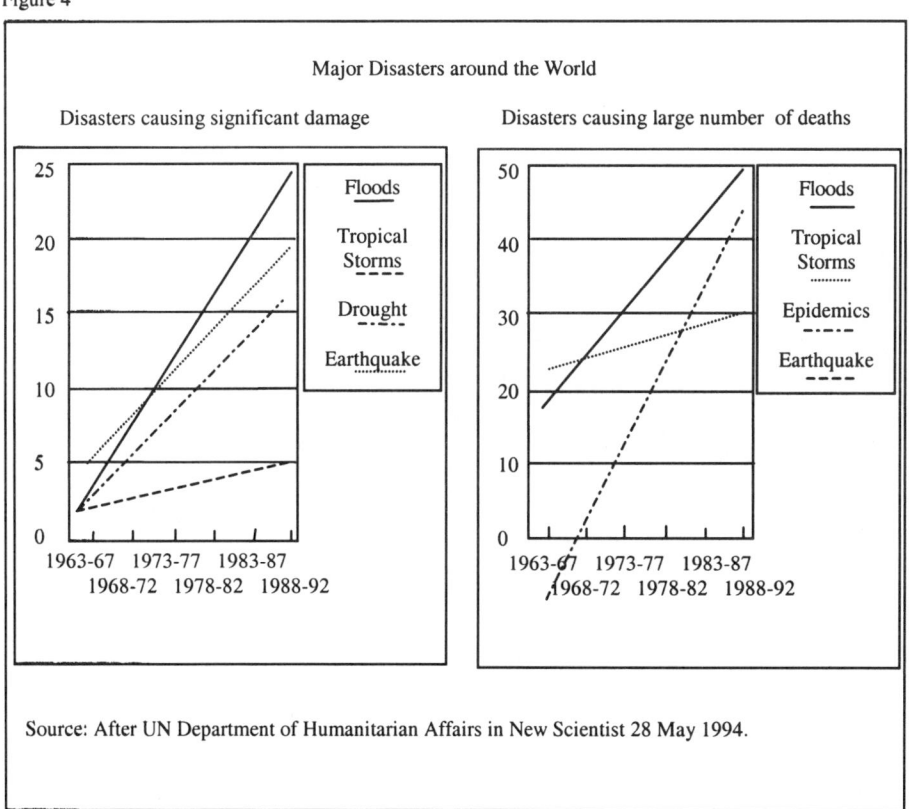

It is a similar story with regard to floods. For example: on the 1st February 1953 a breach in a dike on the North Sea resulted in one of the worst natural catastrophes of our time, flooding large parts of the Netherlands and drowning 2,000 people and 250,000 cattle. Using this example one can determine what is actually understood by the term "natural catastrophe", i.e. a fluctuation in the balance between mankind and nature to the disadvantage of mankind.

Figure5: Breakdown of World Wide Damage due to Natural Catastrophes according to various Criteria for 1993

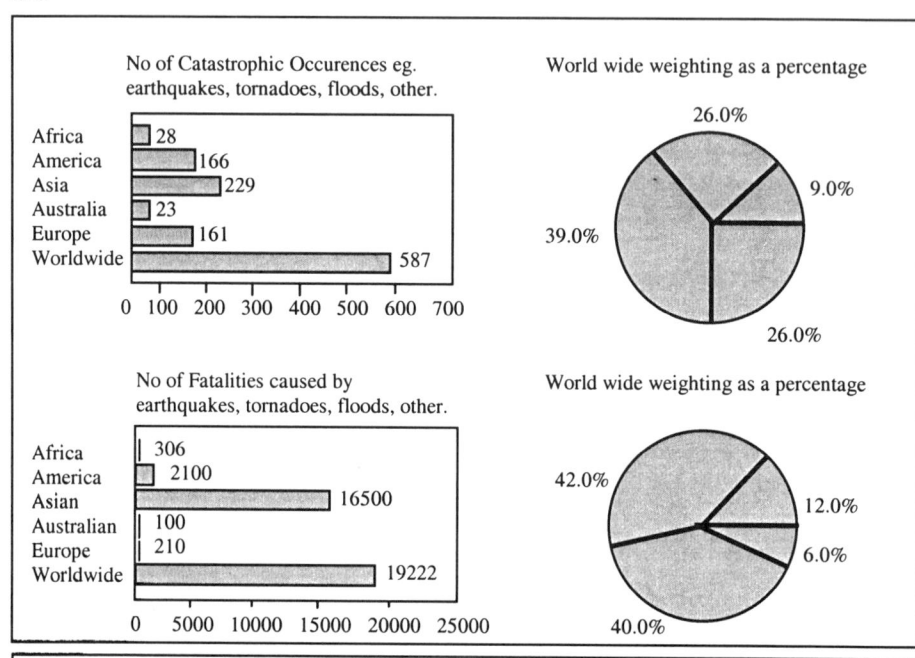

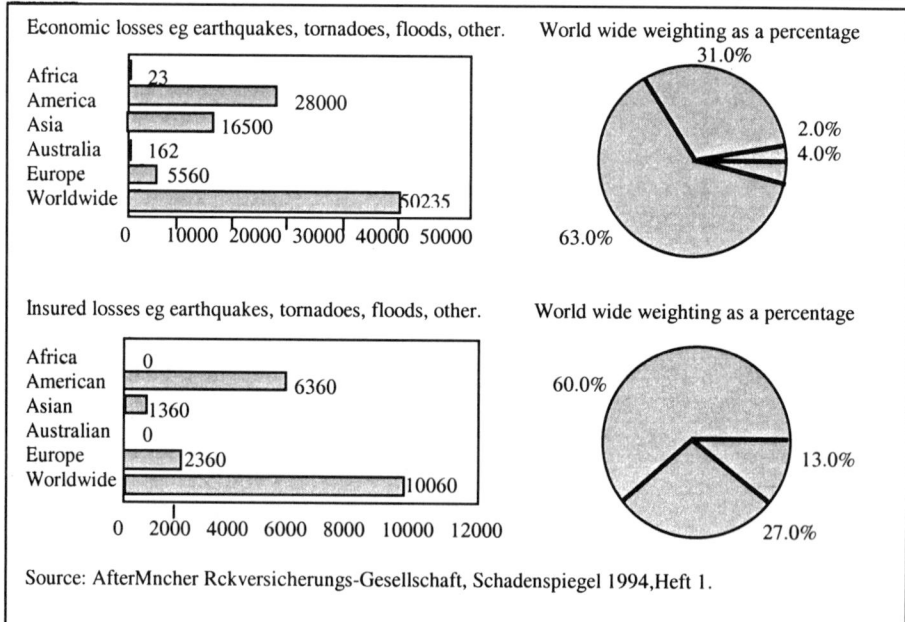

Should the ever pressing issues of the "green-house effect" or the "hole in the ozone layer", both of which could permanently change nature and the environment, be classified in the same way as a flood, a process whereby the sea reconquers her inherent land? (This classical example can already be found in Goethe's "Faust II"! Binswalter, 1994).

A possible change in the climate is a catastrophe largely resulting from mankind's modern industrial activities, which could lead to a breakdown of the whole planet (nature and mankind). Purely natural catastrophes, like for example volcanic eruptions or meteorite showers, have however in the past equally resulted in permanent changes to the atmosphere equating to negative effects for both the climate and living organisms.

Perhaps it does not make sense any longer to differentiate between "man-made" and "natural" risks, as it has been statistically recorded that the majority of natural catastrophes have some kind of disastrous effects on technology and vice-versa (combined natural/technological events) (Showalter et al, 1994).

A further reason as to why differentiation between the kinds of catastrophes could be useful becomes clear from the analysis of the major catastrophes around the world (UN Department of Humanitarian Affairs, 1994).

- Statistics show the number of disasters with regard to the most damage caused in the following order: floods, tropical storms, droughts and earthquakes.
- Statistics show disasters which have taken the greatest number of lives in the order of: droughts, floods, epidemics, tropical storms and earthquakes, (the diagram does not show lives taken through drought, however these are about 25% more than those caused by floods).

It is apparent from this that the criteria for damage greatly influence the statistics. Figure 5 shows a break-down of world-wide damage due to natural catastrophes which varies according to criteria and areas for the year 1993 (Schadenspiegel, 1994). The various weightings between the number of disaster occurrences with losses, the number of fatalities and likewise the economical and likewise insured losses depend upon the complexity of the insurance (parallel to the extent of industrialisation) of areas (very high in Europe and America and low in Africa and Asia).

The long-term trend in the number of natural catastrophes is on the increase (Figure 6). The sum of economic losses, as well as insured losses, is significantly rising (Figure 6). But the development of damage is strongly related to a number of factors in economical development, such as (Nutter, 1994):

- The colonisation of endangered areas, above all of coastal areas, brought about by an increase in the total population in poorer countries as well as a migration of families with high incomes to more scenic areas;
- An increase in insured property values related to the above in coastal regions (number of residences and computers, cars etc. per residence);
- A rapid increase in repair costs;
- An escalated flouting of the building and planning permission regulations (30% of damage would be avoided by abidance of such regulations).

Figure 6: Major Natural Catastrophes

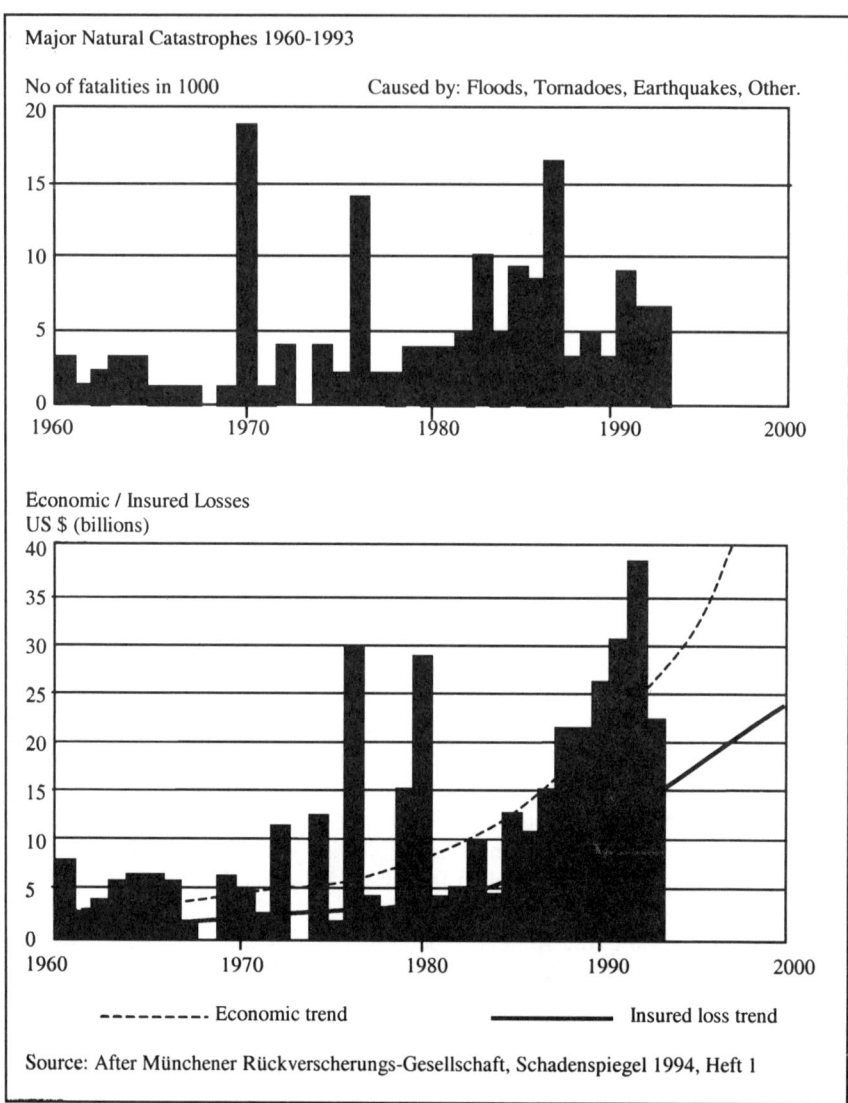

Major Natural Catastrophes 1960-1993

No of fatalities in 1000   Caused by: Floods, Tornadoes, Earthquakes, Other.

Economic / Insured Losses
US $ (billions)

- - - - - - - - - Economic trend   ————————— Insured loss trend

Source: After Münchener Rückverscherungs-Gesellschaft, Schadenspiegel 1994, Heft 1

4. Technical risk management – loss prevention – precautionary principle

THE PRICE OF LOSS PREVENTION

Even with technical risk management, at the beginning one always has to weigh up between risk and profit, and likewise between risks and the costs of risk-minimisation. In this way the near collapse of a bridge on the 'Gotthard-Autoban', which buckled after a storm had swept away one of its pillars, could have been avoided by an increase in the initial building expenses by 1%, either by reinforcing the concrete slab, or by setting the pillar's foundation five metres deeper on solid rock. In cases such as the 'Cypress Freeway' in San Francisco, a multi-storey motorway in an area renowned for its earthquakes, which had been built by flouting the American standards on earthquake resistant constructions, the risk assessment process is replaced by speculation (The motorway collapsed during the major tremor in the 90's, burying a number of vehicles and their passengers).

Table 1: Costs of Damage Prevention Compared to Costs Accrued from Damages.

YEAR	ACCIDENT	REPAIR COSTS	PREVENTION COSTS
1976	Exploding Reactor Seveso	US $ 150 Mio	< US $ 10,000
1981	Collapse Hyatt Regency Hotel Kansas City	US $ 90 Mio	< US $ 1,000
1984	Union Carbide Incident Bhopal	>US $ 200 Mio	<US $ 50,000
1986	Schweizerhalle Fire	US $ 60 Mio	<US $ 100,000
1987	Pont autoroute a Wassen (Un)	150% de Couts Constr.	1% de Couts Constr.

Source: After Zürich Insurance Company: Catastrophe Losses – a Problem for Insurers Only?
Interlaken Symposium 1987.

Table 1 shows the relationship between costs incurred from catastrophes (i.e. insured losses as opposed to economical losses) and the expenditure which would have been necessary for their prevention, drawing reference from a series of well-known catastrophes from previous years. Prevention is basically a question of planning ahead (or proactive management) as opposed to incurring costs end of pipe; or as *John Kletz* put it: "Risk-management is not an additional coat of paint" (*John Kletz* on the 16th January 1989 at the ETH (Swiss Federal Institute of Technology) of Zurich: "A life-cycle engineering approach") (Bernold, 1990). Taking heed from the wisdom of risk-management would in many cases bring about a slower and more sustainable way of managing the economy, e.g. the development of adaptable long-life systems built on a modular basis which can be permanently optimised in the draft and design stages (Figure 7)

WALTER R. STAHEL

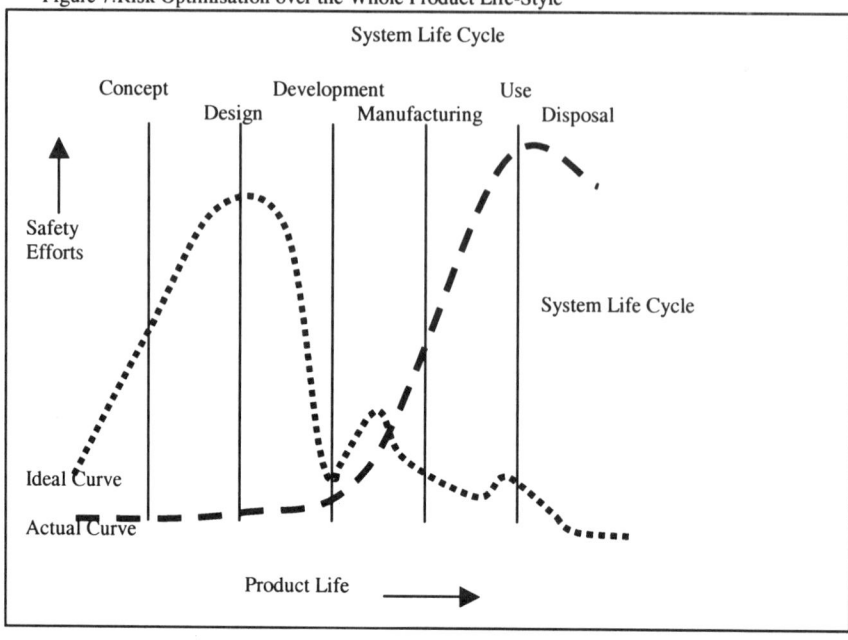

Figure 7:Risk Optimisation over the Whole Product Life-Style

Source: After H.E.Roland / B. Moriarty: System Safety Engineering and Management, J. Wiley & Sons, 1983

The technological risk in itself is principally determined by the relation 'mankind-technology', as well as by technological developments. With regard to the issue *Men and Technology:* in many cases, the additional safety arising from the technological progress of 'engineering-mankind' through the progress-seeking 'user-mankind' is often squashed or even reversed. The 'Titanic', the first 'unsinkable' ship to be built sank on its maiden voyage across the Atlantic due to the fact that the captain really believed it to be unsinkable. The founder had indeed announced this and although it had been very well received by the ears of the public, he had not literally meant it: the 'Titanic' was unsinkable under the supposition that the user (i.e. the captain) would behave as before. (According to the latest technological research, "cost-consciousness" on behalf of the owner was at least as much to blame as the ice-berg itself: the cheap sheet iron used for the hull is extremely recalcitrant at zero temperatures, although people were not aware of this fact at the time. Two sister ships of the 'Titanic' provided long years of service despite their sheet iron hulls).

Modern examples of the "Titanic"-syndrome include "ABS" (anti-skid system) and 'Four-Wheel Drive': these inventions give rise to new 'use-misuse-incentives', comparable with the 'Titanic'. When an advertisement tells the average consumer that a car runs 'as if it were on rails', then he or she will probably buy it in preference to other cars and by the same token probably take to driving it like a locomotive on rails, even with snow or gravel on the ground, the result of which can be easily read in insurance statistics (Amalberti, 1994). If one endeavoured to increase the planned fault tolerance of machines and their protection against misuse by the user, it would help to avoid many accidents.

2.TECHNICAL COMPLEXITY AND INCREASING VULNERABILITY ON A SYSTEMS LEVEL

Pipeline systems in oil fields in the North Sea are a typical example of modern technology, where in digging the utmost care and safety precautions are planned. Many risk problems, however, only arise later on, and develop over longer periods of time, through qualitative and quantitative changes – mostly without it being recognised. Thus the question arises as to how the operation and maintenance of complex technical systems, as well as their individual parts, can be guaranteed without the risks of technological catastrophes of a new kind being subconsciously planned (e.g. the oil-rig 'Piper-Alpha') or without overlooking their creeping emergence.

3. THE LIGHTHOUSE PRINCIPLE, OR SYSTEM IMPROVEMENT BEFORE PRODUCT IPROVEMENT

On 5th June 1988, we learnt that even seemingly harmless systems like a pipeline and a railway bring about 'invisible' risk potentials through the integration of technology and systems. Due to a leak in a liquid gas-pipeline, gas escaped into a hollow through which the Siberian railway track ran: two trains collided in this cloud of gas and brought about a catastrophe with fatalities amounting to approx. 600, and a total loss of both trains. Other examples of technological interactions can be found in the area of 'electromagnetic tolerance': cordless telephones are only partly compatible with our brains, the airbag in a car and the electronics of planes, i.e. flight-control-systems. In addition, every confidentiality in chats over cordless telephones disappear, as Charles and Diana (and many others) have discovered. Even on motorways, the colourful mixture of ecological small cars and 44-tonne trucks that can be found today is in actual fact an undesirable mixture from a risk-management point of view. One solution which would incorporate the goal of risk-minimisation would be to transfer lorry-loads onto the railways.

The EU-directives on the 'consumer policy' issue (product-safety and product-liability) require a 'fail-safe and foolproof' product design. This could best be achieved through in-built redundancy (or reliance) and fault tolerance, which result in an increase in safety. At present, however, statistics for serious accidents (50+ fatalities) show a trend towards an increase in technological catastrophes (Figure 8)

Figure 8: No of Accidents per year with more than 50 Fatalities

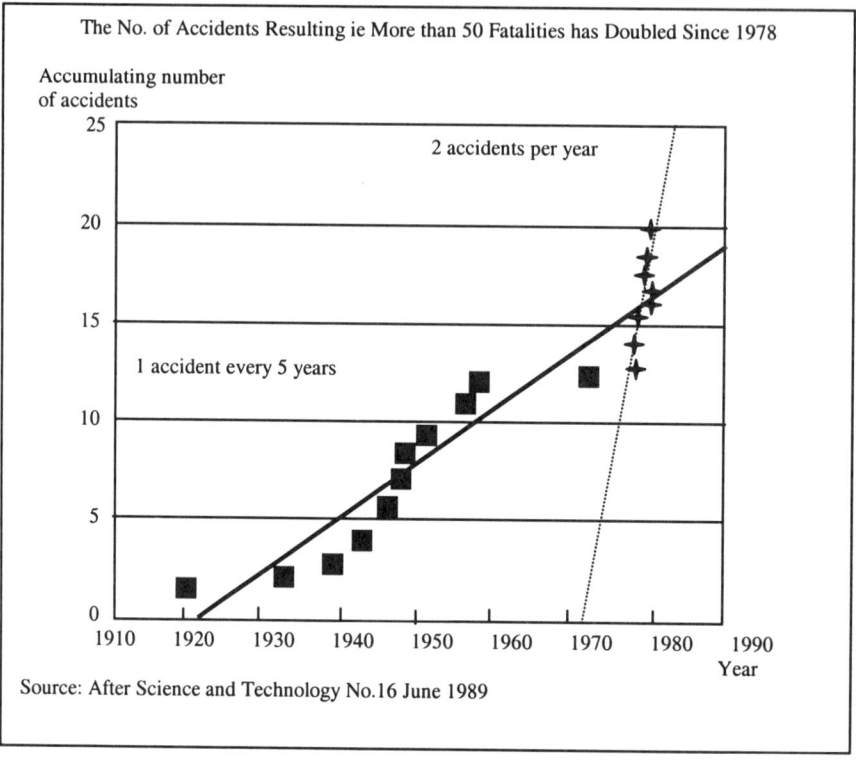

The No. of Accidents Resulting ie More than 50 Fatalities has Doubled Since 1978

Source: After Science and Technology No.16 June 1989

5. The role of insurance in the economic development and the insurability of risks

The insurability of technological risks should be regarded from an insurance as well as an engineering point of view. It is crucial for the work of the engineer that the safety of a product is created within the design phase: risk management and loss prevention are integral components in the design of a product (see diagram 8), not in Manufacturing.

The criteria of insurability restricts the role of insurance as defined by Baruch Berliner and others (Diagram 10) (Berliner, 1982). The mathematical insurability of risks can also be shown in graphical form (Diagram 11), as a function of the frequency and the total possible loss of a catastrophe. Risks appearing in the triangle above the diagonal are non-insurable; risks appearing in the triangle below the diagonal are insurable. In other words: areas with a large number of constantly occurring small accidents, for example motor-vehicle damages, prove to be an ideal insurance area; nuclear power stations, on the other hand, appearing in the filed A1 are in effect insurable. It should therefore be in the interest of both parties (engineers and insurers), to co-operate in the development of technological strategies (which do not include instruction manuals) that enable uninsurable risks to be moved into the lower triangle: the easier it is to insure a risk, the greater the number of insurance policy holders, and as a rule, the lower the premiums – and the smaller the strain on the national budget in the case of a catastrophe. These observations are encouraged by the 'Geneva Association' through its 'M.O.R.E.' programme (Management Of Risks in Engineering).

Figure 9: The Criteria for Insurability

The Criteria for Insurability
1. Degree of probability for a disaster
2. Maximum total amount of disaster
3. Average total amount of disaster
4. Average time span between two disasters
5. Insurance premium
6. Degree of manipulation
7. Undesirable Insurances
8. Legal limitations
9. Insurance cover limitations
Source: After Baruch Berliner: Die Grenzen der Versicherbarkeit von Risiken, Schweizer Rück, Zürich 1982.

Nuclear power stations provide an example of a technology, where for political reasons the mathematics of insurance (and likewise the philosophy of insurability) has been replaced by an international agreement which stipulates a maximum sum of losses for the fully comprehensive cover. The situation in Europe today is now such that the independent insurance of a nuclear power station is greater than its personal liability insurance (Sogh and Fauve, 1991). In addition to this producer liability for the station whilst in operation is explicitly ruled out by international contracts.

This kind of situation can lead to economic distortions (Schmid, 1990), as for example (Swiss) hydroelectric power stations or solar plants have unlimited personal liability. On the other hand, considering the argument put forward in favour of nuclear power, i.e. its cheapness in comparison to expensive solar power (e.g. Photovoltaic energy), it could be said that the insurance industry is playing a possibly decisive role in the choice of new technologies. There is a danger that this example could become the norm for future technologies: it is a topic of discussion in Gene- and biotechnology, as well as in extensive environmental impairment liability. Experts of biotechnology are already saying that the liability insurance premiums cannot be economically born by the mostly small firms, if a compulsory liability insurance without loss limitation should be required by the legislative body. However, these small companies often belong to large concerns, and the reason for their separate existence may well be due to a strategy of 'risk outsourcing'.

Figure 10: Insurable Risks According to Hazard Cause and Hazard Effect

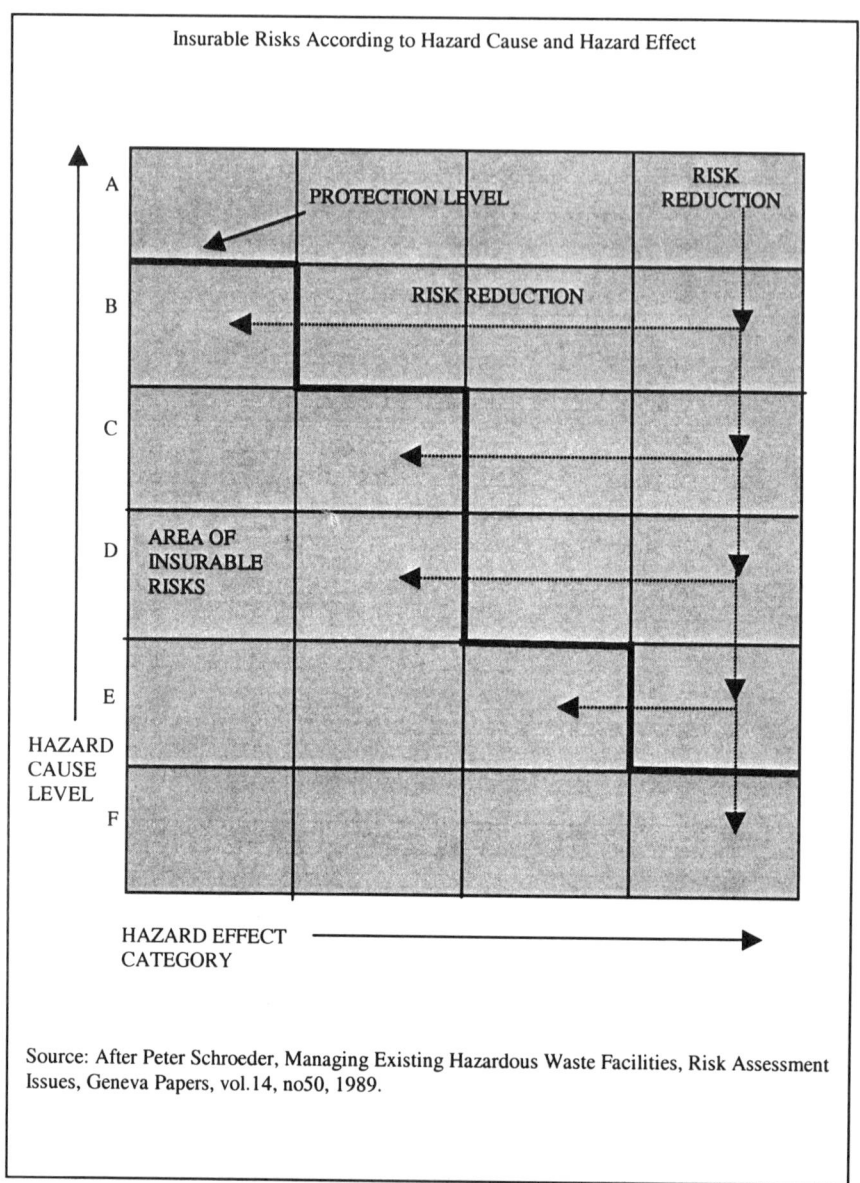

Insurable Risks According to Hazard Cause and Hazard Effect

Source: After Peter Schroeder, Managing Existing Hazardous Waste Facilities, Risk Assessment Issues, Geneva Papers, vol.14, no50, 1989.

Theoretically, insurance is a very good example of how long-term liability costs (e.g. liability for products, environment and waste-management) can, in accordance with the 'polluter principle', and in the same sense of the free market economy, be internalised into the production costs. If the rule for an unlimited liability is not followed, two kinds of distortions emerge. Firstly, large-scale technologies will be given financial preference over small-scale technologies, and secondly, the state will be left to foot the bill at the end of the day. The short-term costs of catastrophe management and of dealing with the immediate consequences of Tschernobyl (in the former USSR alone) were more than 100 billion dollars (WHO Report, 1991). The long-term costs for the human ecology, (i.e. the people in the affected area), however, are much higher: people suffered a long-term destruction of their living environment, and they have to live today as unwanted (and often) outcast) refugees in other parts of the former USSR.

When looking at risk management of large risks it should also be taken into account that a coherence exists between the economies of scale in business (micro-economics) and the dis-economies of risk in macro-economics. (Figure 11). In order to gain some business advantages, catastrophic risks are increasingly being accepted. As long as the last and in additional free insurer is the state (and this is the case wherever no unlimited liability insurance exists), there is little reason for industry to develop alternative technologies focused on waste-prevention and integrated loss- prevention over the whole product-life of goods, from cradle back to cradle.

Figure 11: Balancing and Demarcation Function of the Industrial Insurance in the 'Affluent Risk Society'

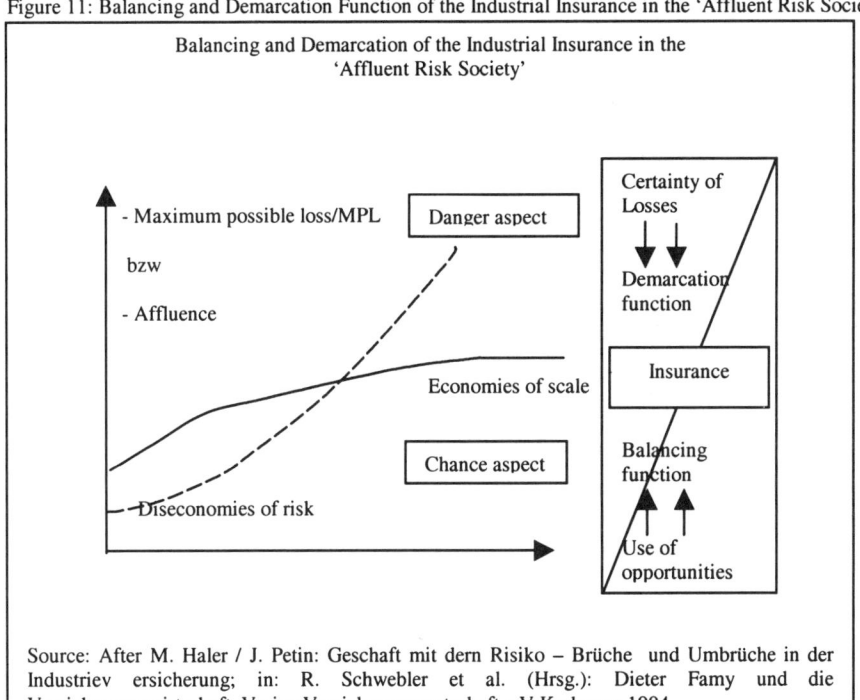

Source: After M. Haler / J. Petin: Geschaft mit dern Risiko – Brüche und Umbrüche in der Industriev ersicherung; in: R. Schwebler et al. (Hrsg.): Dieter Famy und die Versicherungswirtschaft, Veriag Versicherungswrtschaft e.V.Karlsrune 1994

The underlying principle of this apparent failure of the market economy is a lack of understanding for the fundamentally new role of insurance in a highly technological society: in the mid 20th Century, insurance was still seen by many economists as a luxury good playing a secondary economic role, but insurance today has become a prerequisite of technological development in many key areas (Giarini, 1989). Waste-prevention is always loss prevention and vice-versa. This is visible, amongst other things, in the fact that risk management and environmental protection largely use the same terminology. A direct coherence exists between the matters of concern for environmentalists and those of insurers, in that avoidance and prevention, from an economic point of view, almost always have positive outcomes, and in that waste and accidents always include a large risk potential that is difficult to assess. This issue is presently being researched by the EU-commission (DG IX) in an extensive research project with regard to its judicial and economical effects.

6. The objectives of risk management within the State: risk-control - prevention of catastrophes – survival of the nation (national vulnerability)?

For the time being, the 'worst cases' are ruled out in studies on national risk analysis. This creates a possible danger that the state is not setting the optimal priorities for its risk management, that the biggest catastrophes are knowingly accepted – without any measures of prevention. This observation confirms that at the end of the day, risk acceptance is a moral problem, not a technological one.

Figure 12

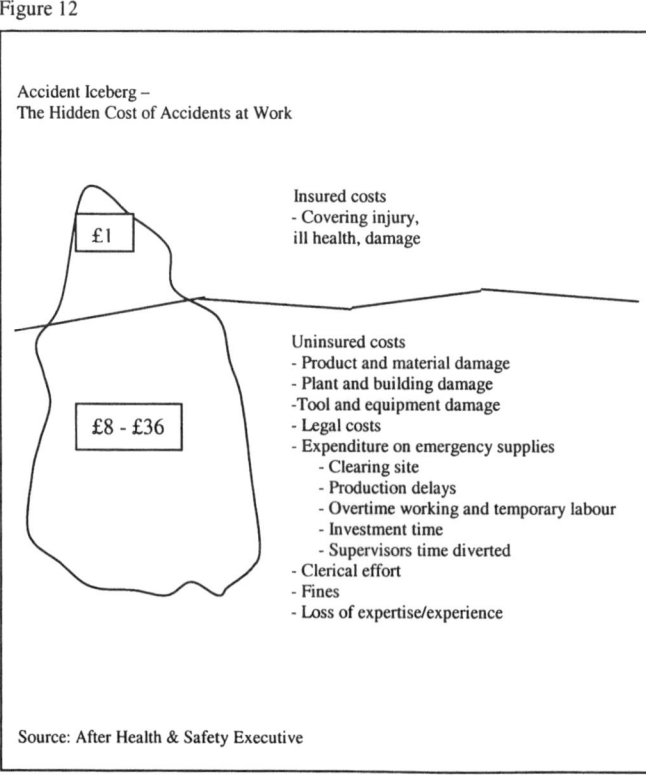

On a business level, it is clear that the risk of a catastrophe which could jeopardise a company's survival is often overlooked, or it is taken for granted that the State will come to the rescue (e.g. Chrysler Corp., German steel industry) A cost analysis of the consequences from the accidents has proven that uninsured (or uninsurable) losses are way above the insurance payments in the case of a disaster (Figure 12): according to an English study, for every pound of insured costs there are between 8 and 36 pounds of uninsured costs! This is due to the fact that insurance only covers the financial consequences of damages, and only partially; whereas many costs are of an immaterial value, such as the loss of a company's good reputation, the loss of market share due to market absence, etc. The extent to which large losses can have catastrophic effects on business is portrayed by the fact that about one quarter of the insured companies that are affected by a major loss do not try to re-establish their business, or go into bankruptcy (famous example: the bankruptcy of the 'White Star Line' after the sinking of the 'Titanic'); over half of those companies affected which survived first do no longer exist three years after the disaster.

On a national level there are several key points that need to be taken into account:

- **The 'accumulation' principle:** Paracelsus already recognised the fact that the difference between poison and medicine depends upon the dosage. There are a number of cases in environmental problems that suffer from accumulation: the CO_2-problem; heavy metals and non-volatile chemicals in the groundwater. Remedial action can only be taken against the dangers arising from accumulation by putting the precautionary principle into practice before a critical threshold is reached, and that is normally before scientific evidence of the danger can be given.

- **The pressure of time:** A series of catastrophes, from the space shuttle 'Challenger' to Tschernobyl and the channel ferry 'Herald of Free Enterprise', can be put down to the pressure of time. The simplest route to loss prevention is through a slowing down of the speed of the consumption of goods and a transition to a more suitable economy, without putting the competitiveness of the economy at stake.

- **The pressure of money**: It takes a lot of courage in the present economy for a manager to stop for example the production of an oil-rig that daily produces 1.5 million dollars of income, or to spend slightly more money to prevent a potential loss in the future – the catastrophe (the explosion of 'Piper-Alpha', or the collapse of the Gotthard-motorway bridge) may after all never have happened. A strategy towards loss prevention is put forward in the form of technological strategies for a transition to a fail-safe and self-curing technology, which is theoretically possible (e.g. built-in disengagement mechanisms). An important contribution can also come from the legislative body, e.g. through the introduction of an extended producer liability 'from cradle back to cradle' (as opposed to the present product liability 'from cradle to the point of sale'). The pressure of money (with potential catastrophic consequences) effects today the running of most infrastructures: in the USA, one street bridge collapses every day, mostly due to poor maintenance, and 180 tonnes of chemicals escape on average per day in 19 chemical incidents. A similar development is foreseeable in Switzerland or any other country, simply due to the fact that the maintenance of the growing volume and increasing age of infrastructures will impose increasingly higher costs, the approval of which is highly unlikely in the age of a tight budget and State budget cuts.

- **The insufficient qualification of personnel in operation and maintenance:** Switzerland recently succeeded in being one of the first industrialised countries, to create a new profession of the maintenance worker. And yet, not least due to financial reasons, there are still insufficiently trained professionals running highly complex technological installations which have been built and designed by highly qualified specialists, catastrophes are thus unavoidable in many cases (e.g. the 1993 accidents at 'Hoechst' or the collision on the Gotthard line of the Swiss railway one Sunday in 1993, when only a trainee was on duty). Even in these cases the precautionary principle is the solution promising the greatest success, as François Ewald quite rightly predicts (Ewald)

7. Conclusion

The trend, regarding both the number of catastrophes and aggravated losses, is increasing for natural phenomena as well as for technological accidents. The 'interaction of systems' (complexity, integration and the resulting vulnerability of technology) will continue to increase unless a conscious change in the objectives of technology occurs, involving a shift to a more sustainable economy.

Certain damage occurrences fall outside the insurance bracket (e.g. the Swiss Railways only liable for accidents due to negligence; heavy goods vehicles from former Eastern Bloc countries carry insufficient or no insurance), or compulsory personal liability insurance laws are violated due to the new kind of poverty: 'unemployment'. In California, one quarter of vehicles are driven without valid insurance, in France it is probably a smaller proportion – but no statistics exist. Due to their vulnerability, many modern technologies are neither compatible with an impoverishment of the population nor with a collapse of social and political structures, as shown by the proliferation of nuclear materials above all in the States of the former USSR – even if it is well known that plutonium is one of the most poisonous substances there is.

At the same time, the insurance industry is finding itself confronted with payments on an unforeseen scale of large losses, as well as a diminishing capacity, which might lead to restrictions in the insurability of risks.

The result of these two developments probably means that in the future the state will have to bear the increasing difference between economic losses and insured losses. These State contributions can take place indirectly in the form of emergency operations carried out by fire-brigades, police or army to fight disasters or clean up afterwards, but also in the form increases in social insurance payments (the economic reintegration of handicapped people in times of high unemployment is hardly successful).

The search for solutions from this cost dilemma should not rule out adjusting present legislation in the sense of applying the precautionary principle, e.g. through a higher personal responsibility of economic actors (extended producer liability), associated with a fostering of innovative ideas (e.g. mandatory insurance cover instead of mandatory standards and legislation).

REFERENCES

Amalberti, R. (1994) Quand l'homme et la machine ne se comprennent plus. In: *Bulletin de liaison de l'Institut Fredrik R. Bull*,Louveciennes.

Berliner, B. (1982) *Limits of insurability of Risks*, Zurich, Swiss Reinsurance Company, Prentice-Hill Inc.,Englewood Cliffs,N.J.

Bernold, T. (1990) Industrial Risk Management: A Life-cycle Engineering Approach, Proceedings of a Conference at the Swiss Federal Institute of Technology, Zurich, In: Elsevier Amsterdam a *Journal of Occupational Accidents*, Vol.13 (1/2)

Binswanger, H.C., (1994) Geld and Magie, Eine Deutung der modernen Wirtschaft anhand von Goethes Faust, In: *Ders. (Hrsg.): Geld and Wachstum, Zur Philosophic und Praxis des Geldes*, Stuttgart.

Ewald, F. (1989) Die Versicherungs-Gesellschaft, In: *Kritische Justiz*, 22 Jahrgang, Heft 4,Baden-Baden.

Frey, P. (1992) Der Risikomarkt – Kontinuitat der Industrieversicherung aus Sicht eines Rueckversicherers, In: *Die Versicherungspraxis hrsg. Vom Deutschen Versicherungsverband*,Sonderdruck Hett 6,Munchen.

Giarini, S. (1989) *The limits to Certainty, Facing Risks in the New Service Economy*, Dordrech Kluwer AcademicPublishers.

Nutter, F.W., (1994) The Role of Government in the United States in Addressing Natural Catastrophes and Environmental Exposure, In: *Geneva Papers*, Vol. 19, No. 72, July, p.244.

Schadenspiegel, (1994) hrsg. *Von der Münchener Rückversicherungs-Gesellschaft*, 37, Jahrgang, Heft 1, München, p.23ff.

Schmid, G. (1990)Rechtsfragen bei Grossrisiken, In *Zeitschrift für Schweizerisches Recht, NF 109* (1990) sowie in den Proceedings des Schweizerischen Juristentages.

Showater, Sands, P. and Myers, M.F. (1994) Natural Disasters in the United States as Release Agents of Oil, Chemicals or Radiological Materials between 1980-1989, Analysis and Recommendations, In *Risk Analysis*, Vol. 14, No.2, p.169ff.

Sogh, G.and Fauve, M. (1991) Compensation for Damages Caused by Nuclear Accidents: A Convention as Insurance, *Etudes et Dossiers* No. 156, July 1991, Geneve, The Geneva Association.

World Health Organisation, (1991) Report on Chernobyl: see *Risk Management Newsletter* No.10 (January 1991), Geneva, The Geneva Association.

Catastrophes Take Their Toll on a Crowded World, In: *New Scientist*, 28 May, p.8.

HISTORY REPEATING ITSELF?

Expertise, barriers to learning and the precautionary principle.

DENIS SMITH
Centre for Risk and Crisis Management
Sheffield University Management School
9 Mappin Street
Sheffield, S1 4DT, UK

JO MCCLOSKEY
School of Marketing
Bristol Business School
University of the West of England
Bristol, BS16 1QY, UK

Introduction

> *"It may not be errors of fact that matter to the general public as much as errors of judgement"*, Porritt (2000 p. 19)

The litany of major crisis events during the last decade alone would suggest that many organisations have poor learning skills with regard to both crisis prevention and disaster containment. A brief consideration of some examples serves to illustrate the problem. The loss of the roll-on, roll-off ferry, the Herald of Free Enterprise at Zeebrugge, raised questions about the culture of the operating company, the role of safety systems within the industry, and even the integrity of the vessel's core design. However, it took the loss of the ferry Estonia to reinforce the view that this class of vessel may have an inherent design flaw. Despite these criticisms and findings, there is still considerable resistance amongst some operators to retrofit stability systems and bulkheads on such ferries, with cost often being the primary reason for such resistance.

Other industrial sectors also display similar characteristics in terms of potential barriers to learning. The nuclear industry, for example, has also shown itself to be prone to the potential for denial after accidents. Following the accident at Chernobyl, Western nuclear operators were quick to reassure concerned publics that the event could not happen to their reactors. The reasons for this were both technical (they did not operate reactors using the Soviet system) and organisational – with the main argument being that

E. Coles, D. Smith and S. Tombs (eds.), Risk Management and Society, 101–124.
© 2000 *Kluwer Academic Publishers. Printed in the Netherlands.*

Western companies were well run, their systems were well maintained and they operated within a strong safety culture. However, the accident at Chernobyl had itself followed on from previous events at Three Mile Island (USA) and Windscale (UK) and it provided further evidence that the potential for accidents was not constrained to a particularly country or organisational culture. However, the process of denial that occurred after Chernobyl, and the accompanying flawed assumption that major accidents could not happen in a well-run Western company, still serves to allow the incubation of latent error to remain unchecked within the industry. The recent accident at Tokamura in Japan (1999) serves to illustrate the point that such accident potential still exists and that errors by both operators and managers can occur even in advanced Western countries.

Evidence also suggests that such problems exist in other sectors. One only has to witness the crash of the high-speed inter-city train close to Paddington, which itself followed on from a similar accident at Southall, to see that similar barriers to learning are in operation in other industry sectors. In this latter case, the effectiveness of the warning systems available to drivers were major factors in the creation of the accident conditions. Why then was action not taken after Southall to improve the integrity of systems defences on board such trains? The fact that the Southall inquiry was delayed by some two years merely served to act as a further contributory factor in creating barriers to learning from the event. The supreme irony is that the Southall inquiry began within days of the Paddington accident.

The question that emerges from an examination of these incidents is: Why do organisations show such poor learning capabilities in the face of experience elsewhere, and why is a considerable body of evidence concerning this failure potential seemingly ignored by managers? It raises a question of whether the process of denial is so strong in organisations that it makes them incapable of seeing the potential for harm within their own systems and decision making processes. Perhaps one reason is that, all too often, the root cause of a catastrophic failure is not adequately identified and consequently, the conditions that served as precursors to the event, are allowed to incubate the potential for further failure elsewhere (see Turner, 1976; 1978; Reason, 1990).

The purpose of this paper is to examine these issues by focusing on policy-making processes within the broad area of the UK's public health policy. Of particular interest here is the apparent failure of successive governments to learn from the problems created by the BSE (Mad Cow Disease) crisis. This is evidenced in part by the perceived muddled approach to policy-making in the field of genetic modification (GM), particularly in terms of GM foods, that is followed by the present Labour government. A central dynamic within this broader analysis will be an examination of the role of the precautionary principle as a means of framing policy decisions and serving as a mechanism for the adoption of a more cautious approach to public health issues. Problems clearly exist in the use of such an approach within policy making, as evidenced by successive examples of failure to prevent environmental damage and risk-based crises from emerging. The precautionary principle is, in itself, invariably ill-defined in both theory and practice and yet remains an important aspect both of the government's policy and their associated media campaigns. It is our contention here, that many of the problems concerning the precautionary principle, stem from the ambiguity that exists around its use within a policy setting and a failure to grasp its potential role as a means of dealing with problems of crisis incubation. We aim to detail here some of the demands that would be imposed by the use of a precautionary approach and question the commitment of the state to a realistic adoption of the demands from such a paradigm shift. An important question that emerges here centres on the role that the precautionary principle might have in reducing the potential for failure within both government and other organisations by surfacing the role of latent managerial error in creating substantial barriers to learning. Before examining the detail of these issues, it is first necessary to set out the key dynamics of the failure process and the manner in which such failure is incubated.

The nature of failure

"People court failure in predictable ways" – Dörner (1996)

The failure processes within organisations occur in complex, often ill-defined ways. There is a widespread recognition that complex technological systems can fail catastrophically and can do so in ways that seem to be embedded within the design of these systems (see Perrow, 1984). Research has suggested that some 70% of such failures occur as a result of human factors rather than technological ones (Turner, 1994). Historical precedent provides us with a myriad of examples to support this view, ranging from high technology failures, through fundamental human blunders to intentional actions and sabotage. In more recent years, there has been a growing awareness of the potential for those latent failures, which are associated with the management process, to occur at every stage of industrial activity, from planning and design through to day-to-day operations (see, for example, Smith, 1995). The process by which latent failure potential becomes incubated within organisations has emerged as an issue of considerable concern within management theory. It has been suggested that the lack of information and explicit knowledge around non-linear systems may ultimately contribute to a sense of denial amongst managers and the possibly rejection of concerns which are expressed by those outside of the (legitimised) expert community (see Smith, 1990; 1993, 2000). A central dynamic of this process has involved the role of knowledge within management and its impact upon corporate success (see amongst others, Boisot, 1999). If knowledge is central to success, then it follows that it must have an important bearing on failure (see Smith, 1995). Research has also sought to explore this dynamic within human factors (see Reason, 1990; 1997), organisational failure (Wilson et al, 1999) and technological risk (Irwin, 1985; Irwin and Wynne, 1996). More recently, concern has been expressed over the impact that scientific expertise can have upon the incubation of disaster potential in such areas as biotechnology and genetic engineering (see for example Green and Yoxen, 1993; Porritt, 2000).

The role of expertise under conditions of uncertainty, invariably assumes a central role in many of these debates, and it is particularly important within risk debates. Science and expertise are increasingly being called upon to help make policy decisions, but this is often done either on the basis of limited information, or in settings within which there is no real understanding of cause and effect relationships. As such, the power of the scientific approach is weakened due to the lack of evidence as well as poor cause and effect relationships between exposure to an "activity" and any damage. An example of the latter problem can be found in the case of the link between BSE in cows and new variant Creutzveldt Jakob Disease[1] (nvCJD) in humans. These concerns should not be seen as a blanket criticism of science and expertise within policy debates. It is clear that expertise is important in helping to support decision-making in those circumstances where the phenomena under consideration are both known and generally well understood. However, expertise is, by definition, domain specific. Once an expert is taken out of his/her knowledge domain then s/he faces inherent problems of knowledge validity. This adds further layers of complexity to the decision making process and may constrain the willingness of experts to consider solutions which are outside of their dominant paradigm. For dynamic, non-linear systems, such problems of knowledge validity assume an even greater significance, as the robust nature of that knowledge domain may be increasingly limited when looked at from a strictly scientific perspective. The reluctance of "experts" to consider the thesis that BSE could jump the species barrier, led both to delays in diagnosing the risk to human meat eaters and an erosion of public trust in those experts and associated policy makers. Clearly, such problems become even more significant when the issues under consideration are not easily theorised or modelled and therefore lie outside of the ability of science to 'prove' cause and effect relationships. Such "trans-scientific" problems (Weinberg, 1972), call into question the traditional role of expertise within the policy making process and raises important challenges to the ways in which we communicate uncertainty to all parties within the debate (see Smith and McCloskey, 1998).

The limitations of expertise

"Assessing risks is simple in principle. It involves identifying hazards or examining what in a particular situation could cause harm or damage and then assessing the likelihood that harm will actually be experienced by a specified population and what the consequences would be (ie the risk). The overall objective is to obtain a view on how to manage the risk or to compare the risk with other risks"- Health and Safety Executive (1996), p. 3.

The first of these challenges centres on the nature of the systems themselves, along with the approaches taken by organisations to deal with their complex and opaque nature. The dynamics of modern systems invariably creates a set of circumstances in which managers find it difficult to cope with issues if decision making, particularly around cause and effect relationships. Decision making is central to an understanding of failure (see Anheier and Moulton, 1999) and, in order to deal with the potential problems associated with poor decision making, organisations attempt to secure stocks of human capital – expressed in terms of skills, knowledge and expertise – to help reduce this risk of failure. This is particularly important in those industries that are using "breakthrough technologies" (see Zucker and Darby, 1999, p. 24) which invariably increases the level of complexity and uncertainty facing the organisation. In addition, these complex systems often display processes of emergence where the interaction between systems elements can give rise to unforeseen events. It is as a function of this complexity that organisations often ensure that they have a well-developed technocracy within their structure. This, in turn, provides them with the knowledge and expertise necessary to deal with external challenges to their activities.

The actions of the tobacco industry, in defending their product, illustrate the possible use of technical expertise in this regard (see, amongst others, Glantz and Balbach, 2000; Jones, 1997; Orey , 1999). Glantz and Balbach (2000) observe, for example, that the so-called "tobacco wars" were typified by a number of important themes that have a relevance to our current discussions. In the first instance, they observe that the tax revenue from tobacco is of importance in shaping political interests. They also point to the tensions that existed between health promotion and health care in the allocation of tax revenues. This Balkanisation of opposition, particularly around financial and technical issues, can be an important dynamic in allowing corporate interests to gain an advantage (see, for example, Smith, 1990). Their second observation concerns the differences in opinion that can arise due to the short-term decision horizons of politicians and the longer-term strategies of those concerned with the protection of public health. This problem has obvious links into a precautionary approach, where the costs of banning an activity are immediate whereas the benefits (if realised) are often only found in the longer term. In this case, the burden of proof becomes an important area of debate. The third point made by Glantz and Balbach (2000) concerns the manner in which

"temporary agreements...become precedents in subsequent budget years because operating programs build their own constituencies" (p. 5).

The creation of an ad hoc policy process embeds considerable difficulties into the policy environment and ensures that subsequent attempts to shift policy are seen by some constituencies as losing ground. The dynamic nature of a precautionary approach will also create difficulties in this regard. Finally, Glantz and Balbach (2000) observe that the degree of openness in policy debates is an important factor in shaping the balance of power between opponents - a point that has been made by other writers in this area (Collingridge and Reeve, 1986; Smith 1990; 1993). This gradient of power within conflicts is extremely important in determining the success of public campaigns against powerful interests but it remains a difficult area to evaluate. For example, Glantz and Balbach (2000) argue that,

"When tobacco policy is left to the inside game played in back rooms, the outcome will favor the existing power structure. The tobacco industry is a tough, experienced inside player who has benefited from the existing power structure. In contrast, the public health groups' power is amplified in public arenas, where tobacco industry's power is sharply curtailed." (p. 5).

This latter point has been developed elsewhere, although a number of authors have taken a different stance on the ability of powerful groups to dominate public debates, especially around risk, though the use of their greater economic power and technical expertise (see, Crenson, 1972; Collingridge and Reeve, 1986; Smith, 1990). The concentration of technical expertise within organisations gives them a considerable advantage within environmental or risk-based conflicts – especially when there are considerable economic benefits to the activity and the costs are ill-defined. In many organisations, this concentration of expertise is located within the research and development function. However, this function has often been cut back as part of a broader policy of re-engineering. The impact that this has on the effectiveness of the role of expert inputs into decision making, conflict resolution and risk management remains an unknown factor. However, even with a significant R&D function, some industries have struggled to cope with the demands imposed upon them by pressure groups and other interested stakeholders in addressing broader public concerns around human and environmental safety.

The second major challenge comes from the broader cultural and organisational dynamics associated with complex activities which may also conspire to create a set of circumstances in which relatively simple and low technology activities may also generate conditions of catastrophic failure. This point is elaborated upon by Tenner (1996) in his discussion of revenge effects, where the technology behaves in ways that were unforeseen

"Technology alone usually doesn't produce a revenge effect. Only when we anchor it in laws, regulations, customs, and habits does an irony reach its full potential" (Tenner, 1996, p.7).

Putting this notion of a revenge effect another way, the emergent properties within systems may create problems that have been largely unforeseen by managers or, more worryingly, may have resulted from the decisions taken by managers themselves. Tenner also raises concerns over management intervention within complex systems. He argues that many attempts at interventions within complex systems may simply move the problem somewhere else, make it more opaque or escalate the consequences of the problem, whether in situ or in another part of the system (Tenner, 1996). Indeed, the view has been expressed that modern systems are often so complex that they may even exceed the ability of many operators to understand them, especially when these systems operate in a degraded mode (see, Smith, 2000). Consequently, any attempts to tinker with the system may ultimately be carried out under conditions of ignorance. Such a veil of ignorance may also apply to those attempts made to transfer scientific principles out of the laboratory and into the real world. In this case, the number of intervening variables increases considerably and creates problems around cause and effect relationships. Indeed, one might argue that the opacity and complexity associated with revenge effects increases considerably within non-linear policy environments.

The limitations of expertise (including management itself), and the weight given to such expert judgements can also be considered to generate revenge potential within organisations. This latent error potential can be seen both in terms of the limitations of expertise around processes of emergence (especially in leading edge technologies) and also in terms of the limitations of management itself. Turner (1976; 1978) argued that the central mechanisms put into place within organisations (expressed as precautionary norms) inevitably reflected the core beliefs, values and assumptions of senior managers and policy makers within the organisation. These core beliefs and values will frame the "regulatory" setting in which the organisation seeks to control its activities. Clearly, such a situation is open to a degree of abuse. Such human involvement can in effect, complicate the maintenance of safety in complex systems. Of particular concern here would be the impact of cultural issues and managerial assumptions upon the organisation's approach towards regulation and control.

A third challenge concerns the organisation's capabilities for communicating any uncertainty about its activities both to interested stakeholders and organisational members alike. Communication processes have been seen as an important dynamic within the literature on both crisis management and organisational failure (see Smith and McCloskey, 1998; Dowie, 1999; Fortune and Peters, 1995), with the view expressed by some, that "secrecy is a common characteristic in crisis situations" (Salter and Levett, 1999, p.174). In cases that involve a higher degree of uncertainty, the manner in which complex issues are communicated to the range of stakeholders, assumes an important role. Again, expertise is important in shaping this process as is the technical nature of the language used by these experts to communicate with the various publics affected by the activities (see Irwin et al 1996).

Finally, and at a level of the organisation's super-ordinate goals, the willingness of managers to move towards a precautionary approach in dealing with issues of risk is also of importance. The precautionary approach requires that organisations challenge their core assumptions and constantly reconsider the limitations of their expertise, especially within emergent, complex, non-linear systems. Clearly, this represents a significant challenge to the ways in which organisations manage their activities. The notion of a precautionary approach has started to attract a considerable degree of attention from policy makers within Western countries. It could be argued that trecent advice from the Department of Health in the UK on the use of mobile phones by children reflects that perspective. Despite the lack of scientific evidence of harm, the government has advised that young children should moderate their use of mobile phones until research shows that there are no adverse health effects. The Stewart Committee

(2000) recommended a precautionary approach be adopted to mobile phone technologies

> "..........until much more detailed and scientifically robust information on any health effects becomes available"... and noted that such a precautionary approach "is not without cost.... But...[they] consider it to be an essential approach at this early stage in our understanding of mobile phone technology and its potential to impact on biological systems and on human health" (p.3).

Such a precautionary approach can be seen as a commendable policy attempt to reduce risk where the evidential basis for safety is vague at best, or weak at worst. Of course it could also be argued that the approach to mobile phones is too limited, and possibly too late given their widespread use. However, the response to the potential risks from mobile phones does raise some interesting questions about what the precautionary approach (expressed in terms of a risk) means within a policy setting and this needs to be explored in some detail. What a precautionary approach does require, is a recognition of the inherent limitations of expertise. This challenge to one of the core elements of the management paradigm may account, in part, for some of the controversy surrounding the principle and this issue now needs to be examined in more detail.

The precautionary principle – theory into practice?

In theory, the precautionary principle offers a potentially useful means of framing a series of debates around such issues as hazard, the role of expertise, risk communication and crisis learning and prevention. At its core, lies the notion that, if there is potential for harm in an activity then a discussion of the acceptability of that risk and the nature and frailty of any evidence should be open to informal public debate. However despite its obvious merits, the use of the principle is shrouded in conflict and disagreement. A key issue has been the fact that the principle has proved difficult to define and, perhaps more significantly, has generated a considerable amount of debate and conflict over its validity. Giddens (1999), for example, has argued that

> "...the precautionary principle isn't always helpful or applicable as a means of coping with problems of risk and responsibility. The precept of 'staying close to nature', or of limiting innovation rather than embracing it, can't always apply. The reason is that the balance of benefits and dangers from scientific and technological advance.....is imponderable" (p. 32)

Jonathon Porritt (2000), a one-time director of Friends of the Earth, observes that the principle:

> "..is still seen by many UK government officials as an extremely unwelcome and scientifically corrupt interloper" (p. 47),

and that many within the scientific community are critical of the precautionary approach because it,

> "appears to threaten the authority of inductive, cause-and-effect-based science, shifting the balance of power between empirical science and other political or social factors" (p. 44).

Indeed, one might argue that the principle's strength can be seen in shifting this balance of power from the "experts" to those who might have to carry the costs associated with such an activity. But what does the precautionary principle mean in policy-making terms and how can its core elements be incorporated into environmental and risk-based decision making? In order to address these concerns, it is necessary to explore the principle's origins and its early interpretations.

The precautionary approach owes much of its origins to policy shifts arising out of the German environmental movement (Bohemer-Christiansen, 199X), although some have argued that its philosophical roots go back to the Hippocratic Oath (Jackson, 1999). At its core lies the notion that sufficient consideration should be given to the potential risks that are inherent in a proposed policy and that these should be set against the benefits that arise from the activity. Herein lies the potential for conflict however, and the principle remains shrouded in controversy. One of the best attempts to define the principle can be found in the so-called Wingspread Statement[2]:

> "When an activity raises threats of harm to human health or the environment, precautionary measures should be taken even if some cause and effect relationships are not fully established scientifically" - Raffensperger and Tickner (1999) p. 8.

In elaborating on the detail of the principle, Raffensperger and Tickner (1999) observe that it was found to have four core elements. The first was the notion that preventative action should be taken before scientific proof of harm and causality is established. Secondly, the risk generators (and not the potential victims) are faced with the burden of proof. Allied to this is the view that it is no longer enough to say there is no evidence of harm bur rather that evidence of safety is required. Thirdly, a range of alternative actions should be considered when evidence of harm materialises. Finally, the decision making process itself must be open and democratic and should include all parties who might be affected by the activity (Raffensperger and Tickner, 1999). O'Riordan and Cameron (1994), in an earlier paper, had envisaged a slightly broader definition of the principle which included, additional elements. These include: the protection of ecological space; concern for the intergenerational impacts of the activity; accepting responsibility for previous damage and reflecting this within the precautionary approach adopted; and ensuring that the cost elements of the approach are recognised and that the burden of such costs is not disproportional. Despite the laudably aims of the precautionary approach, it has attracted criticism from all sides in the environmental debate. Industrial groups are clearly concerned about the implications that a strict interpretation of the principle would have on their competitiveness and profitability. Similarly, the vague nature of the principle also brings with it problems surrounding implementation and enforcement. The principle has even drawn critical comment from environmentalists. In commenting upon the elements outlined by O'Riordan and Cameron (1994), Porritt (2000) argues that each of them can be contested and it is possible to develop "either a weak or a strong formulation" (p. 44) in terms of their implementation. Unless the principle is firmly grounded within a regulatory framework and it is rigorously enforced, it will remain little more than an abstract concept around which debates will occur. In that respect, it can be seem almost akin to an ethical statement about the boundaries of responsibility.

The central issues within the debate involving the precautionary principle can be seen to concern the role of technical expertise and its potential to serve as a barrier to the learning process, thereby, helping to incubate the potential for crises. For decision problems that are high in uncertainty and low in explicit knowledge, reliance upon narrowly defined expertise may serve to prevent an equitable policy outcome due to the limitations of the expert's cognitive schema. This is compounded further when these high levels of uncertainty and lack of knowledge generate considerable hazard potential. The initial decision to use animal protein in the feedstocks for UK cows serves as an example of high uncertainty and considerable hazard in which technical expertise was used to justify the decision. The state's role in policymaking throughout the BSE crisis is also problematic in that it was both promoting the economic interests of industry, whilst also being concerned with the health of the public. One might argue that the latter goal was subjugated by the demands of capital and expertise was used to legitimise that decision.

The BSE crisis

It is now generally accepted that the BSE crisis emerged as consequence of the feeding of herbivore cattle with protein derived from the remains of other animals, notably sheep. Prior to the 1970s, the main source of animal protein in cattle feed had come from fishmeal. However, with the increasing cost of fishmeal, suppliers looked to sheep as an alternative source of protein. Unfortunately, sheep have a tendency to suffer from a brain disorder called scrapie, and an assumption was made that this disorder could not jump the species gap and be passed on to cattle. Invariably, the risks to humans were not considered until the BSE crisis began to emerge in the 1980s

The supposed return of public confidence in the safety of British beef was marked by the announcement by McDonalds on 26th June 1997 that it was lifting the ban on British beef. This was clearly seen as a huge boost to an industry that has been under siege since a government announcement stating a possible link between BSE and the human disease CJD in 1996. This statement followed years of strenuous denial of the risks by both successive Conservative governments (and their agencies) and the meat industry. The emergence and development of this crisis is both long and complex and it is not our intention here to detail the full chronology of events[3]. Instead, this part of the paper deals with the role of expertise in the emergence of the crisis and briefly examines the role of the government's deregulation strategy as a key incubator for the crisis.

To some extent, the government reacted relatively quickly to early signs of the crisis. After the first case of BSE was identified in 1986, the Central Veterinary Laboratory notified government ministers in 1987 who set up a committee, headed by Sir Richard Southwood, to investigate the problem. The committee reported[4] that it was "most unlikely that BSE will have any implications for human health". In 1988, the government legislated to ban the sale of suspect meat and bone meal and introduced a slaughter policy for all BSE infected cattle in an attempt to halt the problem. The BSE Order 1988 also made BSE a notifiable disease, while the government remained adamant that there was no risk of the transfer of BSE to humans. While large numbers of infected cattle were slaughtered, the government argued that the problem was under control. However, research published in March 1996 suggested that at least ten human deaths by CJD were the result of exposure to contaminated beef products. Some ten years after the first acknowledged case of mad cow disease, and following strenuous denial by a number of government ministers, the government was eventually forced to admit that there could be a link between BSE and CJD.

The crisis was probably all the greater as a result of such ministerial denials and their use of supposed scientific evidence and opinion to allay public fears. Such a high profile about-turn was to generate massive repercussions resulting in a European Union ban on British beef, which subsequently spread to other countries, and an emergency meeting of the World Health Organisation was called to consider the ramifications of the problem. The impact on the UK's cattle farming industry was dramatic as the public lost confidence in both the industry's and the government's ability to manage food safety. What became clear was that the government's scientists and regulators had failed both to identify the risk of inter-species transfer of the BSE and arrest its development through the offal ban and cull of infected cattle. All this, despite strong assurances concerning the robustness of state controls. The media began to report on the issue and published the first warnings about the risk of human infection from BSE contaminated beef in 1988. This was largely generated as a result of the work of Professor Richard Lacy at the University of Leeds, although government responded by denying the validity of such claims. Given the government's strong scientific base, and given the varying warnings that were articulated by government and independent scientists, one must question the role of such expertise within the policy and decision making process (see Hogwood and Gunn, 1984).

From the first recognition of the disease in November 1986, it took the government until March 1996 to formally acknowledge that there was a link between

BSE and CJD. Perhaps one major reason for the delay was that such a powerful statement from a government appointed committee served as a barrier against attempts to mount serious challenges to the dominant paradigm of minimum human risk from the disease. Added to this was the significant amount of scientific uncertainty that still surrounds the disease. For example the transmission of the disease between species and its causal agents remains a subject for scientific debate. Whilst there is some consensus that the disease becomes seated in the brain, spleen and spinal cord, the risk factors involved in the generation of CJD from BSE remain uncertain. It does appear, that those who eat infected offal (containing the body parts referred to earlier) were more likely to be at risk although the long incubation periods (five years for BSE and between five and fifteen for CJD) prevent accurate cause and effect relationships from being identified. Similarly, a major barrier to accepting the hazards from BSE stemmed from the dominant view in the biological sciences that the encephalopathies could not reproduce because they did not contain genetic material. The paradigm shift here came from the work of Pruisner (1980) who argued that the misshapen proteins (prions) of encephalopathies would spread by enveloping healthy proteins and converting them to misshapen prions. These unhealthy prions then spread through the spinal cord to the brain almost like a chain reaction. Whilst such an account of the process seems relatively easy to comprehend, the description hides considerable uncertainty in the nature of the transmission. The nature of encephalopathies makes them difficult to study and the prions show a remarkable degree of resilience and are seemingly unaffected by cell enzymes, heat radiation and vaccination.

The mad cow crisis illustrates many of the problems associated with issues of uncertainty and the generation of crises within organisations. The key areas of interest are in the science-policy interface, the dissemination of views concerning risk and uncertainty to the various stakeholders within the problem, the relevance of expertise outside of its domain limits and the extent to which such expertise can be used to inform policy making.

Technical expertise within decision making under uncertainty

The central role of knowledge in risk debates is a matter that needs to be explored further. The utilisation of knowledge within the management process has begun to attract considerable attention within the literature (see, for example, Boisot, 1995). Whilst knowledge is clearly of strategic importance to the competitiveness of organisations, it is also central to the development of policy for the control of risk. However, the processes of risk assessment are largely impenetrable to non-experts and they require that experts serve as mediators in the analytical process. Of particular, concern is the notion that the primacy of (established) expertise over other views, particularly, those of people exposed to the hazards, might lead to a form of social (technocratic) exclusion. Of importance here is the relative weightings given to "expert" and "public" forms of knowledge and the willingness of both groups to engage in dialogue and debate around the relevance of the various forms of knowledge. This is clearly a requirement of a precautionary approach. There have been a number of instances, for example, where a view of the centrality of expertise has compounded the problem. In these cases, the weight given to the views of experts had been given a legitimacy and level of policy influence that circumstances did not justify, especially when considered within a precautionary framework. The reasons for this have been largely due to the high degree of uncertainty that surrounded the scientific arguments used to support policy. In the case of the Canvey Island risk analyses, for example, the first report included a number of 'expert's best estimates' within the calculations because of the absence of quantifiable data (Smith 1990). In this case, the subjective nature of these calculations was not part of any debate between the state and the local publics.

The use of expertise within decision making is justified as an attempt to reduce the extent of the uncertainty inherent within the decision problem. However, the paradox is that if the problem under consideration has a high level of uncertainty within it (especially in terms of *a priori* evidence), then the use of expertise may prove to be something of a constraint (see for example, Sheldon and Smith, 1992). This is especially so when the problem lies outside the accepted boundaries of knowledge and extends beyond the ability of science to prove. Here the use of expertise within the decision- making process may serve to be a limiting factor due primarily to the influence of vested interest and organisational culture, which may assume prime importance in the absence of categoric evidence (Smith, 1990).

This relationship between science and decision/policy making has been viewed by some authors as problematic. Collingridge and Reeve (1985), for example, observe that the relationship between scientific inputs and successful decision making is not always a positive one. This is especially problematic under conditions where uncertainty is high and where scientific concepts and hypotheses are taken outside of strict laboratory conditions and applied in real world settings. In such cases, Collingridge and Reeve, comment that there may be endless scientific debate and, as a result, no resolution of the policy problem under consideration. In such debates, the resolution of policy for risk-based activities may be more of a function of the power of the various stakeholders in the debates, rather than the quality of the scientific arguments articulated by them. This perspective raises some fundamental questions about the nature of decision making for risky activities.

Firstly, to what extent should technical expertise be given precedence over other inputs into decision making in areas of high technical and scientific uncertainty? In a related issue, how do dissident views within the main body of expertise become incorporated into the policy making process, if at all? Secondly, how does decision-making take account of such uncertainty when developing policy for the control of potentially hazardous activities? Finally, how is this process regulated? Can the state occupy the dual role of regulator and promoter of the activity? In addition to these questions, one needs to consider the role of experts' cognitive schema and the limiting effect that these may have in those debates that are trans-scientific. The schema problem here stems from the role of a dominant (organisational) paradigm in creating fundamental cultural barriers that, in turn, prevent a challenge to that paradigm. The case of BSE illustrates this process. For many years, scientists expressed the view that BSE could not jump the species barrier and pass from cows to other animals or, indeed, to humans. This view was strongly held by many within the scientific community and was difficult to shift because of the absence of absolute proof. Any "evidence" (which was invariably weak) was dismissed as scaremongering, despite the fact that it had ultimately passed from sheep (scrapie) to cows (BSE) in the first instance.

Conventional wisdom sees expertise being defined in terms of domain specific knowledge. To qualify for the label of expert, an individual should have sufficient domain-specific knowledge to allow them to make decisions within that domain, with a high probability of a successful outcome. Expertise should, therefore, allow the individual to gain sufficient insight into the nature of a problem to help them solve it. Whilst this should hold true for common problems, it may be a false assumption for those issues in which the problem space is ill-defined, and the issues themselves are trans-scientific. Under these conditions, it is possible that cognitive narrowing may take place, as experts seek to interpret the problem within the rules of their own paradigm. Certain problems, however, may require a paradigm shift in order to cope with the demands of the issues raised. It is held that debates involving TSEs[5] and more recently GM foods, may require such a challenge to the dominant paradigm, in order to prevent the rejection and denial of causal relationships which are informed by the existing paradigm

Regulating genetically modified organisms

One would have expected that the state regulatory system for foodstuffs would have incorporated lessons from the debates and concerns surrounding BSE when they had to consider the approval of genetically modified foods for human consumption. However, it became clear that government officials and the politicians themselves made a series of assumptions concerning the nature of these debates and the extent of public concerns. The attempts to reassure consumers by the Prime Minister Tony Blair had more than a passing resemblance to the infamous Gummer episode, in which the then agriculture minister sought to feed his daughter a beef burger in front of the press. The Daily Mirror ran a headline[6] which was critical of the Prime Minister's stance on eating GM foods (see Porritt, 2000). Such simple reassurances by politicians, who are often perceived as having considerable vested interest in the problem, display a lack of understanding about the role and nature of trust in risk debates. In developing this argument further, our aim in this section is to illustrate how the failure to develop trust can create latent conditions for crisis potential and that this can overcome much of the efforts made towards organisational learning. Within the context of the GM foods debate, the government has, so far, relied heavily upon scientific argument to support their policy stance, despite the fact that there were inevitably quite serious burden of proof issues surrounding the commercial use of GM foods.

Genetically modified foods promised to be a major advance for humanity in terms of providing safe and plentiful food that was free from disease and resilient to pests. Ironically, by comparison to the BSE problem, the precautionary principle has been seen as a major component of government policy on biotechnology from the outset (see, for example, Hill, 1994). Since the first commercial application of genetic modification in 1982 (production of human insulin for diabetes), there has been a requirement on the manufacturers to ensure that the risks from any modification are quantified. This has resulted in the domination of the market by large corporations (see Tait and Levidow, 1992) and in some respects, this makes the task of inspection easier but the problem of regulation and control more difficult. The small number of players within the market for GM foods is easier to track but they have greater power and financial muscle to resist regulation by government. Perhaps more worrying, however, is the ability of large corporate bodies to influence the manner in which analyses are undertaken and the results interpreted (see Smith, 1990). The power of technocratic elites under such conditions of high uncertainty and emergence runs counter to the conventional scientific model of data testing. Here there is little *a priori* data concerning the safety of the genetic modification and so organisations will be dependent on small-scale laboratory tests, which do not deal with the potential real-world problems, due to the highly controlled conditions under which the testing occurs. This becomes even more of a problem when the state adopts more of a non-interventionist stance in terms of regulation and allows for greater corporate self-regulation (see Smith and Tombs, 1994). O'Riordan and Cameron (1994) have argued that a well-regulated regime is essential for a precautionary approach to new technology, as is a knowledgeable set of publics around issues of risk. Given the political environment in which the biotechnology industry developed in the UK, it is surprising that a precautionary approach was taken at all, especially given the government's views on the primacy of scientific knowledge as illustrated in the BSE crisis. However, given that the government had approved the planting of hundreds of acres of GM crop trials in Britain since 1998 and that they also granted licences for the planting of GM maize, soya, tomato and non-animal rennet to be sold in Britain since that time – such an attempt at precaution may seem more like a veneer of legitimacy rather than a policy objective. One might point to the similar shifts in policy followed by the Conservative government over numerous food-related crises during the 1990s to argue that successive governments have failed to embrace the requirements of a precautionary approach. Porritt (2000) points to the problems that exist in the area of food production and management in his observation that:

"Of all government departments in the UK, none has proved more resistant to change in the way we juggle risk and responsibility, nor more obstinate in defending its rights to turn a scientific blind eye on behalf of UK farmers" (p. 47)

Certainly the view of the primacy of science to resolve the dilemma of risk from GM foods continued into successive Conservative governments, and into the Blair administration. Conversely, the environmental lobby was quick to seize upon notions of emergence within the debate, and argued that there was no burden of proof available to justify the widespread, or even, the controlled planting of GM crops. Whether this viewpoint was scientifically correct or not, the environmental lobby won a considerable victory over the government in helping to change the policy of major retailers. In 1999, the top UK supermarkets and food manufacturers announced that they had set up a worldwide consortium to eliminate GM ingredients from all foods. The antecedent for this was a groundbreaking campaign by Iceland, which was followed quickly by other leading UK grocer multiples, to de-list GM ingredients from their own-brand labels. It would appear that the retailers have taken widespread consumer concerns on board and learned the lessons from the BSE crisis, where the demand for beef products declined after the emergence of the scare and the European beef ban was put into force.

These issues also raise a fundamental question concerning the nature and role of organisational learning within such policy debates. Of particular interest here is the question of how learning takes place between organisations, especially when dealing with trans-scientifc problems which have expert mediation at their core.

Organisational learning

The process of organisational learning has attracted considerable attention within the management literature but this has not really translated into the research on risk and crisis management (see Pauchant and Douville, 1993). To-date, most of the research on learning organisations has been grounded (in broad terms) within the strategy and organisational behaviour literature, with some influential research having been undertaken within systems thinking.

Organisational learning can be defined as the processes through which organisations can effectively incorporate the lessons that arise, and the knowledge that accrues from both internal and external sources into their decision making processes. It also involves changing the core beliefs and assumptions of senior decision makers in the process. The move towards continuous improvement is a key element of this activity. Thus, the learning can take place at both the inter- and intra-organisational level and should, if it is to be effective, draw upon lessons from outside of the operating environment of the organisation concerned (Toft and Reynolds 1993; Smith, 1995). Organisational learning can also be considered as an effective means of helping to prevent the development of crisis potential (Smith and Elliott, 2000) – or what Turner (1976; 1978) describes as incubation and Reason (1990; 1998) has termed "latent conditions". A failure to learn (and to challenge core assumptions in the process) will ensure that organisations will continue to incubate these latent conditions. Effective learning will ultimately lead to a challenge to the main core beliefs; values and assumptions held by the organisation and will require that the principal barriers inherent within the organisation will be overcome. These barriers are clearly a double-edged issue. At one level, they exist to prevent unwanted (and negative) changes to the dominant culture of the organisation. When this culture is actively supportive of safety issues then such defences are an important means of ensuring that deviations are not tolerated. However, the downside of such defences occurs when the organisational culture is both resilient to, and not supportive of, a focus on safety. In this case, the organisation may be resistant to any challenges that are made to its deep core values and would only be prepared to negotiate changes around a number of peripheral issues (see, Sabatier, 1987). In many cases, experts are often called upon to provide a justification for the actions of organisations. This search for legitimacy creates an interesting paradox around expert-mediated "learning".

Many organisations seek to provide technical solutions to problems involving risk and yet it has been suggested that such technical issues only account for 30% of major failures (Turner, 1994). Given that some 70% of major failures may arise from human or organisational factors then one might expect that the extent of the interventions to prevent accidents should reflect this balance. It should follow that organisations ought to invest considerable effort into developing an understanding of the manner in which their culture and decision making process can incubate the potential for failure. Invariably, however, these technical solutions are themselves constrained by the cognitive structure of the experts who can propose them. Thus, paradoxically, expertise can incubate the potential for further failure by denying the legitimacy of non-expert assessments of the problem or accounts that do not originate from the same paradigm

The main barriers to learning around the issues of hazard and crisis potential have been outlined by Smith (2000; Smith and Elliott, 2000; Elliott, Smith & McGuiness, 2000)) and are shown in table 1. The key issues that have a relevance to our discussion here centre around the issues of core beliefs and values, the role of corporate responsibility, reconstruction of events and the centrality of expertise. These issues can now be explored by reference to the cases of BSE and GM Foods.

Table 1: Barriers to Learning

Barrier
Rigidity of core beliefs, values and assumptions
Lack of corporate responsibility
Incrementalist approach (failure to deal with emergence)
Reconstruction and projection
Focus on single loop learning
Peripheral inquiry and decoy phenomenon
Centrality of expertise, denial and the disregard of outsiders
Ineffective communication and information difficulties
Cognitive narrowing and fixation (reductionist)
Maladaption, threat minimisation and environmental shifts

Source: Smith (2000)

Conclusions

> "The value of a historical perspective is that it allows for the 'wisdom of hindsight', illuminating matters that were not at all obvious at the time" – Le Fanu (1999), p. 271.

Within the context of our current discussions, it is clearly difficult to argue for an inclusive theory to encompass learning at the level of the state regulatory mechanism. What is clear from the BSE~GMO conflicts, is that there are obvious problems with the use of risk assessment in reconciling public concerns over risk and in terms of developing policy across government departments. McQuaid and Le Guen (1998) observe that the process of risk regulation is, in itself becoming more complex. This is due in part to the complexity of the issues that need to be controlled and the increased concern that is often expressed by public groups, with the result that there is:

> "...growing demand that regulation of risks should take account of the quality of risk as distinct from objective assessment of the quantum of risk". McQuaid and Le Guen (1998), p. 23.

McQuaid and Le Guen (1998) go on to state that:

> "Even using best available science, such risk analyses cannot be undertaken without making a number of assumptions. Parties who do not share the judgement values implicit in those assumptions may well see the outcome of the exercise as invalid, illegitimate or even not pertinent to the problem" (p. 24).

One possible reason why certain forms of scientific data were rejected by the state concerns the apparent reluctance of the UK civil service to employ quantitative data in decision-making (Anand and Forshner, 1995; Hogwood and Gunn, 1984). One interpretation may be that policy makers would be unable to distinguish between the various forms of technical argument and would, therefore, assume that established knowledge (as expressed within the dominant policy paradigm) has the greater legitimacy. McQuaid and Le Guen (1998) point to another issue, namely that liaison between government departments needs strengthening – a point that was clearly an issue in the BSE crisis as MAFF and the Department of Health seemed to have difficulties in co-ordinating their actions around the specified offal ban.

The limitations of expert knowledge have long been considered as contributing to the incubation of failure potential within organisations. Such failures in knowledge, which can be seen as being related to the systems concept of emergent properties, are important in explaining how the conditions for incubation might occur around the use of expertise. However, it does not provide us with insight into the ways in which government departments and other organisations might respond to these conditions.

The conventional wisdom of many managers often envisages the decision-making process as operating under conditions of rationality, especially when informed by technical expertise. For many risks within modern society, that wisdom may be severely misplaced.

The critique of expertise needs to be seen within a much wider discussion of democratic politics, the role of state and the relationship between power and knowledge (see, Gordon, 1991; Foucault, 1991; Fischer, 1986; 1990). Expertise is seen by some as part of a new elite within society, or as Lasch (1995) termed it, a "new aristocracy of brains" (p.6). Unlike previous elites, the is new groups does not carry with it the civic obligations that were previously required of more established elites with the result that this elite has,

> ".....never been so dangerously isolated from its surroundings" (Lasch, 1995, p.4).

This isolation is further compounded by the mobility of these new elites (Lasch, 1995 which may have a significant impact upon the extent of their accountability to local publics – especially where a hazard is geographically defined. Risk becomes entwined with the role of the state as a provider of security for members of society. In dealing with the issue of insurance, Ewald observed that:

> "The technology of risk, in its different epistemological, economic, moral, juridical and political dimensions becomes the principle of a new political and social economy. Insurance becomes social, not just in the sense that new kinds of risk become insurable, but because European societies come to analyze themselves and their problems in terms of the generalized technology of risk" (Ewald, 1991, p.210).

The erosion of the state's role in providing a regulatory mechanism for dealing with risk, and the globalisation of hazard itself, places expertise in a more central position within policy debates. The precautionary principle's challenge to that power may represent a significant shift in government policy providing that it grounds that principle in action rather than leaving it shrouded in rhetoric.

However, if co-ordination between government departments is problematic at present, then it will be even more so if the precautionary approach adopted fully by government. This, combined with the moves towards self regulation and the erosion of the state (see Smith and Tombs, 1995; Rose, 1996), will simply add further confusion to the current muddle surrounding risk policy and may increase the power of vested interests by keeping risk policy debates off the public agenda and in the "back room" of politics. Nevertheless the public relies on governmental agencies to set safe standards. The problem lies in the fact that governments are political entities and regulation requires that they seek guidance from industry experts in establishing and regulating safety standards. This can often prove to be a risk strategy in itself, as the cases of BSE and GM foods clearly demonstrate.

NOTES

[1]. CJD has a well-established pathology and usually affects older members of the population. New variant CJD, on the other hand, affects young people and is believed to be linked to the consumption of BSE infected offal.

[2]. The Wingspread Conference occurred in January 1998 and brought together 35 scientists and activists to discuss the nature and scope of the precautionary principle (Raffensperger and Tickner, 1999).

[3]. A number of reviews of the history of the BSE debates exist. These include, Dealer (1996) and MAFF (1996f).

[4]. The conclusions of this committee were published by the Department of Health under the title "Report of the Working Party on BSE" in 1989.

[5]. Transmittable Spongiform Enapholopathy. This is the collective term for BSE and CJD.

[6]. Daily Mirror 16th February 1999.

REFERENCES

Anand, P. and Forshner, C. (1995) 'Of mad cows and marmosets: From rational choice to organizational behaviour in crisis management', *British Journal of Management*, 6, pp. 221-233.

Anheier and Moulton (1999) 'Organizational Failures, Breakdowns, and Bankruptcies: An Introduction' in Anheier, H. K. (Ed) *When Things go Wrong – Organizational Failures and Breakdowns*. London :Sage. pp 3-14

Boehmer-Christiansen, S. (1994) "The precautionary principle in Germany - enabling government", in O'Riordan, T. and Cameron J. (1994) (Eds) *Interpreting the Precautionary Principle*. London: Earthscan. pp. 31-60.

Boisot, M. H. (1995*) Information space - A framework for learning in organizations, institutions and culture.* London: Routledge.

Butler, D. (1997) 'How BSE crisis forced Europe out of its complacency', *Nature,* 385 (No. 6611), p. 6-7.

Collingridge, D. and Reeve, C. (1986) *Science Speaks to Power.* London: Francis Pinter

Crenson, M.A. (1972) The Un-politics of Pollution: Decision Making in the Cities. Baltimore, MD: Johns Hopkins University Press.

Cutlip, R. (1996) "Current science on transmission of TSE". Paper presented at the conference, Tissue *Distribution, Inactivation and Transmission of Transmissible Spongiform Encephalopathies (TSE) of Animals.* A Symposium Sponsored by Department of Health and Human Services Food and Drug Administration, Center for Veterinary Medicine (CVM) and the United States Department of Agriculture, Animal and Plant Health Inspection Service (APHIS). 13 & 14 May 1996.

http://dairy.umd.edu/varner/cutlip.html Dealer, S. (1996) *Lethal Legacy BSE - the search for the truth.* London: Bloomsbury.

Department of Health (1996a) CJD and public health - Stephen Dorrell Statement. 96/87 20 March 1996.

http://www.coi.gov.uk/coi/depts/GDH/coi6741b.ok Department of Health (1996b) CJD and public health - Statement by the Chief Medical Officer. 96/86. 20 March 1996.

http://www.coi.gov.uk/coi/depts/GDH/coi6738b.ok Department of Health (1996c) CJD and children - Statement by the Chief Medical Officer. 96/91. 25 March 1996

http://www.coi.gov.uk/coi/depts/GDH/coi6903b.ok Department of Health (1996d) CJD and children - Stephen Dorrell Statement. 96/92. 25 March 1996.

http://www.coi.gov.uk/coi/depts/GDH/coi6904b.ok

Department of Health (1996e) Stephen Dorrell's opening statement to joint meeting of Agriculture and Health Select Committees: 27 march 1996. 96/98. 27 March 1996.

http://www.coi.gov.uk/coi/depts/GDH/coi7033b.ok

Deville, A. and Harding, R. (1997) *Applying the Precautionary Principle*. Sydney: The Federation Press.

Dörner, D. (1996) *The logic of failure. Recognizing and avoiding error in complex situations.* Reading, MASS: Persueus Books. (First published in German in 1989).

Dowie, J. (1999) 'Communication for better decisions: not about "risk"', *Health, Risk and Society*, 1(1), pp. 41-53.

Elliott, D., Smith, D. and McGuinness, T. (2000) "Exploring the failure to learn: Crises and the barriers to learning" *Research in Business* (forthcoming)

Ewald, F. (1991) 'Insurance and risk' in, *Governmentality.* Hemel Hempstead: Harvester Wheatsheaf. pp.197-210.

Fischer, F. (1980) *Politics, Values, and Public Policy: The Problem of Methodology.* Boulder: Westview Press.

Fischer, F. (1990) *Technology and the politics of expertise.* Newbury Park: Sage.

Fortune, J. and Peters, G. (1995) *Learning from Failure: The Systems Approach.* Chichester: Wiley.

Le Fanu, (1999) *The Rise and Fall of Modern Medicine.* London: Little, Brown and Company.

Foucault, M. (1991) 'Politics and the study of discourse' in Governmentality. Hemel Hempstead: Harvester Wheatsheaf. pp.53-72.

Garan, H., (1998), "The Human Factor in Industrial Disaster", *Disaster Prevention and Management*, Vol. 7, No. 2, pp92-102.

Gee, H. (1996) 'Genetic link between BSE and CJD', *Nature.*

http://www.nature.com/Nature2/serve?SID=8505&CAT=Corner&PG=Update/update026.html

Giddens, A. (1999) *Runaway World. How globalisation is reshaping our lives.* London: Profile Books.

Glantz, S. A. and Balbach, E.D. (2000) *Tobacco War. Inside the California Battles.* Berkley: University of California Press.

Gordon, C. (1991) 'Governmental rationality: An introduction' in, Burchell, G., Gordon, C. and Miller, P. (Eds.) (1991) *The Foucault Effect. Studies in Governmentality.* Hemel Hempstead: Harvester Wheatsheaf. pp. 1-51.

Green, K. and Yoxen, E. (1993) "Environmental perspectives of biotechnology", in Smith, D. (Ed.) (1993) *Business and the Environment: Implications of the new environmentalism.* London: Paul Chapman Ltd.

Health and Safety Executive (1996) *Use of Risk Assessment within Government Departments. Report prepared by the Interdepartmental Liaison Group on Risk Assessment.* London: HSE Books

Hill, J. (1994) "The precautionary principle and release of Genetically Modified Organisms (GMOs) to the environment", in O'Riordan, T. and Cameron J. (1994) (Eds) *Interpreting the Precautionary Principle.* London: Earthscan. Pp. 172-182.

Hogwood, B.W. and Gunn, L.A. (1984) *Policy Analysis for the Real World.* Oxford: Oxford University Press.

Hueston, W. (1996) "Overview of risk management applied to TSE". Paper presented at the conference, Tissue_*Distribution, Inactivation and Transmission of Transmissible Spongiform Encephalopathies (TSE) of Animals.* A Symposium Sponsored by Department of Health and Human Services Food and Drug Administration, Center for Veterinary Medicine (CVM) and the United States Department of Agriculture, Animal and Plant Health Inspection Service (APHIS). 13 & 14 May 1996.
http://dairy.umd.edu/varner/huston2.html

Irwin, A. (1995) *Citizen Science.* London: Routledge.

Irwin, A. and Wynne, B. (Eds.) (1996) *Misunderstanding Science? The public reconstruction of science and technology.* Cambridge: Cambridge University Press.

Irwin, A., Dale, A. and Smith, D. (1996) 'Science and Hell's Kitchen - The local understanding of hazard issues', in. Irwin, A. and Wynne, B. (Eds.) (1996) *Misunderstanding Science? The public reconstruction of science and technology.* Cambridge: Cambridge University Press. pp. 47-64.

Jackson, W. (1999) 'Foreword' in, Raffensperger, C. and Tinkner, J. (Eds) (1999) *Protecting public health and the environment. Implementing the precautionary principle.* Washington DC: Island Books. pp. xv-xiv.

Jones, R.M. (1997) *Strategic management in a hostile environment. Lessons from the tobacco industry.* Westport: Quorum Books.

Jordan, A. and O'Riordan, T. (1999) 'The precautionary principle in contemporary environmental policy and politics' in, Raffensperger, C. and Tinkner, J. (Eds) (1999) *Protecting public health and the environment. Implementing the precautionary principle.* Washington DC: Island Books. pp. 15-35.

Kimberlin, R. (1996). "Current science on the tissue distribution of TSE". Paper presented at the conference, Tissue_*Distribution, Inactivation and Transmission of Transmissible Spongiform Encephalopathies (TSE) of Animals.* A Symposium Sponsored by Department of Health and Human Services Food and Drug Administration, Center for Veterinary Medicine (CVM) and the United States Department of Agriculture, Animal and Plant Health Inspection Service (APHIS). 13 & 14 May 1996.
http://dairy.umd.edu/varner/kimber.html

Knight, F. (1921) *Risk, Uncertainty and Profit.* Boston: Houghton-Mifflin.

Lacy, R. (1996a) 'How Now Mad Cow?'. *Viva Guide* Number 3.
http://www.veg.org/veg/Orgs/Viva/Guides/madcow.html

Lacy, R. (1996b) 'Bovine spongiform encephalopathy is being maintained by vertical and horizontal transmission', Letters Section, *British Medical Journal*, 312, 20 January 1996.

Lasch, C. (1995) *The revolt of the elites and the betrayal of democracy.* New York: W.W. Norton.

MAFF (1996a) Statement of 20 March 1996 by Agriculture Minister Douglas Hogg to House of Commons.
http://www.maff.gov.uk/animalh/bse/bsestat1.htm

MAFF (1996b) Statement of 25 March 1996 by Agriculture Minister Douglas Hogg to House of Commons.
http://www.maff.gov.uk/animalh/bse/bsestat2.htm

MAFF (1996c) Announcement of 28 March 1996 by Agriculture Minister Douglas Hogg on beef aid.
http://www.maff.gov.uk/animalh/bse/bsestata.htm

MAFF (1996d) Statement of 3 April 1996 by Agriculture Minister Douglas Hogg on tackling the beef crisis.

http://www.maff.gov.uk/animalh/bse/bsestatb.htm
MAFF (1996e) Statement of 16 April 1996 by Agriculture Minister Douglas Hogg to House of Commons
http://www.maff.gov.uk/animalh/bse/bsestat3.htm
MAFF (1996f) 'Programme to eradicate BSE in the United Kingdom'.
http://www.maff.gov.uk/animalh/bse/eradprog/eradprog.htm
MAFF (1996f) 'BSE, a chronology of events'. Appendix 2 of *Bovine Spongiform Encephalopathy in Great Britain: a progress report*, November 1996
http://www.maff.gov.uk/animalh/bse/chron.htm, MAFF (1997a) *BSE: a summary of developments over the past twelve Months*.
http://www.maff.gov.uk/animalh/bse/bseanni.htm MAFF (1997b) UK: *Selective Cull of Cattle Scheme*
http://www.maff.gov.uk/animalh/bse/selcull/sc1.htm#A10
McQuaid, J. and Le Guen, J-M. (1998) 'The use of risk assessment in Government', in Hester, R.E. and Harrison, R.M. (Eds) *Issues in Environmental Science and Technology*. Number 9. Cambridge: Royal Society of Chemistry. pp. 21-36.
Millstone, E. (1994) 'Regulation, innovation and public welfare: The example of the Food Industry', *Technology Analysis and Strategic Management*, 6(3), pp. 329-340.
O'Riordan, T. and Cameron J. (1994) "The history and contemporary significance of the precautionary principle" in O'Riordan, T. and Cameron J. (1994) (Eds) *Interpreting the Precautionary Principle*. London: Earthscan. Pp. 12-30
Orey, M. (1999) *Assuming the risk: The mavericks, the lawyers, and the whistle-blowers who beat big tobacco*. Boston, MASS: Little, Brown and Company.
Pauchant, T. and Douville, R. (1993) 'Recent research in crisis management: a study of 24 authors' publications from 1986 to 1991', *Industrial and Environmental Crisis Quarterly*, 7(1), pp. 43-66.
Porritt, J.(2000) *Playing Safe: Science and the environment*. London: Thames and Hudson.
Raffensperger, C. and Tinkner, J. (1999) 'Introduction: To foresee and forestall' in, Raffensperger, C. and Tinkner, J. (Eds) (1999) *Protecting public health and the environment. Implementing the precautionary principle*. Washington DC: Island Books. pp. 1-12.
Reason, J. T. (1987) 'An interactionist's view of system pathology', in Wise, J.A. and Debons, A. (Eds) *Information Systems: Failure Analysis*. Berlin: Springer-Verlag. Pp. 211-220
Reason, J.T. (1990) *Human Error*. Cambridge: Cambridge University Press.
Rose, N. (1996) "The death of the social? Re-figuring the territory of government", *Economy and Society*, 25(3), pp. 327-356.
Sabatier, P. (1987) 'Knowledge, policy-oriented learning, and Policy change', *Knowledge: Creation, Diffusion, Utilization*. 8(4), pp. 649-692.
Santillo, D., Johnston, P. and Stringer, R. (1999) 'The precautionary principle in practice: A mandate for anticipatory preventative action', in, Raffensperger, C. and Tinkner, J. (Eds) (1999) *Protecting public health and the environment. Implementing the precautionary principle*. Washington DC: Island Books. pp. 36-50.
Schurman, D.L. and Reason, J.T. (1991) 'Systemic Influences of, Management and organisations on safety, in Karwowski, W. and Yates, J.W. (EDS). *Advances in Industrial Ergonomics and Safety III.* London: Taylor and Francis. Pp.693-700.
SEAC (1996a) BSE: statement by Spongiform Encephalopathy Advisory Committee (SEAC) of 20 March 1996
http://www.maff.gov.uk/animalh/bse/seac1.htm
SEAC (1996b) BSE: statement by Spongiform Encephalopathy Advisory Committee (SEAC) of 24 March 1996 http://www.maff.gov.uk/animalh/bse/seac2.htm SEAC (1996c) BSE: statement by Spongiform Encephalopathy Advisory
Committee (SEAC) of 7 June 1996: Recommendations on the handling of waste material from cattle.
http://www.maff.gov.uk/animalh/bse/seac2a.htm
SEAC (1996d) Statement by Spongiform Encephalopathy Advisory Committee (SEAC) on sheep and BSE. 10 July 1996. http://www.maff.gov.uk/animalh/bse/seac3.htm
SEAC (1996e) SEAC statement on maternal transmission of BSE. 29 July 1996
http://www.maff.gov.uk/animalh/bse/seac4.htm
SEAC (1997) Spongiform Encephalopathy Advisory Committee Statement on sheep and BSE. 23 May 1997
http://www.maff.gov.uk/animalh/bse/seac5.htm

Sheldon, T. and Smith, D. (1992) 'Assessing the health effects of waste disposal sites: Issues in risk analysis and some Bayesian conclusions'. In, Clark, M., Smith, D. and Blowers, A. (Eds.) (1992) *Waste location: Spatial aspects of waste management, hazards and disposal.* London: Routledge. Pp. 158-186.

Smith, D. (1990) 'Corporate power and the politics of uncertainty: Risk management at the Canvey Island complex'. *Industrial Crisis Quarterly*, 4(1), pp. 1-26.

Smith, D. (1991)'The Kraken wakes - the political dynamics of the hazardous waste issue' *Industrial Crisis Quarterly* 5(3), pp. 189-207.

Smith, D. (1993) 'The Frankenstein factor - Corporate responsibility and the environment', in Smith, D. (Ed.) (1993) *Business and the Environment: Implications of the new environmentalism.* London: Paul Chapman Ltd. pp.172-189.

Smith, D. (1995) 'The Dark Side of Excellence: Managing Strategic Failures', in Thompson, J. (Ed) (1995) *Handbook of Strategic Management.* London: Butterworth-Heinemann. pp. 161-191.

Smith, D. (1999) Accidents will happen.... but not to us? Safety as a core management function. *Mimeo.* Centre for Risk and Crisis Management Occasional Papers Number 1. University of Sheffield. (www.cracm.com/papers/99.1)

Smith, D. (2000) Living on Factory Row. Issues in risk, public health and the precautionary principle. *Mimeo.* Centre for Risk and Crisis Management Occasional Papers Number 1. University of Sheffield. (www.cracm.com/papers/20.1)

Smith, D. and Elliott, D. (2000) Moving Beyond Denial: Exploring the Barriers to Learning from Crisis. *Mimeo.* Centre for Risk and Crisis Management. University of Sheffield.

Smith, D. and Lloyd, D. (1993) 'Wither objectivity: Technocracy and the social construction of risk', in Cox, R.F. and Watson, I.A. (Eds) (1993) *Engineers and Risk Issues.* Manchester: Safety and Reliability Society. Pp. 2/1-2/18.

Smith, D. and McCloskey, J. (1998) ""Risk Communication and Social Amplification of Public Sector Risk", *Public Money and Management*, 18(4), pp41-50.

Smith, D. and Tombs, S. (1995) 'Self regulation as a control strategy for Major Hazards' *Journal of Management Studies*, 32(5), pp. 619-636.

Tait, J. and Levidow, L. (1992) "Proactive and reactive approaches to risk regulation: the case of biotechnology", *Futures*, pp. 219-231.

Tenner, E. (1996)_*Why things bite back - New technology and the revenge effect.* London: Fourth Estate Limited.

Toft, B. and Reynolds, S. (1994) *Learning from Disasters.* London : Butterworth

Tombs, S. and Smith, D. (1995) 'Corporate responsibility and crisis management: some insights from political and social theory'. *Journal of Contingencies and Crisis Management*, 3(3), pp. 135-148

Turner, B.A. (1976) 'The organizational and interorganizational development of disasters', *Administrative Science Quarterly*, 21, pp. 378-397.

Turner, B.A. (1978) *Man-Made Disasters.* London: Wykeham.

Turner, B.A. (1994) 'Causes of Disaster: Sloppy Management', *British Journal of Management*, 5, pp. 215-219.

Weinberg, A.M. (1972) 'Science and Trans-science'. *Minerva*, 10, pp. 209-222.

Wilson, D., Hickson, D. J. and Miller, S. J. (1999) 'Decision Overreach as a Reason for Failure: How Oreganizations can Overbalance' in Anheier, H. K. (Ed) *When Things go Wrong – Organizational Failures and Breakdowns.London*, Sage. pp 35-49

Zapf, D. and Reason, J.T. (1994) 'Introduction: Human errors and error handling', *Applied Psychology: An International Review*, 43 (4), pp. 427-432.

Zucker, L. G. and Darby, M. R. (1999) 'Costly Information: Firm Transformation, Exit, or Persistent failure' in Anheier, H. K. (Ed) *When Things go Wrong – Organizational Failures and Breakdowns._London*, Sage. pp 17-33

THE SOCIAL CONSTRUCTION AND DECONSTRUCTION OF RISK

HEATHER HÖPFL
Head of School of Operations Analysis and Human Resource
Management
Newcastle Business School
University of Northumbria at Newcastle
Northumberland Building
Newcastle Upon Tyne, NE1 8ST, UK

The Need for a Wider Context

In a paper presented to the 43rd IASS Conference in Rome in 1990, Capt. Heino Caesar, at that time General Manager, Flight Operations, Inspection and Safety Pilot for Lufthansa, identified a number of critical issues relating to air safety which require considered attention. In particular, Caesar drew attention to what he saw as an over-reliance on technical and technological developments in the pursuit of improved air safety at the expense of a more systematic analysis of the organisational context. As an exceptionally experienced pilot, Caesar realised that many of the problems he encountered in his working life arose from the way in which tasks were classified and perceived, that safety was as much a problem of social construction as of technical resolution. As justification for this contention, he cited the evidence that over the previous thirty one years, that is up to 1990, in cases of total losses, 76% of cases were recorded as being the result of errors by cockpit crew.

What this meant and, indeed, how it related to wider organisational issues was rarely considered and despite this statistic, the reports of major air disaster frequently provided little useable information beyond general and summary phrases such as "(the) crew had communications problems". Similarly, Michele Visciola, a Research Psychologist with Aeritalia, has pointed out that "knowledge of pilot errors is fragmented and not supported by a theory" (Visciola, 1990) and John Lauber (1990) of theUS National Transportation Safety Board has commented that "no human error is the result of a single cause or factor". In other words, far more attention needs to be paid to the complexity of human factor failures and the organisational context in which they occur. Caesar points out "that future incidents and accidents must be far more carefully analysed as to human performance factors, to produce tools to develop failure avoidance strategies, and to show how the duties were performed within a social, organisational and cultural context" (Caesar, 1990: 4). This contention is not surprising to organisational theorists who have attempted to give greater emphasis to the wider domain in which aircraft accidents and incidents need to be considered. Moreover, the

E. Coles, D. Smith and S. Tombs (eds.), Risk Management and Society, 125–142.
© 2000 *Kluwer Academic Publishers. Printed in the Netherlands.*

ways in which accidents and incidents are categorised has significant implications for the ways in which the industry and, in particular, aircraft manufacturers come to conceive technological solutions in response to definitions of problems. For example, design considerations might reflect needs created by operational commitments where the latter are inadequately evaluated to assess their contribution to the problem.

A Common Set of Characteristics

The past two decades have seen a series of major disasters affecting such diverse technologies as nuclear installations, chemical plants, oils tankers and ferries, railway networks, oil platforms and, of course, commercial and military aircraft. Despite the obvious differences in the industries involved and their technologies, it has become apparent from the analysis of such disasters that, at a contextual level, there are many common characteristics (Turner, 1978; Reason, 1990). As a result, recent attention has been given to the socio-technical aspects of safety systems to the complexity of the contributory causes in accident analysis, to the multiplicity of ways in which systems can fail, to the predominance of human factor contributions to failure, to perceptual and information difficulties and, not least, to the appreciation of the historical dimension, the fact that disasters often have a long incubation period. This widening of the boundary around safety issues has resulted in a move away from what Toft has described as a "propensity to look for simple causal solutions ...... shaped by the technical concerns of the engineering community" (Toft, 1992) towards a commitment to the recognition of the social and organisational context of incidents and accidents.

The Social Context

The social context of accidents has been receiving more attention in recent years and the significance of tacit assumptions has been seen to have a considerable bearing on the trajectory of disasters. The argument presented here seeks to address some of these issues in order to explore the significance of contextual variables to air safety and to provide an indicative analysis of the social and organisational dynamics of flight safety. British Airways, in the development of the aviation system information system BASIS, took a brave step towards exposing and evaluating the wider context of accidents and incidents in a leading edge approach to safety information. However, in order to see why the British Airways approach to these issues was such a radical one, it is necessary to look at changes which had been and, indeed, continue to take place within the airline as a whole over more than a decade. The reason for this contextual information meets precisely the same need as that for the wider context for safety information. It is easier to understand why British Airways Safety Services were able to produce such an imaginative and different approach to safety when the developments are seen in the context of wider cultural changes within the airline As a whole. Without the wider context of change in the airline, it is unlikely that the paradigm shift which took place in the safety function would have been able to occur.

The Wider Context of Culture Change in British Airways

British Airways has now seen over a decade of sustained change and in June 1996 announced the proposed partnership arrangement with American Airlines which has produced a formidable giant amongst carriers. However, in 1983 British Airways was experiencing considerable problems and a number of external pressures to change. The airline had already made significant reductions in its costs but was finding it increasingly difficult to compete as an airline on the basis of cost alone. Moreover, many costs, such as aviation fuel and airport charges were outside the company's control. This was a time when the company decided to make a commitment to important changes in customer service. Customer service was seen as "the critical factor that could give a competitive edge but could not be easily duplicated" (Bruce, 1987). Colin Marshall, then Chief Executive of British Airways recognised the problems at British Airways as those of an organisation which had become "quite demoralised ...... very demotivated indeed and therefore lacking in customer service ...... was extremely introverted, had really no grasp of what the market place wanted, what the customer wanted" (Young, 1989). Marshall saw the need "to create some motivational vehicle with the employees ...... (to raise) their morale, and in turn .... Customer service" (Young, 1989). Marshall believed in visible management and gave many personal examples of good management and practice as part of his own management style.

Customer Service

In the cost cutting survival strategies of the early 1980's training in British Airways had been cut back. In March 1983, Marshall announced, "we may need to put people through refresher courses to really concentrate on teaching staff how to sell the airline and its services". Eventually, this led to a campaign which became known as Putting the Customer First. Passenger research followed and in November, 1983 an event, Putting People First (PPF) was held. It was felt that Putting People First was a more appropriate term since it embraced both customers and customer service staff in its emphasis on people. The importance of the PPF programme was in what it said about management attitudes and intentions. It represented a considerable investment of time and money on the part of the airline and gave force to the notion of service and a thorough understanding of customer needs. This was to have considerable transformational significance for the way in which airline staff perceived the business and the role they played in it. However, few pilots found much appeal in this message and several British Airways' pilots attending the Flight Safety Foundation Air Safety Seminar in Singapore in1991 commented informally that they were resistant to the idea of being converted into "cuddly pilots". This attitude reflected a concern expressed by flight crew that safety might be compromised in a trade-off between customer care and cost. However, in terms of the work going on in Safety Services at that time, this was far from the case.

Concentrating on Lacks and Deficiencies

After more than a decade of change, British Airways has developed a critical awareness of the organisational implications of developing a service culture. The change sought to promote business awareness at all levels of the organisation and to help staff to understand the functional interdependencies on which strategic outcomes depend. Multidisciplinary teams encouraged collaborative styles of management, and there has been a commitment to management development through higher education and open and bridge building attitude towards the outside world. Over the past fifteen years, BA has gone through a radical transformation and although there are inevitably criticisms which can be made of the style of the changes, the results have been impressive and the company is now an acknowledged leader in its field. The changes opened debate, exposing "what was absent rather than what was there" (Bruce, 1987) and when this notion of a problem defined in terms of absences is extended across the whole operation it provides an insight into some of the more radical philosophical changes which were introduced into safety management. The significance of this is perhaps apparent in relation to the ways in which a privileged position, a style of management and the power to define the safety function were overturned by an ideological change. This change led to a new approach to the way information was regarded and evaluated.

In short, the culture change programme guided the organisation into a new understanding of itself and into new approaches to business practices. The fact that staff had to confront issues related to the way in which they defined themselves and their work produced a range of spaces within which existential issues and matters of corporate identity could be challenged and redefined, albeit that the primary task of re-definition lay with senior management.

A Paradigm Shift in Safety Services

An understanding of the organisational context of change in BA is important to the argument that follows. Culture change in the airline as a whole brought about a greater awareness of the need for a service culture. This involved a paradigm shift in terms of the relationship between airline staff and external customers and also in terms of internal interdependencies. In Safety Services, the change in the culture of the airline away from "a military approach to management" (Bruce, 1987: 25) brought about a greater understanding of the importance of air safety for the public perception of airline companies: a considerable dimension of customer service for an airline operator. Consequently, the need to have a visible commitment to safety issues is part of the way in which airlines have responded to public demands for reassurances about air travel. The ways in which air safety is construed is partly to do with the management of the perception of the risk of air travel. Therefore, to make "safety" a prominent issue was important to the expansion of the notion of customer service. However, to the credit of the Safety Services function, BA made a significant effort not only to change the way in which safety was perceived and the way in which such perceptions were managed but to try to deal with some of the specific problems which arose within the spaces created by the new management philosophy.

As a result, the Safety Services unit in BA made a critical assessment of its role and structuring and decided that a radical change was needed in the system of accident and incident reporting and analysis. Staff saw the need for a fundamental and ideological change in their methods of working. Starting from an inherited forty-seven filing cabinets full of largely redundant and unusable safety data, the department progressed rapidly to a newly designed safety information system which was to provide the active memory of the organisation on matters f safety. The British Airways Safety Information System, BASIS, was designed to maximise data capture and to identify areas of significant risk. It was also designed to provide information regarding the effectiveness of decision-making with a bearing on flight safety and to facilitate rapid distribution of safety information to line managers. Moreover, being designed by experienced end-users of safety information, BASIS incorporates features which offer conspicuous analytical benefits: a risk index, a detailed reference system and a trend analysis function based on operational and technical keywords; patterns of human factor incidents can be analysed and implications for training needs and equipment design and modification can be detected.

End-users Involved in the Design

The evident success of the BASIS system and the interest it generated outside the company (and, indeed, outside the airline industry) attests to the range of practical applications of which the system is capable. In part, this is because BASIS was designed by people with extensive experience of the context in which the system was to be located, with practical working knowledge of incidents, perceptions, technical problems, human factors, the social and organisational context and so on. "The strength of BASIS lies not in the storing of information, but in using it to ask questions about the operation and to provide some answers ...... a practical probing into all the available data with the intention of uncovering the unknown and undesirable" (Holtom, 1991). The statement from Holtom, himself an experienced pilot, indicates a commitment from the company to getting behind surface meanings and to deconstructing taken for granted assumptions.

The ways in which the "unknown and undesirable" might be construed and conceptualised are important in order to explore the dynamic tensions between those things which a safety information system such as BASIS might tackle and, in modification and development, might reasonably incorporate, and those aspects of the system which are destined always to remain outside the scope of precise data capture but which feature significantly in the interpretive domain of the broader system. This appreciation of the dialectics of safety information, between the rational aspects of systems, between those categories which can be used to capture and aggregate data and those which remain elusive, is important from the point of view of making transparent those aspects of safety systems which are irreducible and, therefore, potentially the most threatening. In this respect, the power to construct, define and manipulate information into categories and classes of answer deserves attention. It is important in relation to the construction of the notion of "risk" to consider the power both to define risk and to issue authoritative statements about the nature of risk.

Classification and the Power to Define

The ability to classify information as carrying this or that risk is often associated with the rhetorical position of the organisation which is making such statements. Hence, the capacity to construct statements regarding risk rests on a power relationship within organisations and between organisations and the outside world. In this way, one definition of a situation may carry more significance than another. This may be the result of a hierarchical difference or a difference in apparent expertise. Villagers in Aberfan who voiced their fears about an impending disaster were unable to penetrate the authoritative (and, as time proved, incorrect) definitions of risk which were offered by the National Coal Board. Therefore, where one group has the power to define another group is disempowered. For example, the construction of air safety statistics is regarded as an apparently value-free, neutral and objective means of classifying information about air accident and incident data. However, a simple example of an accident being classified as caused by adverse weather requires further scrutiny. For example, nineteen accidents in the past ten years attributed to bad weather (see IATA statistics) might perhaps have been the result of inadequate training or management pressure to achieve operational targets. Hull loses which are classified as resulting from human factors are categorised in terms of "active failures" where there have been deviations from standard operating procedures, "passive failures" where there has been a failure to act, or "proficiency failures" where there is lack of skill as a contributory cause. However, whereas it is relatively easy to identify what Turner (1978) has termed the "precipitating event" in such accidents, it is far more difficult to determine the history of a particular failure within its organisational context and to understand the contribution of "absences" to the outcome. It is necessary to identify why a flight crew member might fail to act or act inappropriately. However, such answers are usually embedded in the culture of the organisation and its ways of working.

The constructions of definitions and social meanings rests on the use of language and ritualistic behaviour, on territorial and spatial division, on hierarchical arrangements, on theatricality and on the exclusion of certain people from the place where definitions are created. The construction of risk is such that categories of acts and actors are created and then responded to as if such constructions were objective realities. Thus, if adverse weather were thought to be a significant factor in aircraft accidents then the response of manufacturers might be to give more priority to the design of technologies to overcome adverse weather conditions. However, the reason an accident might appear to be caused by bad weather might simply be that management deadlines indicated that a flight be attempted for operational reasons beyond normal safety expectations.

Beyond the Taken for Granted

Reason's work in particular, following earlier work by Turner (1978), has received considerable attention for its emphasis on what is hidden, on "latent failures" (1990: 28) which may only become evident when they occur with a "precipitating event" (Turner, 1978) which causes the system to fail. Moreover, Reason contends that "there is a growing awareness ...... that attempts to discover and remedy these latent failures will achieve greater safety benefits than will localised efforts to minimise active failures" (Reason, 1990: 476-7), for example, in the nuclear industry, failure to perform necessary maintenance activities, that is, latent failure has played a major role in incidents and accidents in nuclear installations (Rasmussen, 1980). Part of the approach to the detection of latent failures must involve an understanding of where the power to define the nature of the activity and its attendant risks actually lies. The need to attend to what is hidden is well understood by theorists working on the nature of risk. However, the extent to which it is possible to expose what is in deconstructionist terms "deferred" by the privileging of specific interpretations is a matter of debate.

There is a need to see behind the rhetoric of safety and to expose the tacit assumptions, privileged constructions and authority to define in order to permit alternative ways of understanding a problem, or the way in which a social situation is defined. The need to give greater attention to data which is not immediately apparent and which does not yield easily to data capture has a growing currency in air safety concerns. In practice, it is difficult to operationalise attention to such factors. What is concealed behind the constructions of appearances, of order and normality, may be to a great extent resistant to attempts to interpret, expose or deconstruct such appearances. The power of such factors, however, is considerable and continually threatens to subvert understandings which organisations may cherish about themselves as, indeed, numerous studies of major disasters have shown (Schwartz, 1991).

An Incident is Reported

The following example offers an illustration of what has referred to as either "Intimidation or Leadership" in the flight deck chain of command (Hurst and Hurst, 1983). The incident, which involved a British Airways scheduled flight to Barcelona, demonstrates a number of features which may be present in a potentially disasterous situation. The flight was flown throughout by the co-pilot, the First Officer. The flight was uneventful and there was no expectation of bad weather. The approach to Barcelona was normal. However, the flight crew were instructed to negotiate a "go-around" due to fog. The Captain was given Gerona as the diversion airfield and accepted a route which was not direct, that is, not in the direction of the airfield. The diversion involved what is termed a "long procedural let down". The fuel situation for the flight was such that, given this alternative route, the aircraft would land with what is technically referred to as "less than reserve fuel". Under such circumstances, regulations require that a "Fuel Emergency" must be declared. However, Barcelona Air Traffic Control was not informed of this situation and no fuel emergency was declared.

In the meantime, the weather at Gerona was deteriorating. The First Officer was flying the aircraft and was aware of the fuel situation. The Captain was commanding the flight and dealing with radio communications with Air Traffic Control. A window in the weather enabled this flight to land as originally planned at Barcelona so fortunately, the aircraft landed safely and uneventfully. However, it landed with less than reserve fuel.

It is a matter of speculation whether there would have been sufficient fuel for a "go-around". The airfield closed seven minutes after the flight landed. The First Officer was aware of the fuel situation but was "unable" to Communicate his fear to the Captain. When the plane landed the First Officer reported later that he had been physically sick. The Captain was completely unaware of the situation. Landing with less than reserve fuel is a situation which requires a formal report. The report indicated simply that the flight had landed with "less than reserve fuel" and a brief outline of the history of the flight.

However, about three weeks after this event, a senior and experienced Captain was on a flight with the same First Officer when they encountered a situation where the same sequence of events began to unfold. However, on this occasion, the Captain requested that Air Traffic Control change the routing. The Captain noticed that the First Officer looked visibly relieved and asked him, "Have you been here before?" Later, in an informal setting, the young First Officer was able to relate his fears. Interestingly, despite the very different management styles of the two Captains, the First Officer explained that on both occasions he had felt unable to voice his concerns at the crucial time.

Some Contextual Issues

This incident demonstrates a number of points. First, it emphasises the power of taken for granted assumptions about the chain of command, of implicit authority and of the difficulties which First Officers have in questioning the Captain's assessment of the situation even when confronted by very different management styles and in potentially life-threatening situations. Secondly, it demonstrates an inability to act on the evaluation of a risk where hierarchy or competence or fear of embarrassment prove more powerful than impending disaster or death. The First Officer in this illustration confided to the senior captain that he had realised that "he would rather die than countermand the orders of a senior officer". It also demonstrates poor job characteristics in relation to the allocation of function between flight crew roles and, fourthly, it shows how rich accounts of incidents may not enter the reporting system. The information which was so potentially rich was excluded from the mandatory report at that time. The complementary account emerged informally because of the sensitivity of an experienced (Management) Captain. Incidents tend to be reported in technical terms at the expense of "rich picture" data as in the example above. Had this incident had more serious consequences, it is the sort of situation which might have been recorded as resulting from "poor crew communication".

Recognising the Organisational Context

Unfortunately, it is not an isolated incident. The literature abounds with examples where first officers have been unable to take control in extremely hazardous situations (Foushee, 1981; Hurst, op.cit, p.169 et seq; Lauber, op.cit; Wiener, 1983, p.91; McKenna, 1990). The Avianca 707 crash, New York January 25th, 1990 provides clear Cockpit Voice Recorder (CVR) evidence that the crew knew that they had insufficient fuel to carry out air traffic control instructions. The account of a human error accident provided by Lauber, a McDonnell Douglas MD-80 operated by North West Airlines which crashed on take-off from Detroit's Metropolitan Wayne County Airport, in which crew failed to run the taxi checklist, makes the point that, "The behaviour of individual pilots never occurs in a vacuum but rather always occurs in an organisational context ...... We will be virtually guaranteeing another accident much like it in the future if we fail to understand that" (1990: 4).

The purpose of these examples is to point out that such situational difficulties cannot be tackled by addressing the issues at the level of the relationship between Captain and First Officer, that is, at the level of role specification and interpersonal interaction. These micro-sociological variables have to be seen in the organisational context in which they occur. The chain of command cannot be seen as an interactional repertoire which can be adjusted by training. The move to a more safety aware culture requires a challenge to the taken for granted assumptions which support existing behaviour. This would need to take account of the formal aspects of flight crew behaviour and the control systems, regulations and procedures: the formal and specified roles of the line of command; the position of flight crew in the organisational structure; power structures and the basis of authority, competence, seniority and experience; and also, how these formal structures and control systems are underpinned and supported by a variety of informal assumptions which are largely taken for granted: the rituals and patterns of behaviour which have their origins in traditions, status differentials and assumptions which establish and re-inforce order; the symbols which establish status. Rank and authority – corporate emblems, the uniforms, livery; and the stories, myths and legends which provide a "rich picture" of the informal aspects of cockpit life. An examination of the cultural basis of cockpit behaviour and an appreciation of the interplay between formal and informal variables is an essential part of any attempt to introduce changes into the status quo whether to the task, the technology, the individuals or their roles. This complex web of interactions, symbols, role repertoires and so on should be seen to have a range of consequences which should not be underestimated and, therefore, any changes which are made which effect flight crew should be attended to with an appreciation of such consequences.

Systems Ergonomics

Traditionally, much of the research in cockpit design and development, ergonomics, interface design, job design and job structuring has been done on a piece-meal basis. Changes were often made retrospectively to correct for design deficiencies or to improve interpersonal skills. There was little evidence of any systematic analysis of the overall context of flight operations. Consequently, whereas there was a considerable body of literature in the human factors area there was little real attempt to integrate studies of human behaviour and performance, cockpit resource management, flight deck ergonomics and so on. In the past, engineering design has concentrated on improving the quality of hardware. Socio-psychological contributions have tended to lag behind. Where they have been tackled, it has tended to be by modification to rectify or adjust for human factors. Social and organisational issues have not been seen to be crucial to the initial design conception. In the past twenty five years there has been a fairly dramatic change of emphasis in design and the emergence of "systems ergonomics" where there is the recognition for the need for human factors to be considered at a much earlier stage of design and development; techniques for the allocation of function between the human being and the machine and the development of off-line selection and training techniques.

Approaches in this area have attempted to deal with many of the problems of classical ergonomics by giving the psychologist a role which is no longer simply remedial. Instead, there is an emphasis on the reasoned allocation of function between user and machine. In practice this commitment has proved much more difficult to effect than the declared commitment of "systems ergonomics" might suggest. Consequently, it is largely in the areas of ergonomics and anthropometry that human factors have tended to be taken into account with the broader aspects of human factors design being considered a marginal concern. Yet even some of the basic ergonomics have proved deficient. There are innumerable well-publicised examples to support this contention within highly technological environments such as the control room design at the Three Mile Island Nuclear power station. Aircraft examples are common, for instance, the positioning of the electronic engine limiters on the Boeing 767. These switch off the engine fuel supply but their proximity to other similar switches meant they could be switched off accidentally. There was a similar problem with the BAC 1-11 where fuel switches have been unintentionally selected instead of adjacent and similar ones. In one modern aircraft, the flight deck flood light switch was originally located close to one that could de-pressurise the aircraft. However, it must be said that the Boeing have attempted over the past ten years to address many of the problems identified above and flight crew have been given more say in the design of their working environment.

Boundaries Too Tightly Drawn

In general, There have been contributions to human factors and human performance in ergonomics in cockpit design, socio-technical theory has contributed to flight crew job design, there is a large amount of accumulated data on the psychological aspects of flight deck behaviour and performance, however, these contributions tend to have a specific range of convenience in theoretical terms. Similar problems have arisen with socio-technical attempts to incluence design issues. Although not part of the original conception of this approach (Trist et al, 1963), in practice socio-technical applications have tended to focus on adjusting the social aspects of the job design to fit around the technology. Furthermore, technology has often produced adverse effects where users have feared unwelcome changes in their patterns of work, experienced de-skilling, new stresses and health problems (Pearce, 1984). In short, for a variety of reasons, design may ignore organisational repercussions and its consequential costs. These may be enormous not only in poor performance but in various forms of resistance to new systems developments, for example, reduced ability to exercise discretion or make decisions, poor response rates to abnormal conditions, poor morale, complacency, de-skilling. This apparent contradiction between the development of increasingly sophisticated hardware and the contribution of human factors to air accidents demonstrates a serious dichotomy in the understanding in which social and technical factors operate in the dynamics of flight deck interaction. All in all, the boundary of what constitutes a human factor problem has been too narrowly drawn.

Many of these concerns have been voiced by flight crew and are being extensively researched. However, even the contribution of cockpit Resource Management (Ruffell Smith; Freeman and Simmon, 1990; Houston, 1991) pays insufficient attention to the organisational dynamics and lacks an appreciation of the wider contextual variables. It could be said that it operates as a sub-set of the management of the organisation within which it operates and, consequently, may never challenge the managerial assumptions on which is based. These contextual variables are extremely important to the way a system works in practice. Freeman and Simmon (1990: 55) provide a valuable framework for understanding the various skill inputs which should feature in a crew resource management programme and they argue convincingly for a more systematic training to promote safety via emphasis on the non-technical aspects of flight deck operations. Hence, it might be desirable to take their argument to a stage further to suggest that crew resource management can only be improved by developing an appreciation of its role in the wider organisational context. An inadequate diagnosis of the cultural context in which any change is necessary is likely to produce various systems repercussions and disequilibrium states some of which will be predictable and other not.

The Historical Context

In a similar way, Turner's (1976) influential work on the causes of disasters argued that large-scale accidents have an "incubation period" in which there are a series of unnoticed events which are likely to run counter to established beliefs about the way that the system operates or that risks are defined. Turner encouraged safety researchers to concern themselves with "the cultural disruption which is produced when anticipated patterns of information fail to materialise" in order to develop an appreciation of the way in which individuals "gradually come to develop and rely on a mistaken view of the world"(1976: 193). "The problem of understanding the origins of disaster is the problem of understanding and accounting for harmful discharges of energy which occur in ways unanticipated by those pursuing orderly goals" (1976: 201).

The incubation period ends when some precipitating event draws attention to the discrepancy between the environment as it is believed to be and the environment as it actually is. This forces into the open the "hidden, ambiguous or anomalous events which have accumulated during the incubation period" (Turner, 1976: 201) producing a sudden shift in information levels. Consequently, Turner argues that sensitivity to information is vital to the prevention of disasters. However, this is more difficult than might at first seem. Some information is completely unknown, some may be known but not fully appreciated, some information may be available to some members of the organisation but not to others, some information may be available but cannot be appreciated within current modes of understanding (1976: 195).

The first case is difficult to deal with: such information may only reveal its significance when some disaster occurs. In the case of information which is available but where its significance is not appreciated, Turner argues that it is necessary to test the understanding which individuals have of their current position. It may be that pressure of work distracts attention from emerging signs of danger; or, that employees distrust the source from which the information is coming; or, that staff are "decoyed" by some aspect of the situation into a failure to perceive the emergent dangers of another aspect of the system; or, because they have difficulty in classifying a phenomenon and may mis-classify it and fail to act or act inappropriately; or, they may have difficulty in separating the information-giving event from the "noise" of other irrelevant information. In the case of information which is available but not in a useable form it is possible to identify any number of different information difficulties, for example, information may be available but hidden amongst other material, similarly warning information may not yield its significance without some sensitivity in the mechanism for assembling, filtering and interpreting it.

A further problem may be that relevant information is distributed between several organisations or parts of one organisation and, hence, its significance may not be appreciated unless by some fortuitous act it is brought together in one person or situation. Similarly, there may be information difficulties associated with the interaction of two or more different systems, each of which when acting independently is safe but when brought into conjunction have in appropriate means for dealing with information at the interface between the separate system. Other examples, may arise in cases where, for instance, prior information is deliberately withheld. There may be a considerable range of behaviours and motives associated with the withholding of information including fear, malice, complacency but the point is that some information will be available within the incubation period but not emerge until after the system has failed. The issue of information for which there are no appropriate categories is an important one. Sometimes individuals are unaware of the extent of their ignorance about the system they are operating, particularly in its wider systems context. However, often it is the case that there is no appropriate channel for the specific or discrepant piece of information to enter the system either because the particular problem is not officially recognised as a hazard or because the existing construction of the situation does not permit the new information to disconcert perceptions. This latter point is significant in that perceptual rigidities may confine attention within an organisation to specific ways of perceiving its task.

Long Run versus Short Run Considerations

Micro-social behaviour which might contribute to accidents needs to be seen in the context of long run variables which are generally more stable. These are the corporate culture, the corporate objectives and the organisation structure. The significance of these contextual variables in shaping and determining the nature of the interaction of the short run variables should not be underestimated. It is beyond the scope of this paper to discuss what influence a corporate culture can have on performance and how culture can be changed (Sathe, 1983; Schein, 1984). However, corporate culture as a system of control where meanings are manufactured and imposed on an organisation clearly does have a bearing on the way in which behaviours and actions are regulated and interpreted. It is possible to argue that corporate values espoused in culture change have a significant role in supporting corporate objectives and that will have consequences for how meaning is construed and managed. Hence some of the issues which are generally taken for granted need to be analysed in order to assess their contribution to the promotion of a safety culture.

Need for a Radical Challenge to Assumptions

The problem for safety management is that it is what is left outside of the apparent rationality of corporate rhetoric and organisational practices which is likely to be more hazardous than those aspects of the system which have been anticipated. This presents considerable difficulties. Clearly, some of the information difficulties discussed above can be dealt with by organisational responses and appropriate systems. However, some information difficulties are much more intricately enmeshed in the social fabric of the organisation and resistant to exposure. Not least, perhaps, is the need for management to radically challenge its own assumptions about its influence on the safety function, specifically in relation to the ways in which meaning is constructed in the organisation. Westrum (1987) has drawn attention to ways in which organisations can promote safer practices and has advocated what he terms a "generative" as opposed to an "calculative" rationality as a means of reducing organisational failure. Many of the features Westrum puts forward, implicitly feature in the way in which British Airways developed its new philosophy for safety management. For instance, Westrum argues that generative organisations should,

1.   encourage system-wide awareness on the part of all members of the system
2.   encourage creative and critical thought
3.   link   interdependent parts of the system
4.   scan the different parts of the system for relevant solution to organisational problems, to be used regardless of their origins
5.   reward system-oriented patterns of thought
6.   avoid overstructuring the organisation
7.   examine mistakes honestly.

Westrums message had important implications for the debate in safety management practices and it is within this context and as an extension of the debate that a concern to activate the generative features of safety management that a concern for organisational learning have come into prominence. In British Airways, this concern has been seen in the way that Safety Services have construed an interpretative environment around its safety information system. A significant part of this environment is rooted in a commitment to the principles of organisational learning. Fundamental to this is the development of a principle of double-loop learning which ties together the relationship between BASIS and its interpretative environment in a continuous iterative process.

Discrete Solutions Not the Answer

Improvements in air safety cannot be achieved on a piece-meal basis. What is being achieved in British Airways has to be seen in the context of changes in the airline as a whole. There have been several attempts by the company to audit the outcomes of the culture change programme and subsequent interventions. Often the message of the change has been conflicting, for example, the importance of staff and staff development as perceived when set in the context of redundancies (Höpfl, Smith and Spencer, 1992) and, despite the impressive gains in financial terms, not all staff have been convinced about the style of the changes (Höpfl, 1993). In part, the problem is that while the culture change in British Airways undoubtedly produced impressive results, the primary objective underlying the changes was the long term viability and profitability of the airline. Clearly, there should be no conflict here with safety and related issues. However, this depends on the coherence of the overview which is taken and how inter-dependencies are construed and traded off against each other. British Airways is fortunate in having David Hyde as Director of Safety, Security and Environment since he brings both broad experience and a broad systems conception of the airlines operations to his remit and made specific appointments to posts in Safety Services which reflected a philosophical commitment to radical change in the safety function (Höpfl and MacGregor, 1993).

Consequently, whereas it is difficult from an external perspective to evaluate the impact of the culture change across the airline as a whole (Höpfl, 1993), it is possible to point to the intentions and achievements within the air safety function. An analysis of the relationship between the corporate culture change and its implications for safety culture is provided elsewhere (Höpfl and MacGregor, 1994; Höpfl, 1994a, 1994b). The changes that have been introduced into Safety Services have had far reaching and beneficial consequences which should promote greater awareness of safety issues across the airline (MacGregor and Höpfl, 1995). To optimise benefits there has to be a complementary analysis/audit of contextual variables, organisational dynamics and change leverage effects on the safety function. What is required is a theoretical framework which reflects the complexity of the organisational dynamics. It is only by achieving a greater appreciation of the contextual variables that it will be possible to attempt to model the relative weighting of human factors in order to improve risk assessment and estimation. This implies developing a sensitivity and responsiveness within the system to the complex, irrational, embedded, conflictual aspects of information which may be permitted to emerge by a commitment to organisational learning and the capacity for exposing itself to radical doubt.

REFERENCES

Ackroyd, S. and Crowdy, P.(1989) Can Culture Be Managed? Working with "Raw" Material: The Case of the English Slaughtermen, *Personnel Review*, 19,5 3-13.

Bate P. (1984) The Impact of Organisational Culture on Approaches to organisational Problem Solving, *Organisation Studies* 1984, 5/1: 43-66.

Blackler, F. and Brown, C. (1986) Alternative Models to guide the Design and Implementation of the New Information Technologies, *Journal of Occupational Psychology*, 59, 287-313.

British Airways, (1987) *BA Management-Staff-Customer Contact Report*

British Airways, (May 1991), *Managing out of a Recession, Business Life*, London: Maxwell.

Bruce, M. (1987) Managing People First – Bringing the Service Concept into British Airways, Industrial and Commercial Training, March/April. Caesar, H.(1990) *Air Transport Development and the Role of Aviation Administration*, Proceedings of the 43rd International Air Safety Seminar (IASS), Rome, 1990.

Caesar, H. (1992) *Living with the Five Million year old Computer*, Proceedings of the 45th International Air Safety Seminar (IASS), Los Angeles, 1992.

Clegg, C. and Corbett, M. (1987) Research and Development into "Humanising" Advanced Manufacturing Technology, in Wall, T. Clegg ,C. and Kemp, N. 1987 *The Human Side of Advanced Manufacturing Technology*, Chichester: John Wiley.

Cummings, T.G. and Salipante, P.F. (1976) Research-Based Strategies for Improving Work Life, Warr, P. (ed) 1976 *Personal Goals and Work Design*, London: John Whiley.

Fernandez, J.W.(1986) *Persuasions and Performances, The Play of Tropes in Culture*, Blooington: Indiana University Press.

Fisher, S. (1984) *Stress and the Perception of Control*, London: Lawrence Erlbaum

Foushee, H.C. (1981) The Role of Communications, Socio-Psychological and Personality Factors in the *Maintenance of Crew Co-ordination*, in proceedings of First Symposium on Aviation Psychology, Aviation Psychology Laboratory, The Ohio State University, Columbus, Ohio, April 1981.

Freeman, C. and Simmon, D. (1990) *Crew resource Management Model*, Proceedings of 43rd IASS, International Air Safety Seminar, Rome, 1990.

Holton, M. (1991) The Basis for Safety Management, *Focus* Nov. 1991, The Flight Safety Committee.

Höpfl, H. Smith, S. and Spencer, S. (1992) Values and Valuations: Corporate Culture and Job Cuts, *Personnel Review* Vol 21 1 24-38 1992.

Höpfl, H. and MacGregor, C. (1992) *Integrating Systems and Safety: Implications for Organisational Learning*, 45th International Air Safety Seminar, Flight Safety Foundation Conference, Long Beach, California, November 1992.

Höpfl, H (1993) Commitments and Conflicts: British Airways, in *Case Studies in Organisational Behaviour*, Legge, K. Clegg, C. and Gower, D. (eds): Paul Chapman Publisher, May 1993.

Höpfl, H. and MacGregor, C. (1993) A Commitment to Change: Safety Management in British Airways, *International Journal of Disaster Prevention and Management*, April 1993.

Höpfl, H. and MacGregor, C. (1994) Safety Management in British Airways, *International Journal of Quality and Reliability*, April 1994.

Höpfl, H. (1994a) Corporate Culture, Safety Culture. *Journal of Disaster Prevention and Management*, August, 1994.

Höpfl, H. (1994b) The Paradoxical Gravity of Planned Organisational Change. *Journal of Change Management*, 7 (5).

Höpfl, H. (1994c) Empowerment and the Managerial Prerogative, *International Journal of Empowerment in Organisations*, November 1994.

Höpfl, H. (1995) Improving Customer Service: The Cultivation of Contempt, Studies in *Cultures, Organisations and Societies*, Mary Jo Hatch and Majken Schlz (eds) I (1).

Houston, R. (1991) Resource Management in the Cockpit, in *Focus on Commercial Aviation Safety*, Winter 1991.

Hurst, R. and Hurst, L. (eds) (1982) Pilot Error, The Human Factors, London: Granada.

IATA (International Transport Association) Report 1992.

Johnson, G. (1987) *Strategic Change and the Management Process*, Oxford: Basil Blackwell.

Lauber, J. (1990) Anatomy of a Human Error Accident, *Flight Safety Digest*, Air New Zealand, September 1990.

Linstead, S.L. and Grafton Small, R. (1992) On Reading Organisational Culture, *Organisation Studies*, 13, 3, 331-335.

Lorsch, J.W. (1985) Managing Culture: The Invisible Barrier to Strategic Change in Kilmann, R.H. Saxton, M.J. Serpa, R. and associates (eds) *Gaining Control of the Corporate Culture*, San Francisco: Jossey Bass.

MacGregor, C. and Höpfl, H. (1995) BASIC Users Conference: The Global Superhighway for Safety Information Systems, *Senior Accident Investigator*, British Airways.

McKenna, J.T. (1990) CVR Shows Avianca Crew Knew Fuel Was Too Low, an abstract in *CRASH*, Cranfield Aviation Safety Centre Headlines, May 1990.

Mumford, E. (1983*) Designing Human Systems*, Manchester: MBS.

Pascale, R. (1985) The paradox of "Corporate Culture": Reconciling Ourselves to Socialisation, *California Management Review*, XXVII, 2: 26-41.

Pearce, B.G. (ed) 1984) *Health Hazards of VDTs?* Chichester: John Wiley.

Perrow, C. (1983) The Organisational Context of Human Factor Engineering, *Administrative Science Quarterly*, 28, 521-541.

Pettigrew, A. (1976) Issues of Change, Warr, P. (ed) 1976 *Personal Goals and Work Design*, London: John Wiley, 245-254.

Rasmussen, J. (1980) What Can Be Learned from Human Error Reports, Duncan, K. and Grunberg ,M. and Wallis, D. (eds*) Changes in Working Life*, London: Whiley.

Ray, C.A. (1986) Corporate Culture: The Last Frontier of Control, *Journal of Management Studies*, 23, 3: 287-297.

Reason, J. (198710 The Chernobyl Errors, *Bulletin of the British Psychlolgical Society*, 1987, 40, 201-206.

Reason, J. (1990) The Contribution of Latent Human Failures to the Breakdown of Complex Systems, Broadbent, D.E. Reason, J. and Baddeley, A. (eds) 1990 *Human Factors in Hazardous Situations*, Oxford: Clarendon Press, 27-36.

Rosenbrock, H. (1981) Engineers and the Work That People Do, Work Research Unit Occasional Paper, no.21, reprinted from the *IEEE Control Systems Magazine*, 1, September 1981.

Ruffell Smith, H.P. (1979) A Simulator Study of the Interaction of Pilot Workload with Errors, Vigilance and Decisions, *NASA Technical Memorandum 78482*, Hurst,R. (op.cit.).

Schein, E.H. (1980) *Organisational Psychology*, Englewood Cliffs, NJ: Prentice Hall.

Schein, E.H. (1984) Coming to a New Awareness of Organisational Culture, *Sloan Management Review*, Winter 1984.

Schein, E.H. (1985) Organisational Culture and Leadership: *Adynamic View*, San Fransico: Jossey Bass.

Schwartz, H. (1991) *The Challenger Disaster: The End of the American Dream*, a paper presented to the SCOS Conference, Copenhagen.

Seaman, C. (1991) *The British Airways Safety Information System*, a presentation to the 44th IASS, Singapore, November 1991.

Smircich, L. and Morgan, G. (1982) Leadership: The Management of Meaning, *Journal of Applied Behavioral Science*, 18 (3) 257-273.

Smircich, L. (1983) Concepts of Culture and Organisational Analysis, *Administrative Science Quarterly*, 28: 339-358.

Tench, W.H. (1985) *Safety Is No Accident*, London: Collins.

Toft B. (1992) *Changing Safety Culture – Decree, Prescription or Learning?* A paper presented at the IRS Hazard Management Seminar, London Business School.

Toft, B. and Turner, B.A. (1987) The Schematic Report Analysis Diagram: a simple aid to learning from large-scale failures, *International CIS Journal: Command and Control, Communication and Information Systems*, Volume 1 (2), April/May.

Trist, E.L. Higgin,G.W. Murray,H. and Pollock,A.B. (1963) *Organisational Choice*, London: Tavistock Publications.

Turner, B. (1978) *Man-Made Disasters*, London: Wykeham.

Turner, B.A. (1989) *How Can We Design A Safe Organisation?* The Second International Conference on Idustrial and Organisational Crisis Management, New York, November 1989.

Van Maanen, J. (1991) The Smile Factory: Work at Disneyland, Frost,P.J. et al (eds) *Reframing Organisational Culture*, New York: Sage.

Visciola, M. (1990) *Pilot Errors: Do We Know Enough?* Proceedings of 43rd International Air Safety Seminar (IASS), Rome,1990.

Waring, A. (1991) Success with Safety Management Systems, *The Health and Safety Practitioner*, September 1991.

Warr, P. (1976) *Personal Goals and Work Design*, London: John Wiley.

Westrum, R. (1987) Management Strategies and Information Failure, 109-127 Wise, J.A. and Debons, A. (eds) *Information Systems failure Analysis, NATO ASI Series f, Computer and Systems Science*, Volume 3 Berline: Springer-Verlag. (see Turner, BA op.cit: 1989.

Wiener, E>L> (1982) Mid-air Collisions: The Accidents, The Systems and The Realpolitik, 101-117 Hurst, R. and Hurst, L. (eds) *Pilot Error, The Human Factors*, London: Granada.

Young, D. (1989) *British Airways: Putting the Customer First*, Ashridge Strategic Management Centre, July.

OPENING PANDORA'S BOX:

Stress at work and its implications for emergency management

DENIS SMITH and DOMINIC ELLIOTT
Sheffield University Management School
9 Mappin Street
Sheffield, S1 4DT, UK

Introduction

> "Few managers can afford to ignore the effects of stress. More
> working days are lost through stress-related illness than ever before. It
> now has a greater impact than days lost through strikes. ... The impact
> of stress is not restricted to illness. It also has a detrimental effect on
> important functions of management such as the effectiveness of
> decision-making, the quality of interpersonal relationships, the
> standard of work, the quality of working life and, of course, ultimately,
> the level of productivity." (Cranwell-Ward, 1990 p.3)

Within the last decade, stress has emerged within the academic literature as one
of the key managerial problems within the areas of human resource management
and occupational psychology. As the opening quotation suggests, stress has both
acute and chronic properties. The worst-case scenario is that stress can
ultimately result in severe illness or even in death (Palumbo and Herbig, 1994).
Even the less dramatic manifestations of the problem can result in serious
consequences by impairing the individual's abilities to cope. The costs of stress-
related illness can be considerable. One estimate suggests that it could cost up to
£1.3 billion, annually, in the UK alone (Summers, 1990). The phenomenon of
stress transcends the private-public sector divide and is manifested across the
range of occupational classifications. Despite the apparently high frequency of
its occurrence, the concept is largely poorly defined and the literature indicates
the existence of a number of paradigmatic camps. Unpacking this concept, and
its associated managerial implications, is akin to opening Pandora's Box (Elliott
and Smith, 1993a). At one level the concept must be examined in order to
provide a greater understanding of the issues although it soon becomes rapidly
apparent that the problem is seemingly trans-scientific in that it goes beyond the
current abilities of science to prove_(Weinberg, 1972). At a cynical level, one

E. Coles, D. Smith and S. Tombs (eds.), Risk Management and Society, 143–163.
© *2000 Kluwer Academic Publishers. Printed in the Netherlands.*

could argue that stress has become the latter day equivalent of a bad back as a means of having paid time off work. Its symptoms are easy to manifest whilst diagnosis and causality are difficult to prove. A more balanced view would suggest that rapid change in organisations - typified by the last 15 years in the UK - create stressors for the workforce, which may become manifested as an apparent inability to cope.

The aim here, however, is to define the boundaries of both the concept and its implications for managing the emergency services with particular emphasis upon the handling of major incidents. In the case of stress, and more specifically PTSD within the emergency services, an important question centres around the potential impairment of their role as rescuers. Given the role of the emergency services in crisis situations and disasters, any impairment of their abilities to effect a rescue or to work in teams may have serious consequences for both colleagues and those being rescued.

Prior to exploring these issues, within the context of emergency and disaster management, it is first necessary to explore the nature of stress and PTSD. Following on from this discussion we will return to the specific context of the emergency services and focus, more specifically, on PTSD and the potential of training and recruitment & selection as intervention strategies. Preliminary data will be presented on the fire service to illustrate the potential that exists in this respect.

The nature of stress

> "The concept of stress is elusive because it is poorly defined. There is no single agreed definition in existence. It is a concept which is familiar to both layman and professional alike; it is understood by all when used in a general context but by very few when a more precise account is required, and this seems to be the central problem" (Cox, 1978 p.1)

Cox's (1978) view is as apt a quarter of a century later as it was when it was first written. Within the academic literature there are various approaches to the study of stress, each with its particular definition of the term. Cranwell-Ward (1990) identified two broad approaches from her study of stress in management: those that related to the *causes* of the stress and those that related to the *effects* induced by it. A third approach is the transactional model, defined below (Cox, 1978; Brown and Campbell, 1994; Fisher, 1986). Each of these approaches needs to be outlined in turn.

Stress as stimulus

Within the stimulus-based responses to stress there is the assumption that the phenomenon is treated

> ".... in terms of the stimulus characteristics of environments which are recognised as disturbing or disruptive in some way. The model used is essentially an engineering one in which external stresses give rise to a stress reaction, or strain, within the individual." (Cox, 1978 pp.12-13)

The importance of environmental stressors on the generation of stress has been recognised since the 1930s and, whilst it has some merits, it has attracted considerable criticism within the literature. These criticisms centre around the difficulty in assuming that the same stimulus will generate the same response; that the stimulus cannot serve as a useful measure of stress across a diverse population and that some stimuli are temporarily and spatially defined (Ross and Altmaier, 1994).

Stress as Response

In contrast, the response variable approach, derived from the work of Seyle (1956), sees the individual responding to any stressor (irrespective of type) in a universal manner, termed the General Adaptation Syndrome. These responses can be seen to occur in three stages. These are, alarm (the immediate reaction to the stressor via Shock and Countershock); resistance; and collapse, and in this context, stress is "quite unequivocally ... the person's ... response to the demands of his environment" (Cox, 1978 p.5). Ross and Altmaier (1994) argue that there are two major groups of criticisms of Seyle's model. The first is that it is difficult to see the response of individuals following a set pattern and that responses to the same stressors may vary. The second criticism is that it is difficult to imagine a common response to different stressors irrespective of their severity and complexity. Seyle's (1956) primary interest was with the nature of an organism's physiological response to extremes of heat and cold. It may be unfair to identify the limitations of this approach to a field of study for which it was never intended.

An Interactional view of Stress

As a consequence of the criticisms made of both models, a third approach to stress has developed in the literature. The transactional or interactional model of stress is developed from the work of Lazarus, Cox and their colleagues

(Lazarus, 1966; Lazarus and Launier, 1978; Lazarus and Folhman, 1984; Cox, 1978; Cox and Mackay, 1976)[1]:

> ".... the physiological and psychological reaction which occurs when people perceive an imbalance between the level of demand placed upon them, and their capability to meet those demands" (Cranwell-Ward, 1990, p.10)

A definition that is close to the one used by Cox (1978) in his observation that:

> "Stress...can only be sensibly defined as a perceptual phenomenon arising from a comparison between the demand on the person and his ability to cope. An imbalance in this mechanism, when coping is important, gives rise to the experience of stress, and to stress response. The latter represent attempts at coping with the source of stress." (p.25)

This interactional model posits that an individual interacts with his or her environment and that whilst the environment can affect the individual, so too can the individual affect the environment. In both cases, there is seen to be an imbalance between the demands placed upon the individual and their perceived abilities to cope with these demands. Both groups of authors emphasise the continuing nature of this process and both emphasise the role of cognitive appraisal in determining the perceived imbalance and therefore stress. For Lazarus et al, the evaluation process occurs in two stages. The first stage, or primary appraisal, occurs when the individual first evaluates the event or demand and focuses upon the negative outcomes that can result from the interaction. The secondary stage occurs when the respondent seeks to evaluate the potential responses that can be made to the negative outcomes that were previously determined.

This paper's concern is with occupational stress. Here the various stressors that impinge upon an individual will transcend the work/social/family divides. The difficulty for researchers lies in isolating the origin of the various stressors and in providing a framework within intervention strategies can be formulated and implemented. To this end, occupational stress can be defined as,

> ".... the interaction of work conditions with characteristics of the worker such that the demands of work exceed the ability of the worker to cope with them. This definition fits the person-environment context ... and allows us to examine the joint contributions to occupational stress of worker characteristics, job conditions and their interactions" (Ross and Altmaier, 1994 p.12)

In this context, stress will occur as a function of the interaction between the individual (expressed in terms of the relationship between perceived abilities and actual abilities) and the resources available to that individual (again both perceived and actual). The extent of these interactions will change over time.

If we accept this transactional model as the basis for stress discussed within this paper, given the identified limitations of alternative explanations, then a number of issues arise which merit further discussion. The first of these centres around the ability of the individual to cope with, or withstand, the impact of stressors placed upon him/herself. The second concerns the various interventions that can take place to either prevent stress building up or to dissipate the effects of that stress once it has occurred. The final issue concerns the likely impacts upon the organisation of long term stress amongst the workforce and its influence upon organisational performance.

Post Traumatic Stress Disorder

Within recent years, the concept of PTSD has emerged within the literature as an extreme manifestation of stress within certain occupational settings. PTSD, as a medical condition, has received considerable support from clinicians (Cullen and Schottenfeld, 1986) and estimates have suggested that the symptoms may become manifest in up to 90% of those individuals who have to deal with a traumatic event. These symptoms include withdrawal, emotional disturbances, sleep disturbance, reliving the trauma and exhaustion (Barton, Braverman and Braverman, 1993; Friedman, Framer and Shearer, 1988). In order to contextualise our field of inquiry the remainder of this paper will draw upon the emergency services in the UK and the fire service in particular.

Stress and the emergency services

The emergency services have a multifarious role. At one extreme they deal with high-energy events which have the potential for causing multiple fatalities; at the other, they undertake routine tasks, procedural drills and spend time waiting to respond to requests for assistance. Each of these activities may create stressors for the individual. For some, the trauma induced by high energy events may create problems in coping. For others, routine shift work or a high administrative load, may cause greater pressure and anxiety. It is this range of stressors that creates a set of unique problems for the emergency services. Of particular importance to this study are the effects of trauma, which result from exposure to the more dramatic events such as multiple fatality accidents, brutal murders or accidents involving children.

Recent events have shown that there is a need to assess the impact that such trauma can have upon emergency personnel (Duckworth, 1988; Elliott & Smith, 1993a; James, 1992; Smith, 1992; Kinchin, 1993). For the emergency services the greatest risk of PTSD is held to be highest after the rescue has been discontinued (Elliott and Smith, 1993a; 1994). The organisation's abilities to provide effective training and counselling in this context is important in ensuring the continued effectiveness of the rescuing function (Smith, 1992). The role of organisational culture in determining the response of management to PTSD is key and evidence indicates that the emergency services suffer from the process of denial in this respect. For example, following the Kegworth aircrash in 1990, a senior officer from one of the fire brigades involved in the rescue denied that any of his staff suffered from PTSD. He claimed that the culture of the organisation allowed for a form of peer counselling, as operational fire fighters discussed incidents quite openly and this dissipated any stress that might build up. The reaction of the Fire Brigades Union to the issue of stress counselling is conditioned by this sort of approach - arguing that any requests for stress counselling may be seen as a sign of weakness and could be used as a means of victimising individual officers. The importance of the emergency services during disaster recovery requires that the initial recruitment and selection, and the post incident support of staff be important components of the human resource strategies of all branches of the emergency services. Sadly, it is our view that current practice falls short in this regard.

The emergency services intervene at the stage of a crisis or disaster, which is typified by high levels of energy, activity and risk (Smith, 1990). Subsequently, emergency services personnel may return to more mundane, routine tasks without the opportunity for discussing the nature of their experiences in a non-threatening manner (Elliott and Smith, 1993a). Consider the following account of a fire-fighter exposed to such a potentially traumatic event.

"When we arrived on the scene, not knowing what to expect, and when we entered the aircraft, we thought it was a transport plane. There was no immediate evidence of seats or passengers. The moaning of the injured soon shattered this initial assessment and, when we managed to get some light into the cabin, we saw the extent of the carnage. Bodies and wreckage were everywhere, it was chaos. We just got on with trying to remove the bodies. The first one that we removed, I thought was dead. Two months later I saw him leave hospital on the TV and I just broke down and cried." - Fire fighter involved in the Kegworth rescue (1992).

The trauma arising from extreme events and disasters is quite apparent. Dealing with multiple fatalities or events involving the death of or serious injury to children can have a detrimental effect upon rescuers. The account of the Kegworth fire-fighter, given above, gives some indication of the main elements facing a rescuer at a major incident. These include a fear of the unknown, surprise at the magnitude of the task facing the individual, initial coping and subsequent trauma if triggered. However, it is not just major events that create stressors for members of the emergency services. Consider the case of an off-duty paramedic who attempted to save the life of a child involved in a fatal road traffic accident:

> "I heard the bang from inside the house and I instinctively ran out to see what happened. My six-year-old son followed me. I saw the child in the road. He was covered in blood and had sustained severe head injuries. I attempted to resuscitate him but knew that it was hopeless. In the evening my son said, 'Dad, I thought that you were clever.' I replied that I was, at which point my son replied, 'You can't be - you couldn't save that boy's life'. At that point I just went to pieces and I was off work for six weeks with stress-related symptoms." - Paramedic (East Midlands, 1996)

Such anecdotal evidence simply provides an insight into the pressures that can face rescue workers that are peculiar to their work.

There are however, a large number of the stressors that they are exposed to may be shared with other workers outside of the service. Cooper & Marshall (1978), for example, identified seven main categories of occupational stressors:

1) Factors intrinsic to the job (physical effort, working conditions, repetition etc)
2) Role based stress (role conflict, role ambiguity etc)
3) Interpersonal relationships with others at work
4) Career development (both in terms of perceived over and under promotion)
5) Organisational structure and culture
6) Extra-organisational stressors, (family etc)
7) The individuals' characteristics.

Research conducted with fire officers, who had significant administrative tasks, identified that all found office work to be generally more stressful than operational work (Elliott, 1991; Hodgkinson & Stewart 1991). From interview data collected from serving officers, it emerged that the reasons for this perceived greater stress included the fact that they joined the fire service to escape the routine of office and factory work (Anon, 1991). Cooper & Marshall's (1978) work indicates that unravelling the complex causes of an individual's experience of stress will be problematic.

There has been a considerable amount of research undertaken in recent years to examine, in a structured manner, the nature of stress within the Services and, more recently, on the likely intervention strategies that can be implemented. Within the emergency services, the greatest amount of research on occupational stress has been undertaken with the police (See for example, Hetherington, 1991; ACPO, 1987), although work with the other two services is increasing. Table 1 illustrates the range of research, focused upon the emergency services and stress. However, because of the dominant paradigm that exists within these organisations, the issues of stress have remained off the political agenda and some senior officers equate stress symptoms with weakness. Whilst such a view may represent an extreme perspective on the problem, it is certainly not unique amongst senior officers although there are signs that values are beginning to change in this respect.

Table 1: Selected research involving the emergency services

Main Stressors Examined	Selected Research Activity		
	Police	Fire Service	Ambulance Service (A&E Service)
Demographic Factors (Generic)	Kroes, 1982 Kroes et al 1974	Douglas et al, 1988	
Management and Organisational Bureaucracy (Eg Shift Work) (Generic)	Cullen, 1989 Brown & Campbell, 1994 ACPO, 1987	Lim Lin Mei, 1990; Ko and Kao, 1993	
Task-Specific Stressors	Hetherington, 1991 Gersons, 1989	Durham et al 1985	James, 1988
Training, Recruitment and Selection	Hetherington, 1991	Elliott and Smith, 1993a; Health and Safety Executive, 1984	Gold et al, 1985
Intervention Strategies (Counselling)	Alkus and Padesky, 1981	Docherty, 1989; Johnston and Kelly, 1988	Duckworth, 1988; Palmer, 1983

The importance of the emergency services in crisis/disaster management is self-evident. Within the context of immediate disaster response, aid and recovery the effective working of rescue teams is of paramount importance. Until recently, the role of stress as an inhibitor of effectiveness within the emergency services has not been given the consideration that it deserves by senior managers. By ignoring the effects of stress, and PTSD in particular, the emergency services run the risk of incubating the potential for crisis within their own organisations. An individual who is suffering the effects of PTSD may well be unable to perform his/her tasks under conditions of rescue and may endanger the lives of colleagues and those requiring rescue. As such, they become akin to resident pathogens within the organisation, which remain dormant until events expose the potential weakness that they create (Reason, 1990).

PTSD and Managerial Intervention

The notion of Post Traumatic Stress Disorder (PTSD) as a potential impairer of individual and thus corporate performance is rapidly becoming a major issue for organisations, which have to deal with major crisis events (see Barton, Braverman and Braverman, 1993). The fire at King's Cross underground station resulted in the award of damages to those traumatised by the event (Frost, 1990). An Association of Chief Police Officers (ACPO) working party reported that much of the literature on PTSD concerned shooting incidents and the police in America. It related to a particular response in the interactionist model and may show both psychological and physiological symptoms. The nausea experienced by the fire fighter at Kegworth may be an example of both a physiological and a psychological response. Thus a psychological response may manifest itself physically. Two basic types of trauma are identified "chronic" and "acute or delayed" (ACPO p2). The literature dealing with the emergency services focuses upon two key variables:

    i) the extent to which the victim is taken by surprise.

    ii) the actual content of the traumatic event itself.

At a basic level, an individual's response to stress appears to be dependent upon two causative factors. First is the nature of the individual with certain personality characteristics modifying the response to stress. This may be influenced by training and other forms of learned behaviour that may arise from experience. The second concerns the nature of the traumatic experience itself and may be influenced by elements such as organisational support. An individual's response to stress may also be influenced by the recovery environment. Clearly an individual's perception of the nature of the experience is key because as Lazarus argues only an individual can describe a particular event as stressful to himself or herself. However, it seems clear that certain situations are more likely than others to lead to a negative "stressful" response, almost irrespective of individual characteristics. Other factors may also serve to modify the experience. For example, strong social support, as indicated by

Green et al (1985) may help the individual to recover quickly. Alternately, previous exposure to some of the hazards encountered may also reduce the perceived threat and thereby reduce the experience of stress (Warr, 1989). The relationship between the psychological make-up of individuals and their stress levels has attracted some attention within the literature (Duckitt and Broll, 1982; Manning, Williams and Wolfe, 1988) although work with the emergency services has been relatively scant given the nature of their activities. Early studies in this area have assessed stress levels amongst operational personnel in all three services (James, 1988; Docherty, 1989; Douglas et. al., 1988; Hetherington, 1991) and pilot research has explored the impact of stress resilience on recruitment and selection within the Fire Service (Elliott and Smith 1993a; 1994).

Table 2: Classification of Post-Traumatic Stress Disorder

	Symptoms
Quality of the Experience	- experience of an event outside of the usual range of experience. This arises from: ~ threat to one's health/life ~ threat to loved ones ~ loss of home/community ~ witness a serious accident/crime
Effects of the Experience	Cluster 1 - re-experience of the trauma ~ recollection of the event ~ distressing dreams ~ reliving the event/nightmares Cluster 2 - avoidance of stimuli associated with trauma ~ thoughts/feelings ~ activities/situations ~ psychogenic amnesia ~ lack of interest ~ detachment/estrangement Cluster 3 - persistent symptoms of increased arousal ~ thoughts/feelings ~ activities/situations ~ psychogenic amnesia ~ lack of interest ~ detachment/estrangement
Duration of the Symptoms	- present for at least one month - delayed onset for up to six months after the trauma

Source: DSM (III) R (para 309.89)

A number of possibilities for human resource managers emerge from this discussion. Effective intervention may range from; the identification of 'hardy' individuals to work in the emergency services; staff development to raise individual capabilities and thereby reduce the perceived threat of a situation (see for example, Meichenbaum, 1985); providing a supportive recovery environment. Where the first is difficult, both politically and practically, and may be interpreted as placing blame upon the individual the second and third recognise management's role in this area and avoid individual blame.

For the emergency services, the effects and duration of the symptoms of PTSD will be critical factors in impairing organisational effectiveness (see table 2 for a clinical definition). Consequently, the remainder of this paper explores these issues with reference to the fire service and assesses, via some preliminary findings, the influence of personality and training upon the experience of stress by new recruits. Although this study places a major emphasis on the individual trainees, the organisational context and culture (see Mintzberg, 1989; Mitroff et al, 1989; Pauchant & Mitroff, 1988; Smith, 1992) also play an important role.

Working on the Edge: Recruiting for the Fire Service

The preliminary study took place at a Fire Service Training College in the North of England. All 30 subjects within the cohort were male and aged between 20 and 30 with a mean age of 25.4 years. The training course was of 13 weeks duration. The college provides three or four training courses per annum for between 90 and 120 recruits in total. The sample for this study, therefore, comprised between 25% and 30% of the average annual total for the college, although within the whole of England and Wales an average of 1,500 recruits are trained each year.

Three tests were issued to each "trainee", 16 PF, GHQ 12, and Hardiness. Cattell's 16 Personality Factor questionnaire (IPAT, 1982) identifies sixteen traits that Cattell argues provide a basis for understanding an individual's behavioural predispositions. The General Health Questionnaire (GHQ) 12 was designed as a self-administered screening test for detecting minor psychiatric disorders among respondents (see Banks et al 1983). The Hardiness Test (see Parkes & Rendall 1988) is a questionnaire-based instrument for exploring three components, commitment, challenge and control, which are held to act as a 'resistance resource, mitigating the adverse affects of stressful life events'. Of these tests only one person failed to return a completed form although two students eventually had to withdraw from the course on health grounds. A stencil for use with the 16 PF test was used to calculate the raw scores in order to allow for a comparison of this sample with data collected from a much wider population. The GHQ 12 and the Self Monitoring Hardiness tests were similarly analysed. A response rate of 90% was achieved.

Among the 28 Fire Service trainees assessed by the GHQ 12 in January, 29% were showing signs of low well being (>2) with nearly 11% identified as experiencing a significantly low level of wellbeing (>6). This was held to have the potential to impair their general health and well being. The mean for the January exercise was 2.54, which was observed to reduce during the subsequent uses of the test to 1.36 in February and to 1.0 in March (Elliot and Smith, 1994). This suggests that either the trainees became better at using strategies to cope with stress, adapted to the use of the instrument or, as their capabilities increased, they perceived a reduced imbalance between their abilities and the demands of the course. The GHQ 12 assessment undertaken in February indicated that 18% of the sample was showing signs of stress although those showing signs of a high level of stress had reduced to 7%. The figures for March showed 14% reporting some experience of stress (>2) but a reduction to 3% of those experiencing significantly high levels of stress (Elliott and Smith, 1994). It should be noted, however, that the mean score fell in each successive month. The sample reduced to 27 during March as a subject who scored highly in the previous two tests >6 and >2 respectively dropped out on health grounds. The mean GHQ 12 scores also fell between the start and end of the training programme. The numbers of trainees scoring above the significant thresholds >2, >6 also fell. The inconsistent correlations between the GHQ 12 test scores and the 16 PF secondary traits are particularly interesting as they suggest that personality is only one modifier of the stress response. Also of interest was the close correlation between a high level of hardiness and a low anxiety score. The positive correlation between hardiness and age also suggests that hardiness may be either learnt or developed with age (Elliott and Smith, 1994).

These results indicate that the training does lead to increased levels of stress, which decline throughout the training programme. Given that the demands of the course remained constant, indeed progressed to more complex and rigorous tasks the development of coping mechanisms provides the most plausible explanation of the reported higher levels of well being. Positive correlations with age, anecdotally linked to long military experience, supported this view.

Other studies (Hetherington 1991, Parkes & Rendall 1988) have suggested that the hardy personality is more resistant to stress. The fact that this was not the case in the study by Elliott and Smith (1994) raises interesting questions regarding the nature of hardiness and its effect as a modifier of the experience of stress. The findings of Manning et al (1988) suggest that potential moderators of stress can be grouped into four categories, which include social resources, constitutional predispositions, personality traits, and others such as economic status. However, they indicated that, "hardiness did not moderate the relationships between stressors and outcomes as reported by others." (Manning et al, 1988, p205)

An alternative explanation is that these findings may indicate that hardiness is learned. A strong correlation between hardiness and age (P= .009) suggests that individuals become more hardy as they grow older, a finding consistent with the view that hardiness is learned.

An earlier study by Lim Lin Mei (1990), of 61 experienced fire fighters found that only 6.56% scored above a threshold comparable to >2. This suggests that levels of stress experienced during training might be considerably higher than those of active fire fighters. This raises an issue of whether high reported levels of stress are peculiar to Fire Service training or whether it is a phenomenon common to other types of residential training. For the new recruit the training course is also an assessment centre, 'fail and you are out'. Unfamiliarity with a situation and the withdrawal of normal levels of family support as at residential training courses may lead to increased levels of reported stress.

An alternative explanation for such discrepancies concerns strategies for coping with stress. When asked about the most useful skills developed during training, nearly 90% of the recruits responded by stating that the experience of team work was of fundamental importance in developing coping strategies. Evidence collected from more experienced fire fighters suggested that stress was coped with most effectively when part of a team (Anon, 1991). These tight knit groups, with their use of black humour, are held to provide fire fighters with a particularly powerful coping mechanism for dealing with stress. Senior Fire Officers reported that after promotion they experienced more adverse reactions than before, not simply because of extra responsibility, but because they often travelled to and from an incident on their own and had lost their sense of team identity (Schofield, 1991). The return journey from the scene of a 'traumatic incident' was identified as a key part of the coping process in which humour and other forms of support would be employed to minimise personal trauma. This suggests that social support derived from work colleagues may play a major role in modifying the response to an experience of stress. The coping strategy adopted through the use of humour and other forms of social support, within a group of close colleagues is not available to the solitary officer.

Discussion

The area of study concerning Fire Service training, stress and PTSD is one, which has not attracted much academic interest until recently. Much of the research already conducted focuses on disaster and not on high levels of occupational stress. It is an area that the Fire Service has itself failed to encourage. This may be due to either a lack of awareness or to the dominant culture of the Fire Service, which prevents its staff from admitting that they do experience stress. Interview data would strongly support the contention that it is the culture of the Fire Service that is largely responsible for the failure in encouraging research into stress and developing intervention strategies as a result. This culture also prevents some brigades from acknowledging that there may be a problem. Recognition of the potential seriousness of this issue is imperative. However, it will not be sufficient for the Fire Service to simply provide a counselling service in the wake of traumatic and more chronic events. As Mitroff et al (1989) suggest, declared intentions by organisations do not necessarily lead to changed managerial actions and a fundamental change in the culture and core assumptions is also required. Pauchant & Mitroff (1988) recommend that organisations should seek to manage organisational and individual anxiety in order to,

> "diminish anxiety through different stress management workshops, specifically targeted to the phenomena of crises." (1988, p62)

For example, some Fire Services employed full time counsellors. In three cases these formed a part of the service personnel unit, which was the only place in which professional women were employed. Visiting the counsellor was not an action supported by the 'macho' cultures of the two Midland and one Northern service visited.

The preliminary results of the fire service study by Elliott and Smith (1994) indicate that certain personality characteristics may modify the stress experience. However, the evidence is not clear and further research is required to explore the relationships between a variety of potential modifying factors. The findings of Manning et al (1988) suggest that the question of whether personality is a modifier of stress is too simple. The interactionist approach seeks to explain this by describing the experience of stress as a complicated interaction between person and environment. This interaction leads to a perceived imbalance between an individual's own resources and the demands that are being made. Consequently, it would suggest that the emergency services should employ a multi-part strategy. Part one concerns the profiling of new recruits during the selection process. Part two is concerned with the training of new staff and in particular exposing trainees, in a controlled manner, to the types of risks, threats and disturbing sights that might be encountered during their emergency service careers. Part three requires the emergency service to provide effective counselling and guidance procedures for staff who may be exposed to 'stressors'. This process must be done in a way that recognises that stress can be considered as an occupational illness rather than a sign of weakness. The fourth part is, arguably, the most important and concerns the creation of a suitable organisational culture.

The focus of this paper has concerned personality and the experience of stress. The literature and the results of the pilot study suggest that levels of stress are experienced by fire fighters which are far greater than those reported by the service. The culture of the Fire Service has been identified as a major factor in preventing the reporting of such experiences. As discussed, individual fire fighters have reported colleagues retiring on the grounds of ill health when retirement has been due to the stresses of the work. Most notable was the reference to the Senior Fire fighter who resigned after the Bradford Fire who reported that he was unable to face up to the risk of such an event again. At one level, the Fire Service cannot afford to lose experienced staff because of the costs of replacing them. At another level, the dangers of an individual succumbing to adverse reactions at the scene of an incident may place colleagues or members of the public in jeopardy. The Fire Service, therefore, needs to determine methods of reducing the risk of individuals experiencing stress. Where this is not possible, support should be provided to ensure that individual fire fighters can cope with the stressors. Courses in stress management or thorough debriefings after an incident are two examples. Caution, however, must be exercised. There is mixed evidence concerning the efficacy of debriefing. Raphael et al (1995), cited in Rick and Briner (2000), suggest that reliving an experience, through debriefing, might harm some people. A greater understanding of the processes which influence any reaction to stress will help the Fire Service to develop strategies to reduce its impact on the Service.

In addition, consideration of suitable personality types will help the Fire Service to recruit and select individuals who are best suited to this type of work. The comments of the experienced fire fighters who found office work more stressful than "active duty" point to inadequacies in the current recruitment strategies. Further study may indicate whether entry at different levels would be useful. In larger brigades the opportunity to employ individuals with specific skills or characteristics might also prove to be useful.

Endpiece

In lieu of conclusions, it is intended to outline a series of managerial strategies that may serve to alleviate the problems of stress within the fire service. Ross and Altmaier (1994) suggested a number of possible workplace interventions that would have relevance for the fire service. These would include, individual counselling for those suffering from the effects of stress either as a formalised session or on an emergency basis following a traumatic event. A second strategy would involve group counselling for a watch, which had been exposed to trauma. Although many senior fire officers might argue that this process of group counselling already exists via the watch, such a support mechanism can only ever be informal and the organisation needs to ensure that formal procedures are in place. The other interventions include outreach work (with the local community), administration and professional development programmes. Given that administrative tasks have been identified as creating stress for firefighters then these latter two strategies would also be important. Additional provision of support by management could also include:

i.   Provision of training to senior staff and other suitable officers to spot the symptoms of psychological distress and to provide immediate support. This strategy is an essential component of the cultural change process that is required within the service.

ii.   Where relevant and appropriately targeted, the provision of compulsory and confidential debriefings with recourse to in-house counsellors (preferably trained both as firefighters and counsellors to remove any bias on the part of operational staff) to ensure that any cultural bias within the organisation does not prevent individuals from seeking help. Again, this has do be done within a supportive organisational climate which recognises that succumbing to stress is not a weakness but a risk associated with the occupation. As Rick and Briner (2000) suggest, organisations should maintain 'robust' processes for evaluating the effectiveness of debriefing as a means of reducing traumatic symptoms.

iii.   The monitoring of individuals who may be potentially at risk from the effects of stress and PTSD. This should follow on from psychological profiling at the recruitment and selection stage and be a commitment from the organisation to ensure the well being of staff.

iv.   Fostering a less 'macho' culture which accepts an adverse response as natural. In addition, the provision of stress management training (see Murphy, 1988) should be an essential part of the initial and advanced training programme for firefighters and should be akin to health promotion within the workplace.

In addition to the management strategies outlined above, it is also possible to identify a series of research issues that need to be explored further but lie beyond the scope of this paper. These can be outlined as:

i)   The identification of those personality factors which may modify the response to stress should be examined further within the fire service. This should involve a much wider study of firefighters at various stages in their careers and, ideally, should include exit interviews with those retiring from the job on health grounds. Such research should seek to determine the specific relationships between different personality types and the various service work environments in causing a stress response.

i.   The experience of stress in other types of professional training should also be examined. In particular this should be undertaken across the emergency services. The levels of stress reported during Fire Service training may simply be a product of training in general rather than specific to this particular programme. At present work is underway which aims to explore this issue further.

ii. The personality types of experienced fire fighters should be examined. The effect of training and career experience on personality is particularly relevant. This would also provide useful data with which to compare evidence collected from Fire Service training.

iii. The behaviour and methods of control used by the Fire Service at times of crisis needs closer examination in order to determine the extent to which its personnel strategies aid it in being prepared to deal with crisis potential. The shift from one configuration to another and the conflicts in cultures between the three services at major events is an area that also requires careful study.

iv. The effectiveness of a single level entry system should also be the subject of further study, again across the three branches of the emergency services. Given that many fire fighters join because they enjoy an active role then one must question whether these the most suitable people for developing corporate strategy and for administering the service. This is especially so when considering the issues of stress management.

v. Following Rick and Briner (2000), a sustained evaluation of the efficacy of post incident debriefing

This paper is inherently speculative and, perhaps over ambitious given the constraints imposed upon it. However, the aim has been to highlight issues rather than to provide simple solutions to complex problems. If it has achieved that it has been successful. We would like to thank those individual fire fighters who have given their time and energy to support our research.

NOTES

1. Both Cox and Mackay and Lazarus and Launier presented their models in 1978. Cox's text makes no reference to the work of Lazarus in this area. Given the fact that both sets of research were published in 1978, this is not surprising. However, Cox and Mackay did present their model at conferences in 1976 although there is no reference to this work being formally published until Cox's book in 1978.

REFERENCES

Association of Chief Police Officers (1987) *Working Party Report on Stress,*
Atherton, P (1990) Private Communication.

Bailey, E. (1993) "Post-traumatic Stress Disorder: A contrasting view" *Disaster Prevention and Management* 2(3) pp.22-25.

Banks, M H et al (1983) The use of the General Health Questionnaire as an indicator of mental health in occupational studies" *Journal of Occupational Psychology,* Vol. 53, 187-94.

Barton, L., Braverman, M. and Braverman, S. (1993) "A comparative analysis of organizational response o traumatic stress among workers in the aftermath of crisis" *Disaster Prevention and Management* 2(1) pp.46-56

Beech J R & Harding L, (Eds.) (1990) *Testing People - A practical guide to psychometrics*; NFER-NELSON publishing Co. Ltd.

Brown, J.M. and Campbell, E.A. (1994) *Stress and Policing: Sources and Strategies.* Chichester: John Wiley and Sons Ltd.

Brown, T J (1988) Stress and the Organisation; Unpublished MA thesis, Leicester Polytechnic, 1988

Burrows, G C, Cox, T, & Simpson, G C (1977) The measurement of stress in a sales training situation; *Journal of Occupational Psychology,* 50, pp45-51.

Chan, K B (1977) Individual differences in reactions to stress and their personality and situational determinants: Some implications for community mental health; "*Society, Science & Medicine,* Vol. 11, pp89-103.

Clews, P J (1990) A study of the attitudes to RAF recruiting practice by officers who have recently completed initial training: Unpublished MSc Dissertation

Cooper, C L & Grimley, P J (1983) Stress Among Police Detectives, *Journal of Occupational Medicine,* Vol. 25, N0. 7, July.

Cooper, C L & Payne, R (1988) *Causes, Coping and Consequences of Stress at Work.* London: John Wiley & Sons.

Cox, T. (1978) *Stress.* London: Macmillan Education Limited.

Cox T (1985) The Nature and Measurement of Stress *Ergonomics,* vol. 28, no. 8 pp 1158-1163.

Cranwell-Ward, J. (1990) *Thriving on Stress.* London: Routledge.

Docherty, R W (1989); Post Disaster Stress in the emergency rescue services; *Fire Brigades Journal,* June 1989.

Douglas, R B et al (1988) A study of stress in West Midlands Firemen, using ambulatory electrocardiograms; *Work & Stress* Vol4 no. 2 pp 309-318

DSM (III)(R) Classification of Post-traumatic Stress Disorder. (Para 309.89). *International Classification of Diseases.* (9th Edition).

Duckitt, J. and Broll, T. (1982) 'Personality factors as moderators of the psychological impact of life stress', *South African Journal of Psychology,* 12(3), pp.76-80.

Duckworth, D.H. (1988) 'Disaster work and Psychological trauma', *Disaster Management,* 1(2) pp. 25-29.

Eades, E., Elliott, D. and Smith, D. (1994) 'Coping with the unthinkable: Post traumatic Stress and the Emergency Services' *Mimeo,* Liverpool John Moores University.

Elliott, D. and Smith, D. (1993a) 'Coping with the sharp end: Recruitment and selection in the Fire Service', *Disaster Management,* 5(1), pp.35-41.

Elliott, D. and Smith, D. (1993b) 'Football stadia disasters in the UK' *Industrial and Environmental Crisis Quarterly,* 7(3).

Elliott, D. and Smith, D. (1994) 'Fitness for purpose: Dealing with stress in Fire Service training' *Mimeo,* Liverpool John Moores University.

Fisher, S. (1986) *Stress and Strategy.* London: Lawrence Erlbaum Associates, Publishers.

Fletcher, B. (C) (1988) "The epidemiology of occupational stress", in Cooper, C.L. and Payne, R. (Eds) (1988) *Causes, coping and consequences of stress at work.* Chichester: John Wiley and Sons Ltd. pp.3-50.

Freeman, M W (1989) Unpublished MSc Dissertation, 1989

Frost, B. (1990) 'Firemen awarded £34,000 for trauma after King's Cross' *The Times*, 19th December 1990 p.3.

Galloway, D, Panckhurst, F, Boswell, C, & Green, K (1984) Mental Health, Absences from work, Stress and Satisfaction in a sample of New Zealand Primary School Teachers; " *Australian and New Zealand Journal of Psychiatry* 18: 359-363.

Goldberg, D P & Hillier, V F (1979) A Scaled version of the General Health Questionnaire; " *Psychological Medicine*, 9, pp 139-145.

Gold, I, Haughey, L, & Barraff, L J (1985) Psychiatric screening in the Emergency Department; *American Journal of Emergency Medicine*, Vol. 3, No. 5,

Health & Safety Executive (1984) *Training for Hazardous Occupations: A case study of the Fire Service*, London: HMSO.

Hendin, H, et al (1983) The influence of precombat personality on post-traumatic stress disorder; *Comprehensive Psychiatry* Vol. 24, No. 6, (November/December).

Henson, A (1991) Private Communication.

Hepworth, S (1980) Moderating factors of the psychological impact of unemployment; *Journal of Occupational Psychology* 1980, 53.

Hetherington, A (1991) *Post Traumatic Stress Disorder in Road Traffic Police*; P.R.S.U. Home Office Report.

Home Office (Fire Department) (1985) *Fire Service Drill Book*, London: HMSO.

Hodgkinson, P E & Stewart, M (1991) *Coping with Catastrophe.* London: Routledge.

IPAT (1985) *Guide to the 16 PF Test*; NFER Nelson, London.

Isles, P. (1992) 'Centres of Excellence? Assessment and development centres, managerial competence, and human resource strategies' *British Journal of Management*, 3, pp79-90.

IPAT (1982) *Guide to the 16 PF Test*, NFER, London: Nelson

James, A (1988) Perceptions of Stress in ambulance personnel; *Work & Stress* Vol. 2,no. 4, 319-326.

James, A. (1992) 'The psychological impact of disaster and the nature of critical incident stress for emergency personnel', *Disaster Prevention and Limitation*, 1(2) pp. 63-69.

Johnston, A J & Kelly, M G (1988) Post Accident/Incident Counselling: Some Exploratory Findings; *Aviation, space, and Environmental Medicine* August 1988.

Kahn R L et al (1964) *Organisational Stress*; John Wiley & Sons, 1964.

Kets de Vries, M. and Miller, D. (1984) *The Neurotic Organization*. San Francisco: Jossey Bass.

Kets de Vries, M. and Miller, D. (1987) *Unstable at the top: Inside the troubled organization*. New York: Mentor.

Kinchin, D. (1993) 'Post-traumatic stress disorder: The consequences' *Disaster Management*, 5(2) pp. 92-94.

Kobasa (1982) 'Commitment and coping in stress over time among lawyers', *Journal of Personality and Social Psychology*, 42, pp.707-717.

Kline, P (1983) *Personality: measurement and theory*; Hutchinson 1983.

Ko, Y. and Kao, H.S.R. (1993) "The effects of paramilitary discipline on the psychology of fire-fighters" *Disaster Prevention and Management* 2(3) pp.26-34

Lazarus, R S (1966) *Psychological Stress and the Coping Process*, McGraw-Hill, New York

Lazarus, R.S. and Folhman, S. (1984) *Stress, Appraisal, and Coping*. New York: Springer.

Lazarus, R.S. and Launier, R. (1978) "Stress-related transactions between person and environment", in Pervin, L.A. and Lewis, M. (Eds) (1978) *Internal and External Determinants of Behavior*. New York: Plenum.

Lim Lin Mei, S (1990) Stress and the Fire Fighter; *Psychological Society Newsletter*, 1990.

Manning, M., Williams, R.F. and Wolfe, D.M. (1988) 'Hardiness and the relationship between stressors and outcomes', *Work and Stress*, 2(3) pp. 205-216.

Matlak, R. (1991) Private Communication.

Meichenbaum, D. (1985) *Stress Inoculation Training*. Elmsford, NY: Pergamon Press.

Mitroff, I I & Pauchant, T. (1988) Crisis Prone versus Crisis Avoiding Organisations. Is your company's culture its own worst enemy in creating crises? ; *Industrial Crisis Quarterly*, 2 (1988) pp 53-63

Mitroff, I I & Pauchant, T. (1989) Do (some) organisations cause their own crises? The cultural profiles of crisis prone vs. crisis- prepared organisations; *Industrial Crisis Quarterly*, 3 (1989) pp 269-283

Mintzberg, H (1989) *Mintzberg on Management*; Free Press.

Monserrate, R. (1993) 'Critical incident stress: A personal experience' *Disaster Management*, 5(2) pp. 89-91.

Murphy, L (1984) Occupational Stress management: A review and appraisal; *Journal of Occupational Psychology*, 57, pp. 1-15.

Murphy, L. (1988) 'Workplace interventions for stress reduction and prevention', in Cooper, C.L. and Payne, R. (Eds) (1988) *Causes, Coping and Consequences of Stress at Work*. Chichester: Wiley. pp. 301-339.

Palmer, C E (1983) A note about paramedics strategies for dealing with death and dying; *Journal of Occupational Psychology*, 56, 83-6.

Palumbo, F.A. and Herbig, P.A. (1994) "Salaryman Sudden Death Syndrome" *Employee Relations* 16(1) pp.54-61.

Parkes, K R & Rendall, D (1988) The Hardy personality and its relationship to extraversion and neuroticism; *Journal of Personality and Individual Differences*, 9, 4, pp 784-790.

Pauchant, T. and Mitroff, I.I. (1992) *Transforming the crisis prone organisation*. San Francisco: Jossey Bass.

Payne, R, Fineman, S, Jackson (1982) An interactionist approach to measuring anxiety at work", *Journal of Occupational Psychology* 55 pp13-25

Payne, R & Firth-Cozens, J (1987) *Stress in Health Professionals*; John Wiley & Sons.

Pervin, L A (1980); *Personality: Theory, Assessment and Research* (3rd Edition); John Wiley & Sons.

Quinn, J B, Mintzberg, H, & James, R M (1988) *The Strategy Process, Concepts, Contexts and Cases*; Prentice Hall International.

Raphael, B, Meldrum, L, McFarlane, A C (1995) Does debriefing after psychological trauma work? *British Medical Journal*, 10 June, Vol 310, 1479-1480

Rhodewalt F & Agustdottir S (1984) On the relationship of Hardiness to Type A behaviour pattern perception of life events versus coping with life events; *Journal of Research in Personality*, 18, pp 212-223.

Rick, J and Briner, R (2000) *Trauma Management vs. Stress Debriefing: What should responsible organisations do?* HSE Books

Richardson, W (1991) Private Communication.

Ross, R.R. and Altmaier, E.M. (1994) *Intervention in Occupational Stress*. London: Sage Publications.

Schofield, S. (1991) Private Communication.

Seyle, H. (1956) *The Stress of Life*. New York: McGraw-Hill.

Sharlin and Mor Barak (1983) 'Intervention phases following a disaster' *Disasters*

Smith, D. (1990) 'Beyond contingency planning: Towards a model of crisis management', *Industrial Crisis Quarterly*, 4(4), pp. 263-275.

Smith, D (1991) Organisational Response to Crisis: The Case of the Kegworth aircrash; *Crisis Management Working Papers*, Leicester Business School.

Smith, D. (1992) 'The Kegworth aircrash: A crisis in three phases?' *Disaster Management*, 4(2) pp. 63-72.

Smith, M (1988) Stress and the NHS Middle Manager; Unpublished MA Dissertation, Leicester Polytechnic.

Snyder, M (1974) Self-Monitoring of Expressive Behaviour; *Journal of Personality and Social Psychology* Vol. 30, 326-337.

Stiff, J B (1989) Air Disasters- rescue, post incident stress, airport fire service evaluation and special training; *Fire Engineers Journal*, June 1989

Summers, D. (1990) 'Testing for stress in the workplace' *Financial Times*, 23 November cited in McKenna, E. (1994) *Business Psychology and Organisational Behaviour*. Hove: Lawrence Erlbaum Associates, Publishers.

Sutherland, V J, & Cooper, C L (1990) *Understanding Stress*; Chapman and Hall.

Taylor, A J W & Frazer D C H (1982) The stress of post disaster body handling and victim identification work; *Journal of Human Stress*, pp 4- 11.

Walsh M (1989); *Disasters: current planning and recent experience*; Edward Arnold, 1989.

Warr, P (1990) The measurement of wellbeing and other aspects of mental health; *Journal of Occupational Psychology* 1990, 63 pp 193-210

Weinberg, A.A. (1972) *Science and trans-science* Miner

QUESTIONS OF RISK AND REGULATION:

Hegemony, governance and the US chemical industry

FRANK PEARCE,
Queens University,
Kingston
Ontario, Canada

STEVE TOMBS,
Liverpool John Moores University
School of Law, Social Work & Social Policy
Josephine Butler House
1 Myrtle Street
Liverpool, L7 4DN, UK

Introduction

If, since the second world war, accelerated rates of technological development (and, in particular, the growth of 'big science' in the OECD countries) has fulfilled the Enlightenment promise of progress, at the same moment it has called it into question. Developments in the chemical and nuclear industries have helped magnify material wealth but, as the new social movements argued, and recent disasters have demonstrated, they have also created a society where we have an increased susceptibility to large scale disasters and chronic illnesses (Beck 1992a: 1992b). The response of the hegemonic social groups has been to significantly modify the ways in which production takes place, and the ways in which it is regulated.   In some ways, at least, within these countries, it is now undertaken more safely than previously, if never as safe as government and corporate propagandists claim.   However, the move to produce more basic chemicals in less developed countries where there is usually less regulation both displaces risk by exporting hazard (Ives 1985), and, at the same time, may well, in the long run, contribute to a general lowering of standards in OECD countries too (Pearce and Tombs 1994; Pearce and Snider 1995). In this chapter we will relate the changes in the ways in which chemical production is organised within the U.S. to these changes and to changes in its mode of governance. Overall, we want to argue that there has been effected 'a passive revolution' (Gramsci 1971: 119-120; Sassoon 1987: 204-217) in that these changes have occurred through procedures, and through discourses, that marginally modify but crucially sustain both the overall conception of group and societal interests held by the groups comprising the power block and, relatedly, the pre-existing modes of hegemonic dominance. Yet, because of a continu-

E. Coles, D. Smith and S. Tombs (eds.), Risk Management and Society, 165–187.
© 2000 *Kluwer Academic Publishers. Printed in the Netherlands.*

ous reconfiguring of social relations and social forces within the American and global context, this new 'settlement' is inevitably unstable and recuperable.

Hazardous Production and Modes of Governance

STRUCTURAL CHANGES IN THE AMERICAN CHEMICAL INDUSTRY

The American chemical industry that emerged in the late nineteenth century was first primarily involved in the production of such industrial chemicals as alkalis, acids, dyestuffs, explosives, and metallic alloys and then, after a tremendous boost provided by the first world war, diversified into such products as paints and lacquers, man-made fibres, electrochemicals, pesticides and pharmaceuticals (Chandler 1990: 146-188). Although organic chemicals had always played a key role in the industry, the addition of oil and natural gas feedstocks to those of wood, tar and coal happened early in the U.S. and led to the development of polymers and plastics - some 1 billion pounds of synthetic organic chemicals were produced in 1940. The boost to this production associated with the war effort, most importantly the programmes to create synthetic rubber and high octane aviation fuel (Aftalion 1991: 214-215), helped generate major new feedstocks and a proliferation of new chemicals; by 1950 30 billion pounds of organic chemicals were being produced, which had risen to some 300 billion pounds by 1976 (Barnett 1994: 14). The years up to 1973 were, for the chemical industry (as with most American industries) generally extremely prosperous; but already there were signs of the problems that would become only too evident in the 1980s. Internationally, the chemical industry had matured, in that its markets were relatively saturated and static - major new breakthroughs were few and far between. Competition had increased - the British and French chemical industries had adjusted to the postwar status quo and were now holding their own, the German chemical industry had risen from the ashes and Japan and other non-European industries were developing rapidly. This competition was met in all countries by an increase in investment leading, in turn, to a crisis of overcapacity.

At the same time in different countries there were somewhat uneven pressures from environmental groups to change the modus operandi of the industry. In the U.S. the response of the chemical industry was to disengage its production both upstream and downstream and to shift to specialist chemicals. It also diversified into other unrelated areas thereby creating conglomerates, less dependent on any particular market, and into speculating in financial markets. In the 1980s, U.S. chemical companies were pushed to (a relatively new) shortermism by increasing dominance by finance oriented executives (Fligstein 1990), corporate raiding, leveraged buyouts etc.. By the end of that decade, even the largest US chemicals companies were forced into dramatic restructuring as a consequence of over-capacity, to some extent independent of the US recession (Chemistry & Industry, 1991: 524).

Our focus in this chapter, then, is upon the formal political role of the state in terms of regulation, and the translation of, and relationships between, this and (risk-producing and risk-avoiding) activities within the US chemical industry. Clearly, neither focus can be addressed adequately without reference to wider social forces and relationships. Thus our concerns are partly with government, referring to state activity within the political domain, much more with changing forms of governance, as the 'conduct of conduct', which extends our analysis beyond the narrowly defined political domain, as well as, necessarily, with the inter-relationships between these two functions.

Our particular empirical concern is with the predominant ways in which, in the last thirty years, there have been changes in the ways in which companies within the American chemical industry have managed the hazards and risks associated with the use, production, distribution, and disposal of hazardous chemical substances. These practices are clearly related, in part, to the regulatory processes to which they are subject, and the past thirty years have seen the passage, (sometimes tardy) implementation of, and synergy between, such Federal Acts as the 1963 Food Drug and Cosmetic Act, the 1969 National Environmental Policy Act, the 1970 Clean Air Act, the 1970 Occupational Safety and Health Act, the 1972 Federal Water Pollution Act, the 1974 Safe Drinking Water Act, the 1976 Resource Conservation and Recovery Act (RCRA), the 1976 Toxic Substances Control ACT (TSCA), the 1978 Federal Insecticide, Fungicide and Rodenticide Act, the 1980 Comprehensive Environmental Response, Compensation, and Liability Act (CERCLA), the 1984 RCRA Amendment Act and the 1986 Superfund Amendment and Reauthorization Act (SARA), including in title 7, EPCRA - Emergency Planning and Community Right to Know Act. Thus, in the 1990s, many companies are now supposed to engage in more rigorous site-management, by developing inventories of hazardous substances and ensuring that they are stored in a safe manner, by developing safer production processes, by aiming at source reduction and the recovery and reuse of materials, and by on-site disposal and neutralisation of hazardous wastes. Safety and evacuation plans must be developed, and workers, local communities and government authorities informed of potential dangers, from point or moving sources, including routine discharges into the ecosphere. Land use is now more controlled and production facilities are inspected and certified. Damage already inflicted on the ecosphere is assessed and dealt with, to some extent, according to the 'polluter pays' principle. This new regulatory regime involves regulatory agencies, some market based solutions and the various uses of the legal system. Most commentators agree that chemical production in the US is now safer than in the 1950s, while many recognise that there remain important improvements that need to be effected.

Despite real improvements, it was perhaps inevitable that the emergent systems of regulation towards risk mitigation have failed to match the specific hopes of any group for their programmes had first to be 'translated' into the political rationalities and programmes of government. Government in its concern with 'the wealth, health and happiness of the population' must follow the procedures of consultation, encouragement, and use the carrot and stick to develop alignments that will create a new viable regime, where the "translatability between the moralities, epistemologies and idioms of political power, and the government of a specific problem space, establishes a mutuality between what is desirable and what can be made possible through the calculated activities of political forces" (Rose and Miller 1992: 182). Now, since a key element of such regulation has been the increasingly active role of the state it is necessary to relate the particular kinds of legislation attempted (and their relative success and failure) to the complex and often contradictory economic and political constraints within which they are developed and utilised. If there is legislation that holds corporations liable for harms to public health and the environment this imposes substantial costs on industry, the absorption of which may impair capital accumulation, the engine of growth in liberal capitalism. At the same time an absence of governmental actions that address certain harm to public health and the environment may threaten the legitimacy of the system as a whole (Barnett 1992: 105) [1].

THE AMERICAN CHEMICAL INDUSTRY AND DISCOURSES OF GOVERNANCE

We can identify, during this century at least three discourses and technologies of governance [2]. It should be emphasised that although there is chronological element to these, they cannot be understood in terms of a linear emergence and transcendence. Rather, they must also be viewed as co-existent, and as more or less dominant at particular periods; in other words, these discourses need to be understood in the context of hegemonic struggles. For this reason, while the following sections impose some chronological order upon the phenomena which we are discussing, the complexity of such phenomena renders a strictly chronological discussion both undesirable and impossible.

The 'first' discourse of governance, a corporate liberal discourse (Weinstein 1968; Lustig 1982; Sklar 1988), emerged during the first half of this century. Political authorities aimed to circumscribe severely the extent to which the state interfered with the ways in which the production and disposal of dangerous chemicals took place, delegating responsibility for its organisation to the technical expertise, self- interest, and sense of social responsibility of those operating the large chemical companies. It was believed that efficient production and social progress depended upon the activities of such large companies. They developed sophisticated and safe industrial processes that made products that were benign if used properly.

Inevitably waste products were generated but new uses for these were continuously found (see Aftalion 1991), while any remaining excess could be disposed of into an unperturbable ecosphere. If this was particularly environmentally destructive, this damage would be subsequently remedied by new technologies. This discourse is essentially one of progress - of increases in scientific knowledge, leading to increasing conquest of nature and control of the world, development of new materials, all leading to continuous improvements in 'the wealth, health and happiness of the population'. Industry experts, of course, knew (and made some allowances for the fact) that they were often dealing with dangerous processes which could cause significant harm to plant workers and surrounding communities. However, within this discourse, it was damage caused by explosions and by the immediately observable effects of corrosive and poisonous chemicals that were the major concern. A certain number of disasters and accidents were the unavoidable side effects of progress but injuries to human beings could be dealt with by the courts or through workers' compensation (routine accidents were easily actuariseable). Furthermore, quite extensive state intervention in the economy was accepted, for example, in the area of economic regulation and in the context of defence procurement. This discourse played a key role in creating and regulating the particular intensive mode of capitalist accumulation developed in the U.S. (Aglietta 1979; Jessop 1990: 174). However, we would stress that Fordism in the U.S. was always an aspect of a dual economy internally, while dependent upon favourable external economic relations (Pearce and Snider 1995) and, indeed, transnational processes also (Holman 1993). When the latter conditions of existence became problematised in the 1970s (Gill and Law 1993), there was a strong impetus to reconfigure, but not totally abandon (Woodiwiss 1993), the corporate liberal mode of governance.

Corporate liberal arrangements, however, had earlier been problematised by medical professionals and trade unionists, concerned with the chronic side effects of chemical and nuclear products and production processes on the health of local residents, consumers and workers, on animals, and on the ecosystem. These issues first became 'newsworthy' at the beginning of the 1960s, with the publication of Rachel Carson's book, Silent Spring (1962); then, the radical movements associated with civil rights, nuclear war and the Vietnam War, the consumer movement with its concern with the price, quality and safety of consumer goods, and union involvement with occupational safety and health issues, all prepared the way for debates and legislation concerning water quality, air quality, nuclear power and the production, use and disposal of toxic chemicals. In the early 1970s environmentalism per se became a major focus of American public consciousness.

Having 'measured the real against the ideal and found it wanting' (Rose and Miller 1992: 181), intellectuals, activist trade unionists and environmentalists developed ways of redescribing the real, so that programmes of action were developed which garnered enough support to influence the legislature. Those subscribing to this radical liberal discourse believed that many existing Federal agencies were captured by the industries that they were meant to regulate (Fellmeth 1970; Turner 1970). On the one hand, they argued for more activist interventionist regulatory agencies. On the other hand, they argued for: independent public-interest science; greater empowerment of employees, communities and individual consumers, who were, or might well be, exposed to dangerous chemicals; and legal regimes that were conducive to the compensation of actual or potential victims. Most fundamentally their aim was to force corporate capitalism to pay for past irresponsible conduct, and to force large corporations not only to internalise externalities but also to actually take seriously the interests of other stakeholders. Aided to no small extent by a series of highly visible industrial disasters (at Flixborough, U.K. in 1974, Seveso, Italy in 1976, Love Canal, U.S.A. in 1978, Three Mile Island, U.S. in 1979, Bhopal, India and San Juan Ixhuatepec in 1984, Mexico in 1984, and Chernobyl, Ukraine and Basel, Switzerland in 1986), those subscribing to this discourse were able to play key roles in the passage of much legislation. Nevertheless, they have remained highly critical of how capitalism is regulated, how victims are compensated, how the environment is repaired, and how the costs of these are distributed (Hofrichter, 1993).

There had also emerged in the 1970s and 1980s a neo-liberal 'enterprise' discourse (Miller and Rose 1992; Adams and Brock 1991). This was also critical of Federal agencies (Stigler 1971, Peltzman 1976; Weaver 1978; Weidenbaum and Fina 1978), though this criticism was on the grounds that a deregulated economy, opened up to market forces, would not require regulators to act as 'policemen', merely as advisors

(Bardach and Kagan, 1982), since regulation would be provided by the market and the legal system. General tendencies towards innovation, characteristic of high technology sectors like the chemical industry, had been further stimulated by a growing global awareness of environmental matters. Once these issues were identified industry could and did respond responsibly. For example, the U.S. Chemical Manufacturers Association "Responsible Care Initiative" and its role in implementing' legislation on 'Your Right To Know about what chemicals are being produced and used in your neighbourhoods' (New York Times Magazine, April 15, 1991) can be seen as testimony to this. Progress had been achieved, and would continue to be achieved, when scope was given to the skills and innovatory capacity of corporate management to produce a continuous improvement in managerial skills, new technologies and new risk management techniques. Built into this perspective, however, was some kind of recognition of the limits of scientific knowledge and the need to acknowledge ignorance and uncertainty. On the other hand, there was a general belief that, overall, the wealthier the society the safer the society. The rational response to environmental problems was for the state to help industry develop new standards of conduct and for the courts to develop new - but rational - rules of liability. While some state initiatives were seen as constructive - for example, clarifying the rules for responsibility for disposing of hazardous waste safely and setting up a competitive market under the RCRA in 1976 - many were deemed clearly irrational, most notably aspects of the Superfund Acts of 1980 and 1986. Over all, critics argued that greater attention should be paid to both the opportunity costs and transaction costs of policies pursued [3], and there should be a very definite refusal of their opponents implicit utopia of a no-risk society (Aharoni 1981; Douglas and Wildavsky 1982). Indeed, this discourse stresses both the inevitability of uncertainty because of a necessary incompleteness of knowledge (Hayek 1949) and a positive conception of risk-taking as potentially productive and beneficial as well as harmful (cf. Douglas 1993). Hence, there is a negative view of risk averse behaviour and a positive evaluation of venture capital. This discourse, of course, has not been restricted to the U.S., but has also informed the reasoning of the IMF, World Bank and OECD (Cox 1993: 266), and thereby of many national governments.

Of interest here are the answers to the such questions as: what events and processes occasioned these reactions; to what extent were these actions justified; how viable were they and what were their solutions; to what extent were these discourses adopted to guide conduct and to what extent were they used selectively and/or as a gloss for other kinds of activities? To explore these questions we need to examine the changing nature of the regulation of the American chemical industry, and of practices and discourses within the industry, with respect to various hazards associated with its production activities.

The Emergence of Risk Assessment

INSURANCE AND RISK

While the post-war expansion in production brought immense benefit to the US chemical industry, it entailed enormous costs which were disproportionately borne by its workers and by local communities. As these, and other pro-regulatory groups, began to accumulate evidence regarding occupational ill-health, strategies of denial through a refusal to acknowledge such evidence became increasingly difficult for employers, the courts and regulatory agencies to sustain, certainly from the late 1960s. This period saw a reluctant recognition of the industrial causes of a range of diseases and a more general awareness that chemicals - in pharmaceutical products, in food, in the air, in the water, in toxic waste sites, and from chemical or nuclear disasters - negatively affect the health of people irrespective of their class, race or gender.

The long-standing efforts of the chemical industry to refuse to recognise that many chronic illnesses were induced either through occupational or more general exposure to toxic chemicals is explicable given that such illnesses and diseases, once legally acknowledged, constitute an actuarial nightmare. They involve mass victims, multiple injurers, fuzzy loss, multicollinearity (complex causal chains) and latency. They occur infrequently, often involve strict liability and joint and several liability, they depend upon the vagaries of individuals deciding to make claims, and occurrences cannot be predicted but at best estimated using the very imprecise procedures of risk analysis (Katzman 1989: 133).

Since 1973, insurers have proved unwilling to provide Comprehensive General Liability policies to the Chemical Industry without significant exclusions and although there was a subsequent attempt to cover these with an Environmental Impairment Policy this did not work to the satisfaction of the insurance companies (Katzman 1989: 131-134). Indeed, much of the recent crisis in the world insurance market, most notably at Lloyds, has been associated with environmental and disaster insurance (Raphael 1994: 150-172). Insurance companies have also become increasingly concerned about the safety of chemical plants (Smith, 1985), while the spectacular expansion - both in the scale and geographical spread of production - of the international chemical industries in the post second world war period means that the prospect of a high consequence event has increased (King, 1990),

In the context of their responsibility for first party site clean up and joint and several liability for other sites, many insurance companies are now at loggerheads with their chemical industry clients (Barnett 1994: 41), differing with the latter over what has been insured, accusing them of failing to fulfil their legally defined environmental obligations and even, on occasion, of clandestinely dumping hazardous waste (cf. New York Times 1992 May 17, p. F 11). It is hardly surprising that insurance cover has become increasingly difficult to secure, and is often dependent on external inspections of facilities by insurance companies. As a result, many firms and groups of firms are beginning to organise self-insurance, or at least are attempting to do so.

For Priese, recent increases in liability should be understood in terms of a new utilitarianism concerned with the most effective means of reducing both the general risk of injury and the effects of injury in society as a whole. This is achieved by using civil damage judgments as a public policy instrument 'for internalizing costs to the parties that generate them' (Priese 1993: 208), where generation means any contribution by such a party to an increase in the risk of loss faced by any other party. Here, since 'all actions can be arrayed at some point upon the risk contribution continuum, sharp moral distinctions lose moment' (ibid.: 215).

Undoubtedly the greatest increase in liability, both in number of cases and in size of awards, has been for corporate, professional and governmental defendants. These increase, however have had less to do with incentives to increase safety than with injury compensation insurance through tort law. But this is both a socially inefficient and inequitable way of providing such insurance (Galanter 1994). Because benefits tend and operational costs tend to be higher, and because insurers are less able to differentiate among the insured population to reduce insurance costs, the provision of insurance through an adversarial tort system is at least five to ten times as expensive as first-party insurance (Priest, 1993: 225; Galanter, 1994). Because all consumers end up paying the same additional amount for the cost of liability insurance but the wealthy get larger damage settlements this does not work to redistribute income downwards. The general historical verdict upon tort as a compensation device is that it has been, at best, 'uneven in performance'; moreover, evidence on the extent to which it has proven effective in controlling risk is highly equivocal (Galanter, 1994). Furthermore, fear of liability suits has meant that a number of the goods and services have been withdrawn from the market and there has been a reluctance to offer new products.

On the other hand, in tort cases, where injury is caused by something of which the plaintiffs were unaware while it was affecting them, but which would be dreaded by most people, courts and juries have recently, and increasingly, tended to reject the 'minimal' accounting based upon the calculation of the probability of risks, cost/benefit analysis and risk analysis, and have instead applied a more comprehensive - and, crucially, less predictable - social concept of responsibility and reasonable risk including a concern with human autonomy and victim compensation (Sanders 1992: 76).

Useful here is Michael Baram's discussion of the challenges posed for the chemical industry and its insurers by what he refers to as an emerging 'moral code' established by the common law of 'toxic torts' (Baram, 1987: 421). Baram charts four key adaptations within US tort law which have 'increased the economic vulnerability of industrial firms using hazardous materials' (ibid.), and he places these in the context of an increasing inability of traditional regulatory and corporate risk management methods to respond to the impression made upon 'public perceptions' and 'public consciousness' by the acute and chronic health, safety and environmental effects caused by hazardous technologies (ibid.: 416). Moreover, in introducing these adaptations he (perhaps unwittingly) summarises the nature of the challenges that these pose for the hegemony of an existent form of governance:

> "The challenge for industry and its insurers is to develop the ability to overtake increasing risk and economic vulnerability, particularly for firms doing business in the USA, where vulnerability is greatest, and bring about the control and abatement of these dynamic forces" (ibid.: 417).

First, many state statutes of limitations have been revised by the 'discovery' rule, which permits the bringing of tort actions within a period of time from the discovery of the tortuous act, rather than from the date of the act itself, so that the period during which tort actions can be brought is considerably extended (ibid.: 422). Second, many states have adopted the doctrine of strict liability over tests of negligence where toxic chemicals are involved (ibid.: 422-424). Third, certain states have replaced the requirement to establish which among multiple parties caused a particular injurious effect with a 'joint and several liability' (Baram, 1987: 424). Finally, Baram notes shifts in some US states towards a weakening of the burden of the proof of causation (ibid.: 425-6). Alongside this shift he notes the progressive potential of both Right-to-Know legislation and the results of biological monitoring and medical surveillance in terms of plaintiffs being able to gather the evidence necessary to show causation (ibid.). We shall consider briefly the potential significance of Rights-to-Know below. On biological monitoring and medical surveillance, it is worth noting that the optimism expressed by Baram regarding the provision of key information for counter-hegemonic groups through the use of such techniques is hardly, at present, being realised. As Draper has documented thoroughly, these techniques have been usurped by corporations themselves, and have provided a means of exclusionary and discriminatory practices in the relative distribution of what have come to be recognised as more and less healthy jobs (Draper, 1991; see also Robinson, 1991). The net effect of the use of such techniques in the US chemical industry has been to exacerbate what have been called, in a different context, existent 'structures of vulnerability' (Nichols, 1986). As we shall note below, however, these particular consequences of the use of the biological monitoring and medical surveillance reveal less about any inherent natures of these techniques, and much more about the balance of social forces within which they have been developed and deployed.

Thus the growing awareness of the dangerous nature of modern production processes, expressed through increased litigation, environmental movements and news of dramatic disasters combined to make it impossible for the corporate executives, scientists and technologists and state administrators to retain legitimacy without making changes. Their strategy has been to claim that all of these are the unavoidable side effects of progress, that they were unexpected but that they can be controlled or ameliorated. They are merely a new episode in a long running saga. It is business as usual.

RISK MANAGEMENT VIA RISK ASSESSMENT?

Companies (and regulators) have made increasing use of techniques of risk analysis, safety assessment, consequence analysis, hazard operability studies, hazard analysis, and so on, which were initially developed to satisfy the demands of those insuring hazardous technologies, this remaining their primary rationale (Kletz, 1990: 281). The most inclusive of these techniques is risk assessment. This generates data which is then used to define the health effects of individual or population exposures to hazardous materials and/or situations (National Research Council 1983). Risk assessment consists of four stages (Cutter 1993: 38), where each are is reduced to a series of techniques whereby risks are represented as identifiable, quantifiable, predictable and calculable. Thus risk assessment as a process remains wholly subsumed within the discourses of (Western) scientific rationality. Yet it remains no surprise that each of the particular stages of risk assessment are subject to controversy. In their present form, risk assessments necessarily involve rough and ready methodologies, meaning that whichever particular sets of assumptions are chosen, widely divergent results are produced.

It needs to be emphasised more generally that in addition to what appear to be essentially internal disputes regarding the range of quantitative techniques which constitute risk analysis, there are a further series of disputes over the subsequent uses to which such data are put. That is, numerical estimates do not simply translate themselves into particular kinds of policy decisions. There are various and competing strategies regarding the ways in which such a 'translation' is effected. Cost-benefit analysis, often favoured by the chemical industry, requires assigning a numerical value to the quality of life is not only inappropriate but provides insuperable problem for such analyses. Attempts such as those of Leonard and Zeckhauser (1986) to acknowledge that "some social values will never fit in a cost-benefit framework and will have to be treated as 'additional considerations' in coming to a final decision" are necessarily ad hoc and unpersuasive (see Teuber, 1993, Wynne, 1989).

Under these circumstances, the deployment of risk analysis to develop safety strategies is necessarily limited. Indeed, the public rhetoric and the estimates presented by risk analysts are frequently pacifying, although sometimes alarmist, and based upon poorly founded claims that scientific methods are the appropriate basis to make decisions about what constitutes socially acceptable risk.

While techniques of risk assessment have been predominantly developed and deployed within the context of a technocratic scientism, these techniques have, more recently, been seized upon by deregulationists, who have urged a more effective regulatory strategy which relies upon market based solutions and the recognition of self-interest on the part of corporations. Indeed, a critique of the burden of over- regulation was a central argument in the re-emergence of free market political forces in the US in the latter part of the 1970s. Yet as these groups manoeuvred towards formal political power, it was clear, given the nature and weight of counter-hegemonic discourses around risk, that their arguments for deregulation could not simply be translated, in policy terms, into a complete withdrawal of regulatory oversight and general state abstention from any concern with the private production of hazard and risk. Risk assessment, then, provides a useful weapon for those who would otherwise pose almost moral objections to social regulation by the state of private business; moreover, and unfortunately, as regulatory capacities became undermined once these neo-liberal forces assumed hegemony, social liberals also put ever greater faith into such techniques, seeing in these a better-than-nothing alternative in the face of encroaching deregulation. Thus the use of risk assessment, and other forms of quantitative argument (such as accident and incident statistics) were central to experiments under the Reagan administrations in the removal of OSHA inspections from certain companies. Through the development of risk management programmes, based upon effective techniques of risk assessment, chemicals companies were able to 'demonstrate' - that is, on paper - that they were competent and to be trusted in effective safety and health management. Subsequently, however, a combination of events and phenomena have proven a mode of governance within which such forms of risk assessment are central to be an unstable one. The emergence of deregulatory arguments and the neo-liberal hegemonic discourse in the US can be explored briefly in relation to one arena of sources of hazards and risk, namely workplaces.

Prior to the passage of the Occupational Safety and Health Act in 1970 legislation of conditions endangering the health of workers was the responsibility of individual states and relatively ineffective (Rosner and Markowitz 1989: 95). In the 1960s, such legislation had been initially proposed by radical discontented rank and file workers and by activist trade union leaders (Donnelly 1982:18-19; Szasz 1984:105) but it was Richard Nixon who eventually (and opportunistically) supported and passed a comparatively weak Occupational Safety and Health Act. The new Occupational Safety and Health Administration was under-resourced, uncritically adopted pre-existing business generated standards (Stellman and Daum 1973: 9-11) and its inspections and enforcement activities, initially at least, were directed mainly against small businesses (Szasz 1984: 108).

Nevertheless, in many ways a product of a radical liberal discourse, the Act created a general employer duty "to furnish to each of his employees employment and a place of employment which are free from recognized hazards that are causing or are likely to cause death or serious physical harm to his employees" (Sec. 5a). The OSHAct gave some participative rights to employees, and in legitimating occupational health and safety concerns provided a mobilising basis for trade unions (Noble 1986). It also granted considerable powers to OSHA inspectors. These nevertheless usually failed to secure the imposition of maximum penalties, while many early citations were for easily observable but relatively trivial violations. Furthermore, every observed violation had to be recorded and, less happily, there was an obligation to make a formal finding of guilt. This latter provision removed from the inspector the kind of discretion, and negotiating power, that, contra Kagan (1984), is routine in most police work. Moreover, these practices helped provide reasons (and justifications or excuses) for the climate of opposition to OSHA which developed into the vocal deregulatory movement (see Wilson 1985).

Although the deregulationists' arguments did not, at first, affect what happened within the production process and did not have much influence on OSHA, they were much more significant at the general political level. During the Carter administration, the President himself, Congress and the Supreme Court all acted to constrain OSHA's activities and organised labour lost political and economic power. Reasonable questions about the relative effectiveness of different regulatory strategies were subtly redefined as issues resolved through the application of very narrowly construed cost-benefit analyses.

When Reagan came to power the Office of Management and Budget, a crucial and antagonistic hegemonic apparatus (Mahon 1977, 1979), gained effective control over many regulatory agencies. A new unambiguously pro-business Secretary of Labor and a pro-business OSHA director were appointed. The budget, number of inspectors, inspections and follow up inspections were all cut, and worker's rights and inputs curtailed (Calavita 1983: 441-443; Navarro 1983: 523).

It was argued that the OSHAct had little impact on job risk levels. The activities of OSHA were unlikely to change the behaviour of those who own and/or control businesses: first, because it was ineffective in imposing sanctions (Smith 1976: 63-64); second, because it imposed excessively high uniform standards, and demanded a re-engineering of workplaces where employers more rationally used situationally specific standards, and provided less restrictive (and cheaper) personal protection devices.

Drawing on Becker's (1975) human capital concepts, Viscusi, argued that occupational safety and health was best regarded as a marketised good, bought and sold as part of the wage bargain, best regulated by labour and other markets and by the civil courts (Viscusi 1983: 35). Then, in the absence of market rigidities produced by governmental or trade union activities, the forces of supply and demand can determine the level and kinds of risk faced by individual workers, the level of investment in occupational health and safety, and the approaches taken to control hazards. Thus, while some market imperfections of course need to be remedied (Smith 1982: 330-334; Chelius 1977: 63-69), the most socially efficient health and safety standards will be realised when employers find the level of expenditure on safety which is equal to the expenditure incurred by accidents. Improving the market mechanisms thereby allows for the dismantling of OSHA and the restoration of the private system that dominated health and safety before 1970, albeit with a somewhat reformed workers' compensation program (Chelius 1977). These ideas were not merely theoretical but informed much OSHA practice in the 1980s when Viscusi became a consultant for it. It should be clear how contemporary forms of risk assessment could thus be deployed within this general neo-liberal discourse, and used to justify some degree of regulatory withdrawal. Companies were best left to determine the nature and levels of risk that they were creating, and it was in their own interests to seek effective management of these.

As has been argued at length elsewhere (Pearce 1995; Noble 1986), this representation of what determines occupational safety and health is deeply flawed. In reality, the general level of occupational safety and health is determined by the interests of capital, by its power relative to workers, by the extent to which it dictates state policies, and by the specific ways in which the mode of regulation articulates together market relations, legal instances and regulatory agencies. This is particularly clear in the case of workers' exposure to toxic chemicals. Thus, there we find that: information on the dangers of chemicals is rarely even generated or is inadequate; 'safe' exposure levels, if set, are often problematic as are the monitoring of actual exposure; monitoring of workers' health is undertaken by companies who control the information that they gather. Thus, if a problem is identified the tendency is not to develop safe working environments, rather, those individuals who are putatively most likely to suffer from exposure are excluded from these particular occupations. There is no evidence that workers choose this solution, one based upon the freedom of employers not to contract with certain individuals, rather than, keeping their options open by rendering workplaces safe for anyone to work in, the solution preferred by trade unions.

This certainly does not mean, moreover, that individuals are always excluded from occupations dangerous to their health nor does it mean that those individuals who actually work with dangerous chemicals are not also at risk from them. Many chemicals can also affect men's reproductive capacity, and while more than 100,000 jobs are currently denied women of child bearing age because they entail using risky chemicals, there remain as many as 20 million women's jobs (that is, where industry is dependent on women workers) that equally entail toxic risks (Draper 1991: 71-73. These considerations suggest not some failure of these programmes but rather that both these concepts of marketisation and the strategies of exclusion can be made to function in relation to other strategic goals - viz. against equal rights legislation for women (and also incidentally for African-Americans of either gender). In fact, monitoring and screening procedures are not inherently positive or negative from the point of view of workers. Similar procedures using similar technology, when under the control of employees and their unions can have a worker empowering and life enhancing import (Draper 1991: 138-139).

Reconstructing a Hegemonic Mode of Governance?

It is important to note that while neo-liberalism clearly forged a hegemony through a representational and actual incorporation of a range of social forces, neo-liberal forces were also quite explicit in their marginalisation and exclusion of certain social groups from this hegemony (trades unionists, welfare recipents, various ethnic minoroties). There is a real sense in which neo-liberal hegemony was actually only partial, to the extent that neo-liberal forces actually eschewed the construction of consent in favour of dominance over some groups, even if the latter was often couched in terms of a 'simple' unfettering of the free market. This may well indicate an inherent fragility within this mode of governance. And the neo-liberal hegemony that had been constructed through the latter half of the 1970s and the early 1980s in the US was subject to potentially over-bearing counter-pressures in the latter half of that decade.

The 1980s witnessed both heightened public opposition to the apparent hubris of science-technology where this conjoined with corporate capital, such opposition being organised around a re-emergence of environmental issues. Two events in particular added a qualitatively new impetus to this opposition, and forced the chemical industry to begin to act in ways which were consistent with its own implicit questionings regarding an almost total prior faith in science-technology. It would, of course, be false to claim that the disasters to be discussed below were in any simple way 'causal' of shifting practices and attitudes towards risk on the part of the chemical industry. Indeed, what is clear is that these need to be placed in the context of the kind of general opposition noted immediately above, so that it is recognised that disasters can potentially be used as opportunities to organise around "resonances already present in social discourse" (Hadden, 1994: 108). Crucial to the realisation of such potential is the absence or presence, and nature, of the social forces that might seek to take advantage of these, such forces being most effective where they can attach themselves to existent activities, and where they have some tradition in, or experience of, political action (ibid.).

In 1984 there occurred three major chemicals disasters, at Cubatao (Brazil), Ixhuatapec (Mexico) and Bhopal. Cubatao and Ixhuatapec cannot have been without some reverberations in the US, given their relative proximity, at least in North American perceptions. But Bhopal was clearly the major disaster, not only in the obvious sense of the massive and unprecedented loss of life and chronic illness associated with that gas leak, but also because it raised fundamental questions concerning the operations of one of the largest US-based multinational companies in a host nation; and a host nation, at that, which could hardly be dismissed as backward (even if such a dismissal was precisely one of the post-disaster tactics employed by senior Carbide officials; Pearce and Tombs, 1993).

Following the disaster, the US chemical industry was plunged further into the insurance crisis that was noted above (Smith, 1986). Indeed,

> "This insurance crisis was already under way before Bhopal but was intensified by the safety crisis Bhopal produced, as well uncertainty over the full extent of UCC's liability in Bhopal. Before Bhopal, the environmental impact liability insurance market had practically evaporated. The sudden and accidental pollution insurance market had reduced individual companies' coverage from £300 million to £50 million, with even this lower figure unavailable to large chemical companies" (Jones, 1988: 198).

Moreover, the trends in the use of tort law, also referred to above, were crucial in exacerbating this crisis. As Jones again notes, the increasing number and levels of court awards won against chemicals companies is indicated somewhat by the fact that "the number of multi- million dollar verdicts rose from one in 1961 to 251 in 1982 and 401 in 1984" (ibid.). Further compounding this sense of crisis in the US chemical industry was the fact chemicals companies there witnessed the near collapse of Union Carbide in the face of a hostile takeover bid following the disaster (Lepkowski, 1994).

Thus for many chemicals corporations in the US the possibility of a Bhopal-type disaster in an industrialised economy, or in the US itself, was at once real, yet unthinkable (not least economically). In public, however, representatives of the industry sought to reassure workers and publics in classic technocratic fashion, claiming that 'it couldn't happen here', that US plants were technologically better and, in frequent and often quite explicit resorts to racism, operated by well qualified Americans rather than backward Indians (Pearce and Tombs, 1993).

Federal OSHA continued to use "records inspections" as late as 1987. Under this policy, inspections of operations were halted where records showed that employer to have fewer than average injuries and illnesses. Indeed, on the basis of a good safety record, UCC's plant at Institute had been exempt from OSHA inspections as one specific element of more general moves towards deregulation in the US. There is no denying that UCC as a company had (latterly, at least) what appeared to be a relatively good safety record in the US. But this counted for nothing when aldicarb oxime, a compound with methyl isocyanate as one of its constituents, leaked from the Institute plant in August 1985. Following this incident, OSHA sent several teams of inspectors to conduct a 'wall-to-wall' inspection - and this led to 221 charges of 'wilful violations', though OSHA and UCC eventually reached a settlement which saw the company fined just over $4,400 for five serious violations on the agreement that the others would be corrected. Of particular interest is the fact that US Labour Secretary Brock revealed that had UCC actually kept accurate records (that is, rather than those on the basis of which they were exempt from inspection), then their accident record would have been 'substantially higher' than the US chemical industry average (Jones, 1988 : 163-186, Pearce, 1990).

Bhopal provided the context in which Institute would be received. Yet in many respects it seems that it was the latter that provided the final impetus for many chemicals corporations to calculate that it was in their own economic interests to develop more effective systems of safety and health protection. The particular inadequacies of probabilistic techniques such as risk assessment to deal effectively with low probability/high consequence situations, such as Bhopal or Chernobyl, was implicitly accepted, as many operators of major or high hazard sites shifted towards their risk management strategies to encompass a recognition of worst case scenarios. Thus in 1987 Baram could write that there were clear signs of a trend amongst the operators of hazardous facilities from what he calls a 'rudimentary' form of engineering risk analysis (Baram, 1987: 427), to an approach wherein risk is viewed as 'an ongoing operational problem involving full consideration of personnel turnover, training, human factors, and other aspects of its capability for ongoing control' (ibid.: 432).

It is interesting that although these shifts have occurred in practice, they tend not to be explicitly recognised in public rhetoric on the part of the industry. The reason is clear. For these are no formal nor procedural shifts. The increasing use of a worst case scenarios is a recognition that a cataclysmic event such as Bhopal can occur, a likelihood that was effectively denied by previous forms of statistical calculations of risk which would posit its probability as one in so many million years that this meant nothing to lay-people in terms of a real threat. The move towards a more encompassing approach to risk management entails attention to managerial and organisational systems; and in this is the possibility of a recognition of the need to transcend technocratic forms of management and accept some participation from a range of 'outside' groups in effective risk management - namely, workers, local communities, and other potentially counter-hegemonic groups.

If these changes have not been made explicit, their one most tangible effect has been the development of emergency procedures in the form of evacuation plans, ongoing 'risk education', and various other forms of contingency planning (Jones, 1988). When one places these shifts in practice alongside other chemical industry initiatives during the same period, such as 'Responsible Care' and claims towards greater disclosure of information, then these changes are best viewed as both pre-emptive and as attempts to regain legitimacy. That is, taking these together, there is evidence here of a real threat to the chemical industry, of possibly punitive forms of regulation being forced upon local and national states, and of an attempt to reconstruct a new hegemonic discourse in the context of overwhelming threats to the previous one. This process can be understood through Gramsci's concept of a passive revolution.

Importantly, partly through reluctant recognition of their own limitations, partly through new legal requirements regarding Rights-to-Know and Hazard Communication, and partly through the need to involve workers and local communities in emergency procedures (which has its origins in both law and the demands of insurers), this new form of governance is partially constructed in a way that is inclusive of (some) formerly 'outside' groups. Thus in this very attempt to construct a new hegemony crucial ground is ceded in a reluctant recognition (however implicitly) that previous reliance upon both scientific-technological capacities and rationalities was an inadequate and, to some extent, a duplicitous, one. And with this more or less public recognition there emerges a space within which workers, local communities, and other counter-hegemonic groups can contest the ways in which risks are defined and managed. The exploitation of this space is facilitated by the techniques (such as rights to know, worker and community involvement in risk management, health monitoring and biological surveillance) which have been developed within the hegemonic mode of governance itself. That is, the emerging mode of governance is one that is open to contestation and struggle, the outcomes of which are far from settled.

There is no necessity that these techniques will be exploited. For example, rights-to-know can be more or less progressive, more or less open to corporate manipulation. Moreover, even where hazard and risk information is genuinely made more accessible, we reject the simplistic assumption that this in any way constitutes some form of 'empowerment'. Yet there is created at least the potential for worker and citizen involvement in hazard prevention through involvement in decision-making in local plants, and instances of this have been described recently (Hadden, 1994). Hadden's work also indicates that workers and local communities, once they have won a level of involvement in plant organisation, tend to develop more radical demands, both at local level and beyond, through alliances with other pro-regulatory forces.

Conclusion and Discussion

In this chapter, we have considered the management and regulation of risks and hazards associated with the US chemical industry, and have attempted to characterise these in terms of hegemonic modes of governance. This exercise is useful, since it allows an appreciation of the limits of the possible in terms of the regulation and management of such risks, both historically, and in the present. Moreover, it indicates the significance of competing and allied social groups in defining, contesting, and exploiting those limits. While arguing that there have been clear shifts in the hegemonic modes of discourse, these shifts have involved a range of social groups and forces, some of which have been counter-hegemonic, some of which have been integral to various hegemonic groups. Thus industrialists, science- technologists, regulators, insurers, lawyers, workers, local communities, and various national activist groups, have struggled over contested and changing ideological and material terrain, sometimes consciously but at other times not, sometimes pre-emptively but at other times clearly reactively, and from positions of varying strengths and weaknesses, these strengths and weaknesses determined by conditions some of which may have been amenable to control, others of which escaped control, and perhaps even predictability.

For us, it is clearly the case that the contemporary regulation of major hazards associated with the production, use, storage, distribution and the disposal of the by-products of the chemical industry is infinitely better than it was thirty years ago. It is equally clear, however, that these responses to hazardous production on the part of the chemical industry have not been primarily motivated by any new humanitarianism but, rather, represent pragmatic responses to various pressures. Indeed, we have argued that recent responses by chemicals companies, and by local and national states, to a combination of almost unprecedented pressures have at least entailed a recognition of a serious challenge to neo-liberal hegemony, at most provoked attempts to effect the construction of a new, hegemonic mode of governance through a process akin to what Gramsci termed a 'passive revolution'. This 'new' hegemony is a precarious one, not least because it entails a recognition of the limitations of scientific rationality.

While the conclusions of this chapter are somewhat tentative, and highlight areas of a future research agenda, they do provide important insights into the nature of hegemony in the context of the United States. Most importantly, it is clear that there are complex relationships between the hegemonic ideolgies and the ways in which the power bloc actually operates. For example, whilst maintaing a representation of all citizens as of equal value, and emphasising meritocratic and individualistic factors as determinants of individual fates, marginalsed groups have, and continue to be, either rendered invisible and/or subjected to disproportionate levels of risk; indeed, there is a real sense in which some social groups have never been included within any hegemonic order (Hall, cited in Simon, 1988: 110). However, as particular moments have revealed the extent to which large numbers of the US population find themselves included within marginalised groups, distrust in the ability or willingness of the state to attend to the social welfare (here discussed in relation to risk) of a majority of citizens has developed, so that there has developed and is reinforced a popular cynicism. This is clearly illustrated through even a schematic overview of popular responses and opposition to industrial activities which are recognised as entailing high levels of risk. Such cynicism makes the sustaining of hegemony highly problematic. Indeed, there must now be a real question as to whether or not there has been a shift within the US social order from hegemony, with its emphasis on an 'ethico-political' leadership and 'organised concerns' to what Gramsci called 'supremacy', where a power block rules based upon domination over a fragmented population (Gramsci, 1971: 276).

Further theoretical implications follow from the analysis presented in this chapter. The recent work of Ulrich Beck, as well as that of Rose, Miller et al., while not referred to in any detail in this chapter, provide obvious theoretical backcloths against which some of our arguments might be considered.

A key virtue of Ulrich Beck's work is, as we have indicated, his highlighting of qualitative discontinuities between a modernist industrialisation and a reflexively modernist risk society, and between the ways in which these are regulated. Nonetheless, the preceding analysis also suggests that Beck has over-stated his case. New ways in which risks might be regulated are, certainly, essential; but to some extent, as we have indicated, these have been realised. Moreover, the struggles towards their realisation have made use of, and been furthered by, the existence of some traditional forms of regulation. In other words, although we have not explicitly argued here, but as has been detailed at length elsewhere (Pearce and Tombs, 1990, 1991), regulatory techniques based upon deterrence and punitive sanctioning strategies are crucial, so that these are necessary, though not sufficient, conditions of more effective forms of regulation of hazardous production. While we have considered shifts in regulatory strategy here, these need to be considered alongside these more traditional techniques, which in our view remain no more or less suitable than was previously the case.

Further, while particular techniques of governance - such as tort law, risk assessment, biological monitoring - may not be neutral, nor are their possibilities exhausted by their historical uses. Thus, simply because certain techniques have historically been used to distribute certain costs and benefits in particular ways does not mean that their possibilities are thereby exhausted, and that they cannot be redefined in order that such risks and benefits can be redistributed. Risk assessment or tort law, for example, are open to deployment for different ends. These ends are not, of course, 'open' in any infinite nor uncontested fashion. Again, as we have indicated, the historical uses of such techniques is only to be understood within the context of a more general hegemonic mode of governance, which itself is a constant site of lesser or greater struggle, resistance and counter- hegemonic pressures. In this we are perhaps closer to O'Malley (1992) than to Rose and Miller (Miller and Rose 1990) (Rose and Miller 1992) (Rose 1993; Rose 1994).

We follow Beck's argument concerning the challenges that science and scientific institutions face in an age that he characterises as 'reflexive modernity'. Nonetheless, this chapter also indicates that the institutional power of science-technology grants it an enormous adaptability and durability, to the extent that it has historically reconfigured and reconstituted its internal discourses, as well as its relationships with the state and corporate capital, so that even in what Beck calls a 'Risk Society' there is no reason to under-estimate the power of science-technology. Where we find ourselves in opposition to Beck's thesis, is in his idealist understanding of scientific rationality (which, therefore, leads him into idealist prescriptions for effective opposition). As Rustin has observed recently, absent from Beck's analysis 'is the concept of capital itself' (Rustin, 1994: 11), so that his critique becomes

> "directed towards "techno-scientific rationality", not to the institutional power of capital, as if he thinks that it is the mode of scientific thinking itself rather than its sponsoring corporate agencies which is the decisive agent of change" (ibid.: 9).

Relatedly, we see that common to the work of Beck on the one hand, and theorists such as Rose and Miller on the other, is a virtual absence of any recognition of class relations in a Marxist (or, indeed, any other) sense. Even if men and women who, in the context of this chapter, are engaged in struggles against toxic capital, do not define this as a class struggle, this does not detract from the fact that such struggles are precisley that. That is, they are struggles to a large extent framed by the ownership and non-ownership of the means of (toxic) production. Further, they are struggles in which the state and law are deeply imbricated and implicated, are certainly no class-neutral arbiters, but nonetheless are also sites of struggle within which counter-hegemonic groups can win gains.

Finally, it remains for us to take issue with one popular 'truism', one which Beck himself represents, but which happens, in fact, to be only partially true. It seems to us clearly to be the case that the unifying power of issues around toxic capital, especially where these are posed in terms of environmental degradation or protection, is partly explained by the fact that the risks associated with the chemical industry constitute a threat to all, and ultimately to the eco-system upon which we all depend. Yet to accept this is not to accept that all are affected equally. The distribution of hazards and risks is not only a differential one, but it is one which is organised around traditional cleavages of class, ethnicity and gender. And that fact makes some very 'modernist' forms of analysis necessary in understanding the nature of contemporary modes of chemicals production, their regulation, how this might be improved, and how we might identify the key agents in forcing any such improvements.

NOTES

[1] A similar structural marxist approach has been used by McNamee (1987) to discuss the relation between the American state and the chemical industry and by Horwitz (1986; 1989) to make sense of the general deregulation movement in the U.S.

[2] These discourses are quite similar to, but were constructed quite independently from, what Dake and Wildavsky identify as the three dominant 'cultural biases' in the U.S. - 'hierarchy, individualism, and egalitarianism' (Dake and Wildavsky 1993: 47). Although these authors write as if cultural and political factors are in a competitive relationship, there own data suggests that they are rather in a complementary relationship with each other.

[3] With some reason. In Superfund settlements, very large industrial corporations are estimated to face transaction costs, equal to about 21 per cent of their total settlement costs and insurance companies 88 percent (of $410 million of $470 million in 1989) (Acton and Dixon 1992).

REFERENCES

Acton, J.P and Dixon, L.S. (1992) *Superfund and Transaction Costs: The Experience of Insurers and Very Large Industrial Firms*, Santa Monica: Rand Corporation.

Adams, W. and Brock, J. (1986) *The Bigness Complex: Industry, Labor and Government in the American Economy*, New York: Pantheon: 76-78.

Adams, W. and Brock, J. (1991) *Antitrust Economics on Trial: A Dialogue on the New Laissez-Faire*, Princeton: Princeton University Press.

Aftalion, F. (1991) *A History of the International Chemical Industry*, Philadelphia: Pennsylvania University Press.

Aharoni, Y. (1991) *The No-Risk Society*, Chatham, NJ: Chatham House.

Aglietta, M. (1979) *A Theory of Capitalist Regulation*, London: New Left Books.

Baram, M. (1987) 'Chemical Industry Hazards: liability, insurance and the role of risk analysis', in Kleindorfer, P.R. and Kunreuther, H.C., eds., *Insuring and Managing Hazardous Risks: From Seveso to Bhopal and Beyond*, Springer Verlag, 415-442.

Bardach, P. and Kagan, R. (1982) *Going by the Book: The Problem of Regulatory Unreasonableness*, Philadelphia: Temple University Press.

Barnett, H. (1982) *'The Production of Corporate Crime in Corporate Capitalism'* in P. Wickham & T. Dailey (eds) *White Collar and Economic Crime*, Lexington: Lexington Books: 157-170.

Barnett, H. (1992) 'Hazardous Waste, Distributional Conflict and a Trilogy of Failures', *Journal of Human Justice*, 3, (2), 93-110.

Barnett, H. (1994) *Toxic Debts and the Superfund Dilemma*, Chapel Hill: The University of North Carolina Press.

Beck, U. (1992a) 'From Industrial Society to the Risk Society: Questions of Survival,Social Structural and Ecological Enlightenment', *Theory, Culture and Society*, Volume 9.

Beck, U. (1992b) *Risk Society: Towards a New Modernity*, London: Sage.

Becker, G. C. (1975) *Human Capital*, Chicago: Chicago University Press.

Calavita, K. (1983) 'The Demise of the Occupational Safety and Health Administration: a case study in symbolic action', *Social Problems*, 30, 4, pp. 437-448.

Carson, R. (1962) *Silent Spring*, New York: Fawcett Crest.

Chandler, A.D. (1990) Scale and Scope: The Dynamics of Industrial Capitalism, Cambridge: The Belknap Press of Harvard University Press.

Chelius, J.R. (1977) *Workplace Safety and Health*, Washington DC: American Enterprise Institute.

Cox, R. (1993) 'Structural Issues of Global Governance: implications for Europe', in Gill, S., ed., *Gramsci, Historical Materialism and International Relations*, Cambridge: Cambridge University Press.

Crooks, H. (1993) Giants of Garbage, Toronto: Lorrimer.

Cutter, S. (1993) *Living with Risk*, London: Edwin Arnold.

Dake, K. and Wildavsky, A. (1993) 'Theories of Risk Perception: Who Fears What and Why?', in Burger, E.J., ed., *Risk*, Ann Arbor: University of Michigan Press.

Donnelly, P. (1982) 'The Origins of the Occupational Safety and Health Act of 1970', *Social Problems*, 30, (1), 13-25.

Douglas, M. (1993) 'Risk as a Forensic Resource', in Burger, E.J., ed, *Risk*, Ann Arbor: University of Michigan Press.

Douglas, M. and Wildavsky, A. (1982) *Risk and Culture*, Berkeley: University of California Press.

Draper, E. (1991) *Risky Business. Genetic Testing and Exclusionary Practices in The Hazardous Workplace*, Cambridge: Cambridge University Press.

Fellmeth, R. C. (1970) *The Interstate Commerce Commission: The Public Interests and the ICC*, New York: Grossman.

Fine, B. (1990) "Scaling the Commanding Heights of Public Enterprise Economics", *Cambridge Journal of Economics*, 14.

Fligstein, N. (1990) *The Transformation of Corporate Control*, Cambridge, MA: Harvard University Press.

Galanter, M. (1994) 'The Transnational Traffic in Legal Remedies', in Jasanoff, S., ed., *Learning From Disaster. Risk Management After Bhopal*, Philadelphia: University of Pennsylvania Press, 133-157.

Gill S. and Law, D. (1993) 'Global Hegemony and the Structural Power of Capital', in Gill, S., ed., Gramsci, *Historical Materialism and International Relations*, Cambridge: Cambridge University Press.

Gramsci, A. (1971) *Selections from the Prison Notebooks of Antonio Gramsci*. Translated by Hoare, Q. and Nowell Smith, G., New York: International Publishers; London: Lawrence and Wisconsin.

Hadden, S. (1994) 'Citizen Participation in Environmental Policy-Making', in Jasanoff, S., ed., *Learning From Disaster. Risk Management After Bhopal*, Philadelphia: University of Pennsylvania Press.

Hayek, F. (1949) *Individualism and Society*, London: Routledge and Kegan Paul.

Hilgartner, S. (1992) 'The Social Construction of Risk Objects', in Short Jr., J. F. and Clarke, L., *Organizations, Uncertainty and Risk*, Boulder: Westview Press.

Hofrichter, R., ed. (1993) Toxic Struggles. The Theory and Practice of Environmental Justice, Philadelphia, PA: New Society Publishers.

Hohfeld, W. (1913) 'Some Fundamental Legal Conceptions as Applied in Legal Reasoning', *Yale Law Journal*, 23.

Holman, O. (1993) 'Internationalisation and Democratisation: Southern Europe, Latin America and the World Economic Order' in Gill, S., ed., Gramsci, *Historical Materialism and International Relations*, Cambridge: Cambridge University Press.

Horwitz, R.B. (1986) 'Understanding Deregulation', *Theory and Society*, Vol. 15.

Horwitz, R. B. (1989) *The Irony of Regulatory Reform*, New York: Oxford University Press.

Ives, J., ed. (1985) *The Export of Hazard*, London: Routledge & Kegan Paul.

Jessop, B. (1990) *State Theory: Putting Capitalist States in Their Place*, Cambridge: Polity Press.

Jones, T. (1988) *Corporate Killing: Bhopals Will Happen*, London: Free Association Books.

Kagan, R., (1984) 'On Regulatory Inspectorates and Police', in Hawkins, K., and Thomas, J., eds. (1984) *Enforcing Regulation*.

Katzman, M. T. (1989) 'Pollution Liability Insurance as a Mechanism for Managing Chemical Risks', in Schnare, D. W. and Killingsworth, M.J and Palmer, J.S. (1992) *Ecospeak: Rhetoric and Environmental Politics in America*, Carbondale and Edwardsville: Southern Illinois University Press.

King, R. (1990) *Safety in the Process Industries*, London: Butterworth-Heinemann.

Kletz, T. (1990) *Critical Aspects of Safety and Loss Prevention*, London: Butterworths.

Lepkowski, W. (1994) 'The Restructuring of Union Carbide', in Jasanoff, S., ed., *Learning From Disaster. Risk Management After Bhopal*, Philadelphia: University of Pennsylvania Press, 22-43.

Lustig, R.J. (1982) *Corporate Liberalism: The Origins of Modern American Political Theory, 1890-1920*, Berkeley: University of California Press.

Lyotard, J-F. (1984) *The Postmodern Condition: A Report on Knowledge*, Minneapolis: University of Minnesota Press.

Mahon, R. (1977) 'Canadian Public Policy: the unequal structure of representation', in Panitch, L., ed., *The Canadian State: Political Economy and Political Power*, Toronto: University of Toronto Press

Mahon, R. (1979) 'Regulatory Agencies: Captive Agents or Hegemonic Apparatuses', *Studies in Political Economy*, 1, (1).

McNamee, S.J. (1987) "Du Pont - State Relations", *Social Problems*, 34, (1), February.

Miller, P. and Rose, N. (1990) *'Governing Economic Life'*, *Economy and Society*, 19, (1).

National Research Council (1983) *Risk Assessment in the Federal Government: managing the process*, Washington DC: National Academy Press.

Navarro, V. (1983) 'The Determinants of Social Policy, A Case Study: Regulating Health and Safety at the Workplace in Sweden', *International Journal of Health Services*, 13.

Nichols, T. (1986) 'Industrial Injuries in British Manufacturing in the 1980s: a commentary on Wright's article', *Sociological Review*, (2).

Noble, C. (1986) *Liberalism at Work: The Rise and Fall of OSHA*, Philadelphia: Temple University Press.

O'Malley, P. (1992) 'Risk, Power and Crime Prevention', *Economy and Society*, 21, (3).

Pearce, F. (1995) 'Controlling, Reforming or Reconstructing the Corporation and its Economy: Remedies for Corporate Antisocial Conduct within US Capitalism", in P. Stenning (ed) *Essays on Accountability*, Oxford: Oxford University Press.

Pearce, F. and Snider, L. (1995) 'Regulating Capitalism', in Pearce, F. and Snider, L., eds., *Corporate Crime: Ethics, Law and the State*, Toronto: University of Toronto Press.

Pearce, F. and Tombs, S. (1989) 'Bhopal: Union Carbide and the Hubris of the Capitalist Technocracy' *Social Justice*, 16, (2).

Pearce, F. and Tombs, S. (1990) 'Ideology, Hegemony and Empiricism: Compliance Theories of regulation', *British Journal Of Criminology* (Autumn).

Pearce, F. and Tombs, S. (1993) 'US Capital versus the Third World: Union Carbide and Bhopal', in Pearce, F. and Woodiwiss, M., eds., *Global Crime Connections*, London: Macmillan, pp. 187-211.

Peltzman, Sam (1976) 'Toward a More General Theory of Regulation', *The Journal of Law and Economics*, 19.

Priest, G.L. (1993) 'The New Legal Structure of Risk Control', in Burger, E.J., ed., *Risk*, Ann Arbor: University of Michigan Press.

Raphael, A. (1994) *Ultimate Risk: The Inside Story of the Lloyd's* Catastrophe, London: Bantam Press.

Robinson, J.C. (1991) *Toil and Toxics. Workplace Struggles and Political Strategies for* Occupational Health, Berkeley: University of California Press.

Rose, N. (1993) 'Government, Authority and Expertise in Advanced Liberalism', *Economy and Society*, 22, (3).

Rose, N. and Miller. P. (1992) 'Political Power beyond the State: Problematics of Government', *British Journal of Sociology*, 43, (2).

Rosner, D. and Markowitz, G. (1989) Dying for Work: Safety and Health in *Twentieth America*, Bloomington: Indiana University Press.

Rustin, M. (1994) 'Incomplete Modernity: Ulrich Beck's Risk Society', *Radical Philosophy*, 67, Summer, 3-12.

Sanders, J. (1992) 'Firm Risk Management in the Face of Product Liability Rules', in Short, Jr. J. F. and Clarke, L., *Organizations, Uncertainty and Risk*, Boulder: Westview Press.

Sassoon, A.S., (1987) *Gramsci's Politics*, Minneapolis: University of Minnesota Press.

Sklar, M. (1988) *The Corporate Reconstruction of American Capitalism 1890-1916*, New York: Cambridge University Press.

Smith, M.A., ed. (1985) *The Chemical Industry after Bhopal*, London: IBC Technical Services Ltd.

Smith, R.S. (1976) *The Occupational Safety and Health Act: Its Goals and Achievements*, Washington: American Enterprise Institute.

Smith, R.S. (1982) 'Protecting Workers' Health and Safety', in Robert W. Poole Jr., ed., *Instead of Regulation*, Lexington, Mass.: Lexington Books.

Stellman, J. and Daum, S. (1973) *Work Is Dangerous to Your Health*, New York: Vintage Books.

Stigler, G. J. (1971) 'The Theory of Economic Regulation', *Bell Journal of Economics and Managerial Science*, no. 2 (Spring).

Szasz, A. (1984) 'Industrial Resistance to Occupational Safety and Health Legislation: 1971-1981', *Social Problems*, 32, (2).

Teuber, A. (1993) 'Justifying Risk', in Burger, E.J., ed., *Risk*, Ann Arbor: University of Michigan Press.

Tinker, T., Lehman, C., Neimark, M. (1988) 'Bookkeeping for Capitalism: The Mystery of Accounting for Unequal Exchange', in Mosco, V. and Wasco, J., *The Political Economy of Information*, Madison: The University of Wisconsin Press.

Turner, J. (1970) *The Chemical Feast*, New York: Grossman.

Viscusi, W.K. (1983) *Risk by Choice: Regulating Health and Safety in the Workplace*, Cambridge, Mass.: Harvard University Press.

Weaver, P. (1978) 'Regulation, Social Policy and Class Conflict', in D.P. Jacobs, *Regulating Business: The Search for an Optimum*, San Francisco: Institute for Contemporary Studies.

Weidenbaum, M. and de Fina, R. (1978) *The Costs of Federal Regulation of Economic Activity*. AEI Reprint no. 88, Washington, D.C.: American Enterprise Institute.

Weinstein, J. (1968) *The Corporate Ideal in the Liberal State*, Boston: Beacon Press.

Wilson, G.K. (1985) *The Politics of Safety and Health*, Oxford: Clarendon Press.

Wilson, H.T. (1976) The American Ideology: Science, Technology and Organization as Modes of Rationality in *Advanced Industrial Societies*, London: Routledge and Kegan Paul.

Woodiwiss, A. (1993) *Postmodernity USA*, London: Sage.

Wynne, B. (1989) 'Frameworks of Rationality in Risk Management: towards the testing of naive sociology', in Brown, J. ed., *Environmental Threats: perception, analysis and management*, London: Belhaven/ESRC, 33-47.

INJURY, DEATH AND THE DEREGULATION FETISH:

The politics of occupational safety regulation in UK manufacturing industries

STEVE TOMBS
Liverpool John Moores University
School of Law, Social Work & Social Policy
Josephine Butler House
1, Myrtle Street
Liverpool L7 4DN, UK

Introduction

In May 1994 the Health and Safety Commission (HSC) publicly reported upon its year long review into deregulation and UK health and safety legislation. The response of TUC secretary John Monks to the report was clearly one of relief. It should, he said, "mark the death of the deregulation fetish in health and safety" (Guardian, 25/5/94).

This reaction was understandable but, as this chapter will demonstrate, it is both premature and somewhat superficial. The deregulation 'fetish', as Monks has called it, has not gone, and will not go, away. This fetish - and that is precisely what it is, as my dictionary defines this term as "any object, activity etc. to which one is excessively or irrationally devoted" (Collins English Dictionary) - provides a clear link between the Conservative administrations of Thatcher and Major; it sits at the heart of a changed political terrain, and the 'new' hegemonic political discourse; and it is linked with injury and death on an horrendous scale in British industry.

The aim of this chapter is to examine some of the more recent developments in the social and political environments within which the deregulation fetish is crucial, but of which it remains only one element. It demonstrates that the deregulation fetish, as part of a broader assault on the legitimacy of the external regulation of business activity, will not go away since its effects are already being felt in the context of the regulation of occupational safety in the UK.

The body of the chapter is divided into four main sections. The first outlines recent trends in recorded injuries in UK workplaces, with particular reference to manufacturing industries. [1] A second section charts the nature and effects of the social and political contexts of the work of the HSC, its Executive (HSE) and the factory inspectorates (HMFI) in the 1980s, arguing that while Thatcher Governments withdrew from any direct deregulatory assault on occupational safety, what transpired was a gradual but continual undermining of the ability of these agencies to fulfil their mandated functions. Thirdly, the chapter charts, and assesses, the nature and effects of a new politics of deregulation in the

189

E. Coles, D. Smith and S. Tombs (eds.), Risk Management and Society, 189–206.
© 2000 *Kluwer Academic Publishers. Printed in the Netherlands.*

UK. This new politics is then related, in a fourth section, to UK Governmental opposition to EU influence in domestic social policy, which it is argued stood in a symbiotic relationship with the re-emergence of a more sustained deregulatory discourse in the UK.

It must be emphasised that while the focus of this chapter is the external regulation of occupational safety in the UK, this is not to imply that 'effective' standards of occupational safety can be guaranteed by the state. There are a whole series of influences upon injury and death rates which extend far beyond regulatory agencies and the law. Moreover, these need to be increasingly understood not just at workplace, sectoral, local and national levels, but also in both international and global terms. Thus an understanding of changing levels of occupational safety requires attention to factors outside the scope of this chapter. Not least amongst these factors are the following: the relative powers of capital and labour at various levels; the nature and extent of pro-regulatory or oppositional forces; levels of, and trends in, unemployment; trends in labour markets and employment patterns (notably attempts to imitate models of flexibility claimed to exist in the US and Japan); the nature of, and changes in, contractual arrangements and methods of payment; and the introduction of new technologies and new forms of work organisation, particluarly in terms of the impact of these upon workers' skills and functions. There have been some recent attempts to conceptualise the range and relative importance of such factors in terms of a overarching framework (Carson and Henenberg, 1989, Dawson et al., 1988, Moore, 1991, Nichols, 1990, Pearce and Tombs, 1994), although what is gained in terms of general analytical clarity tends inevitably to be at the expense of detail regarding particular sets of factors.

In this chapter, therefore, a more focused approach is adopted. Notwithstanding the range of factors identified, it remains the case that the law and its enforcement does have real effects on levels of deaths and injuries at work, so that external regulation is a necessary though not sufficient condition of effective standards of occupational safety (Pearce and Tombs, 1990).

Recent Trends in Injuries in Manufacturing Industries

Previous work has reviewed data on injury rates in UK manufacturing industries in order to isolate any trends from the early 1970s, that is, from the time of the Robens Report (Robens, 1972) and its consolidation in the subsequent Health and Safety at Work Act (HSW Act) 1974 (Tombs, 1990). Two trends became apparent. First, that from 1974 up to 1980, there occurred a clear, if far from uniform or simple, downward trend in the incidence rates for 'total reported accidents' in the manufacturing sector. Second, that in the years between 1981-1985, there occurred an even clearer upward trend in the incidence rates of fatal and major injuries in manufacturing industries. Although the bases of the data within the two periods differ, the change in trends is clear: between 1974 and 1980 there occurred a 20% reduction in total reported accidents, while between 1981 and 1985 there was a 30% increase in fatal and major injuries. The first half of the 1980s undoubtedly witnessed a deterioration in safety protection for employees in UK manufacturing industries (see also Barrett and James, 1988, Dawson et al., 1988, James, 1993).

Such a dramatic change in trends is perhaps not surprising given that the two periods which surround the watershed year of 1980 - the second half of the 1970s and the first half of the 1980s - were characterised by very different political economies in the UK.

Following what might be called the high-point of corporatism in the UK, which, for all its limitations, was an era when workers gained rights and exercised some voice, not least of all in the area of occupational safety, the demise of an increasingly discredited Labour government marked a key turning point in the recent political economy of the UK. This was significant with respect to occupational safety as it was in so many other areas of British life. Thus with the election of the first Thatcher government in 1979, committed to removing the 'burdens' of social regulation from business, the HSE was subject to funding cuts during the first two-thirds of the 1980s, accompanied by enormous increases in its workload. The result was that there emerged a state of de facto deregulation in the UK. It is instructive that in this period of regulatory weakness, coupled with the backcloth of a trade union movement weakened by recession, successive pieces of employment legislation, mass unemployment and governments prioritising managements 'right' to manage, many employers took the opportunity to reduce expenditure on capital investment, remove safety personnel, cut back on basic maintenance, and used the economic and new legislative climate to bypass trades unions, where these existed. These factors clearly had deleterious effects upon safety performance (Freeman, 1988, Grunberg, 1986, Stow, 1988).

Now, since the criteria by which data is gathered were altered again in 1986/87, it is not possible simply to trace trends from 1981 to the present; however, it is possible to examine the period from 1986 onwards, and draw comparisons with the trends highlighted above. In looking at 1986-1993, what emerges is actually a rather imprecise picture.

Across **all** industries during this period, fatality rates - while dropping (HSC, 1993a, 1993b) - merely reflected changes in the employment structure, so that such rates are said to have reached a "plateau" (HSC, 1992: 78). Further, this pattern of stasis is repeated in the data for non-fatal injury rates, which "are virtually the same as we might have expected given the employment changes in the last six years", so that during this period there had, within these rates, "been no significant changes" (HSC, 1992: 80). At the same time, areas of economic activity previously thought of as relatively 'safe' appear to have become more dangerous as employment within them has grown (while that in manufacturing has, of course, declined). For example, the 1991-92 Annual Report of the Local Authorities to the HSC details how, since 1986-87, the over three-day injury rate had increased consistently and substantially in the service sector as a whole, to the point where a whole range of work in this sector was thought to be as risky as manufacturing, even on the basis of recorded, but substantially under-reported, data (Health and Safety Information Bulletin, henceforth HSIB, 212: 6).

More specifically within manufacturing industries, the picture is a less even one (HSC, 1993a, 1993b). Fatality rates did drop over this period, though there seems to be some official prevarication over whether or not this decline can be **wholly** explained in terms of the shift within manufacturing from employment in high towards less hazardous occupations (HSC, 1992: 83, HSIB, 223: 2). The same period was also one of unevenness in the direction of non-fatal injury rates, so that an examination of the data on major injuries and over three-day injuries shows no clear trends. Certainly there were particularly high rates recorded for the year 1989-1990. More generally, the HSC concluded in 1992 that

> "there is little evidence of a general decline in the non-fatal injury rates in this sector. For particular manufacturing classes, the injury rate trends move in different directions..." (HSC, 1992: 83).

In other words, while the increase in injuries of the early eighties may have been reversed, the longer term trend which this increase had interrupted - namely a decline in the rates of injuries of various types of severity - has not been re-established.

To these observations might be added clearer, and more worrying, conclusions regarding trends in injury rates for non-employees during this period. Between 1986-1993 there occurred a clear increase in injuries for non-employees in the manufacturing sector (HSC, 1993b) - a group which is becoming more significant as many corporations undergo reorganisation along various lines (often encapsulated, somewhat misleadingly, under the umbrella term of flexibility). Such workers face greater levels of risk since they are more likely to be non-unionised, on short-term or fixed contracts, and/or working on the basis of extremely tight tenders. Official comment on such data provides a disturbing comparison; for example, by 1990-91, "the fatal injury rate for self-employed people in this sector ... stood at double the employee rate" (Employment Gazette, 1992: 14-15).

Broadly, then, the period since the latter half of the eighties has at best seen the state of occupational safety, as measured in injury rates, remain static. Moreover, this picture of stasis can obscure the worrying, and clearer, trends in rates of injuries for the self-employed, a group becoming more numerically significant (Hughes, 1992). What is clear is that such data offer no room for complacency. Indeed, even the HSC has noted recently that it has become increasingly clear that injury rates would not have improved since the mid-1980s without structural changes in employment, adding that any real progress requires improvements in safety management (HSIB, 212). Finally, echoing some of the points highlighted above, the HSC has warned that the "positive effects" of structural changes could be offset by a series of factors including the "fragmentation" of firms through the growth of self-employment and small employers, and privatisation (ibid.; HSC, 1994).

Finally, let us be clear about the scale of death and injury from which such trends are extrapolated. In 1992/93, provisional (and thus under-estimated) figures record over 420 work-related fatalities, almost 30,000 major injuries, and in excess of 140,000 'over three-day injuries' (HSC, 1993a, 1993b). Moreover, as the supplement to the 1990 Labour Force Survey indicated, less than a third of non-fatal injuries at work are actually reported (Stevens, 1992). Figures on work-related illnesses are far more horrific - for example, about 700,000 people were absent from work due to work-related illness in 1989/90 (HSC, 1993b: 23-4) - though less well recorded. If trends in injury data do not offer room for complacency, nor does the scale of the carnage.

Deregulation or a Slow Strangling ?

From 1979 onwards, with the election of the first Thatcher government, there unfolded a programme of economic and institutional reform, a programme which at times was sufficiently radical, and seemingly coherent, to herald the emergence of the (albeit contested) label 'Thatcherism'. A key element of Thatcherism was a fundamental belief in the virtues of the 'free' market and private economic institutions; the "commercial enterprise" became a model for reform of a whole range of institutions (Keat, 1991; Savage and Robins, 1990). Many areas of economic activity passed from the public to the private sector, while many were deregulated. Accompanying these material changes were a series of ideological assaults upon 'big' government, bureaucracy, state interference, regulation, and so on. The themes are familiar to most. Nichols and O'Connell-Davidson have commented upon this period thus:

> "During the 1980s, political economy in Britain was shaped by the new right, a trend which was facilitated by the weakening of labour in relation to capital. Privatisation was seen as playing a key role in this weakening of labour, ideologically and organisationally, which accounts for the almost feverish rhetoric which has sometimes accompanied it ... The political rhetoric of privatisation not only celebrates the virtues of the free market, it also denigrates the public sector in general .. " (Nichols and O'Connell-Davidson, 1993: 707-708; and see Samson, 1994: 91 and passim).

During this period there did not occur, in the UK any sustained or direct governmental assault on occupational safety legislation as occurred, for example, in the USA (Claybrook, 1984, Pearce and Tombs, 1994). While plans for changes to existing regulation were mooted, even detailed in Government White Papers in 1985 and 1986, these were in fact abandoned in the face of opposition, not least from some employers themselves, the latter indicating that they did not perceive existing regulations as excessively burdensome (Cmnd. 9571, Cmnd. 9794, Health and Safety at Work, April, 1986). However, of real significance during the 1980s were clear shifts in governmental policy towards the HSE, and a creation by governments of a particular climate in which regulators were to operate.

In the first half of the 1980s there was effected a virtual emasculation of the HSE and in particular the factory inspectorate in terms of the roles prescribed for these agencies by the Robens' system of self-regulation and the HSW Act. It has been argued elsewhere that this was one key contributory factor to the increase in injury rates in that period. Indeed, what could be discerned from the fragments of data available on the work of the inspectorate was the increasing significance of something akin to a 'vicious cycle of regulatory non-enforcement' (Tombs, 1990: 333-7): the factory inspectorate became increasingly unable to secure and oversee compliance with safety obligations, and to enforce the law, where necessary, in a "quick", "rigorous" and "effective" fashion Robens (Robens, 1972: 80).   In the second half of the 1980s, the political-economic context within which occupational safety is managed and regulated was altered, largely through a process of what Snider has termed 'public crises' and 'agitation' (Snider, 1991). Thus, notwithstanding significant deregulatory pressures at this time, a series of events and struggles combined to produce some pressures for stricter regulation, and indeed criminalisation, of occupational safety (Pearce and Tombs, 1994).

Probably more important as a catalyst here than the half decade of rising injury rates were a series of highly publicised disasters and subsequent inquiries and reports that occurred so frequently in UK in the latter half of the decade (see Harrison, 1992, Slapper, 1993, Wells, 1993). In most of these, *"the relevant companies"* were *"inculpated by the evidence (and with some official enquiry report) in contributing in some significant way to the cause of death"* (Slapper, 1993).

Surrounding these events were a series of political struggles by a range of groups including trades unions, 'Hazards' groups, and victims groups such as the Herald Families Association, the Piper Alpha Families Support Group, and Disaster Action. The disasters acted as a catalyst for new groupings and new demands, and provided a wider and receptive audience for more longstanding activists and activities. Many of those involved in struggles following these disasters were already active in relation to occupational safety, organising around other 'mundane' sites of employee death and injury, perhaps most notably in the construction industry. These disasters, as is frequently the case, provided opportunities to organise around "resonances already present in social discourse" (Hadden, 1994: 108). **[2]**

Other factors had combined to make this period one in which demands regarding occupational safety could be more effectively articulated. The most strident anti-trades unionism of the Thatcher governments seemed to have dimmed, while a changed economic situation also saw workers' movements increase in confidence somewhat. This period also witnessed: the protracted spectacle of legal hearings over the Zeebrugge disaster; renewed concern over environmental degradation; a recognition of the **public** harm that industrial activities caused; and vulnerability on these issues of Conservative governments, given their implication in many of the disasters of this period (notably in transport and in oil production). This was a time in which issues of worker and public safety became highly politicised. It is unsurprising that the HSC/E received increased funding from 1988/89, even if this may have been more symbolic than real (Dalton, 1991: 15).

While it was written of this period that "the notion of corporate manslaughter .. [entered] .. popular vocabulary" (Guardian, 26 August, 1988), other legal and popular developments contributed to a tentative criminalisation of offences under social regulation. Amongst these were: the first ever custodial sentence under the HSW Act in 1987; the first ever disqualification of a company director under s37 of the HSW Act in June 1992; attempts to bring cases of manslaughter against individual company directors (Bergman, 1991: 23); attempts (albeit unsuccessful) to introduce a Corporate Accountability Bill into parliament; an agreement between the Metropolitan Police and the Union of Constructional and Allied Trades and Technicians (UCATT), ensuring that deaths on construction sites in London will be investigated by the police; and calls by senior regulators themselves for deterrent levels of monetary sanctions against convicted corporate safety offenders, coupled with some relatively large, though still exceptional, fines actually imposed (although these remained miniscule in the context of the turnover and profits of the companies involved).

This changed political-economic conjuncture, and the activity and gains within it, provoked a response on the part of the Major Government; since 1992 a rather different scenario has emerged. For more recent years in the UK have witnessed a resurgence of deregulatory rhetoric and practices of the kind that appeared, and then seemed to disappear, in the middle of the previous decade. This resurgence of deregulation as an explicit political theme needs further to be understood in the context of a more consistent theme of 'privatisation'.

Many of the economic and institutional reforms set in train by Thatcher came to be understood under the umbrella term of 'privatisation'. This one term provided a shorthand for a whole series of quite distinct phenomena, including deregulation, denationalisation, contracting-out, the 'Next Steps' programme of civil service reform, and, more latterly, Compulsory Competitive Tendering (CCT) and market testing. While these recent aspects derive from the Major administrations, it is important to recognise that they are consistent with initiatives of previous Thatcher governments.

Civil service reform in general, and market testing of civil service work in particular, now constitute key elements of contemporary political discourse (Atkinson, 1990); and the Civil Service, of course, includes the HSE and the factory inspectorate. The term 'market testing' refers specifically to the restructuring of central government services, and is based upon the same principles as CCT, namely that "competition in the open market, relying upon a process of tendering, provides the best guarantee of value for money" (Napier, 1993: 2). To gain some idea of the extent of such testing, it has been reported that "most" of Whitehall's work could be put out to tender; the Government has announced a ten-year programme of civil service reform; and it has initially targeted 'testing' 1.5 billion pounds sterling of civil service work in the 'free' market (The Times, November 1992). While the precise nature and extent of market testing remain subject to political and legal controversy, there is no doubt that the principle will lead to the privatisation of some of the aspects of the work of regulatory agencies. The HSE and factory inspectorates will not be immune from these developments. Some HSE functions have been subject to market-testing, and this practice is to be expanded and developed (HSC, 1993c, HSIB, 212). The privatisation of some HSE functions is certain to follow, though at present it remains to be seen precisely what elements of their work will be first affected (IPMS, 1993).

Within this general context, the work of the HSC/E has been subject to various forms of Governmental scrutiny regarding efficiency in recent years; at the same time, the HSC has made successive "gains" in "efficiency". For example, following a Treasury report published in September 1992, the HSE exceeded Treasury expectations of "efficiency gains" equivalent to 1.5% of running costs in 1992-93 'saving' 1.74%, this following savings of 1.73% and 1.54% in the previous two years (Dawson, 1994: 14); and these years were ones in which the HSC/E continued to assume a vast range of new responsibilities. Notwithstanding these facts, the enforcement work of the HSE was subject to further review, in 1994, by the National Audit Office (National Audit Office, 1994).

This latter report was critical of enforcement inconsistencies across and within regions on the part of HSE inspectorates.

Prior to the publication of the National Audit Report, but following the HSE's announcements of successive 'efficiency gains', the HSC/E learnt that its budget was to be cut by the Government. Thus, at the end of 1993 the Public Expenditure Survey revealed reductions in grant-in-aid to the HSC/E in the order of a 2.6% reduction for 1994/95, and a 5% cut for 1995/96; these figures represent a reduction of £15 million over two years. The immediate reaction of John Rimington, Director-General of the HSE, was to speak of the need to "pare or defer some activities" (HSIB, 218: 2). The need to abandon some work would further undermine the ability of the HSE to perform its inspection functions in a way that would match even its own targets (IPMS, 1993). Writing to HSE staff in December 1993, Rimington outlined the probable need to reduce staffing through recruitment restrictions and early retirement. As he noted, early retirement would inevitably "exacerbate the current lack of experience within inspectorates" (HSIB, 218: 2). This is consistent with the reference in the 1992/93 Annual Report to "imbalances" amongst field inspectors, where it was noted of the factory inspectorate that over 35% of factory inspectors in the field had less than 5 years service, while 27% were undergoing preliminary training (HSC, 1993a). Little wonder that the National Audit Office found enforcement inconsistencies; but its recommendations that enforcement should be conducted by "well-trained, experienced" inspectors is hardly made more realistic by the Government's funding reductions, nor the HSC/E's responses to these.

What this discussion indicates is not any direct form of deregulation, but a (more subtle?) gradual erosion of the ability of the HSE to do that with which it is charged. The Major administrations seem now to be following the course of the first two Thatcher Governments of the 1980s, namely engaging in a series of moves by which large parts of economic activity become, in a de facto sense, deregulated, as inspectorates become increasingly unable to perform the functions with which they are formally charged. This gradual erosion is, ironically, almost furthered by the HSC/E which seem continually keen to meet, even surpass, government expectations. The details of the 'debates' over deregulation per se (see below) provide an excellent example of this supine attitude on the part of regulators. The response of the HSE to the Report of the National Audit Office - to develop plans for yet another re-organisation in its Field Operations Division - is another example. This is not to deny that this defensiveness is partly explicable given the ideological climate that exists around 'regulation'. Of course, what this also demonstrates is the inherent weakness of the current regulatory system in the UK.

The weakness of this tripartite form of self-regulation is further indicated by even a cursory look at the other 'partner' in the formal structure, namely the trades unions. Dalton (1992) has argued that the unions did not relinquish their efforts in respect of occupational safety issues during their most difficult times of the 1980s, and indeed such efforts were developed through new alliances and initiatives. Moreover, it might be argued that trades unions have recovered somewhat in the early 1990s in the UK. However, other government moves have been taken to reduce the influence of these organisations in safety matters. Most notable, here, was the Government's announcement of its intention to phase out its grant to Trades Unions for training by April 1996; one-third of this grant is used by the

TUC to part-fund its health and safety training activities (HSIB, 209). Coming on top of the Government's requirements of trades unions over the "check off" system, the consequences of this phasing out of a grant are likely to be exacerbated. These financial attacks on trades unions are clear indications that the Major governments are no less committed than their immediate predecessors to reducing the power of organised labour (Miller and Steele, 1993: 233-34), even if the adopted tactics are rather different.

As indicated above, to be fully understood these material changes must be viewed in the context of a more general ideological climate; thus it has been noted that 'privatisation', in its various guises, is less an industrial strategy, more a series of "climate-setting measures" (Massey, 1992: 493). One of the key links between Major and Thatcher, the 90s and the 80s, is the common thread of distrust of government activity in the economic sphere - that is, a distrust of regulation, of law, of 'bureaucracy'. Public sector activity remains equated with inefficiency; private economic activity with value for money. Fifteen years of such rhetoric means that it seeps into the national consciousness (Samson, 1994: 91-94).

Indeed, under the Major administrations, curbing the scope and practices of governmental agencies has become closely intertwined with a whole plethora of 'Citizens' Charters'. The aim of civil service reform is apparently to develop a more efficient and flexible service for taxpayers. (Positions maintained, with no apparent irony, while 'Quangos' proliferate). Other groups, notably the civil service unions (who hardly have a tradition of radicalism), see such moves as highly damaging in terms of morale, effectiveness in a sense other than simply commercial efficiency, and a threat to existing standards of impartiality (IPMS, 1993). These fears are not exactly groundless - where its effects have been examined rather than assumed, the shift from public to private ownership has not had revolutionising effects in terms of efficiency, and is perceived by employees and managers in such companies as having had some very detrimental consequences (Nichols and O'Connell-Davidson, 1993). Where public bodies have been subjected to the rigors of the 'enterprise culture', one consequence has been increased risks to worker and public safety (Moore, 1991, Richardson, 1993).

Of course, it is important to be clear that none of these developments mean (as such) the wholesale privatisation of the HSE or its inspectorates. Yet as the record of Right-wing think-tanks in the past twenty years indicates, the ability to think the unthinkable can lay the ideological groundwork for what might once have seemed impossible 'reforms'. Indeed, privatisation discourse in the UK has consistently involved reference not simply to the inefficiency of public agencies, but also to almost moral objections to their existence. Put simply, calls for reform of the HSE and its inspectorates, where these come from the right of the political spectrum, are generally couched in language which rejects the legitimacy of the external regulation of business. The confidence of regulators, and indeed their very existence, has been consistently undermined in the UK, even where they have escaped certain forms of material attack. This undermining can also be furthered in relatively 'minor' ways - for example, the decision of the Government to make the position of Chair a 'part-time' one, and to appoint someone with apparently no safety or health expertise (HSIB, 215), is another small dent in the legitimacy of this regulatory body.

The combination of the material and ideological changes sketched out above further undermines the ability of the HSE and inspectorates to do that with which they are legally charged. The 'vicious cycle' outlined above may have different elements, but it is still of relevance in understanding regulatory ineffectiveness in the UK in the 1990s. The 'slow strangling' continues. To these changes, however, must be added a much more concrete, threatening, though entirely derivative, set of developments, which can be considered under the umbrella term 'deregulation'.

The Rise and Fall of the Deregulation Fetish?

Late in 1993 the HSE produced a Consultative Document which proposed a series of deregulatory changes to - that is, the repeal of - existing health and safety legislation. Although the consensus at the time was that the proposed changes were in themselves minor, it was noted that to achieve them might require an amendment of the HSW Act, more specifically to Sub-section 1(1) of the HSW Act, which prevents stand alone repeals or revocations. It is no surprise, then, to read of this proposed change that it "could represent the thin edge of a deregulatory wedge", which would "facilitate a more concerted attack on safety legislation" (HSIB, 216: 3).

It is important to bear in mind that these proposals, and most importantly the mooted amendment to the HSW Act, emerged from within the HSE itself. Thus the change was, formally, independent of the DTI's general deregulatory review, and the work of its seven task forces.

The latter produced their final report in January 1994. The key recommendation - reform of the HSW Act through amendment to Sub-section 1(1), which stood as an "obstacle to reform of H&S legislation since it prevents 'stand alone' repeals or revocations" (Business Deregulation Task Forces, 1994: 14) - coincided with that of the HSE's earlier Consultative Document. Yet only the Task Forces recommendations on the implementation of the overly prescriptive requirements of EU legislation prompted criticism from the HSE - not due to any substantial disagreement, but simply borne of the fact that the HSE has "no choice but to fully implement" such legislation (HSIB, 216).

Despite this lack of HSC/E opposition, the Task Forces' report was hardly uncontroversial. Among the more worrying sections were insistences that: the HSE "thoroughly reassess the real costs and burdens" imposed by regulation against "expected real benefits" (Business Deregulation Task Forces, 1994: 14); the HSE offer guidance which limits the need and scope of risk assessments (ibid.: 15); the HSE should encourage "practical, sensible implementation of safety procedures rather than bureaucratic form-filling" (ibid.); "rationalisation" of health and safety legislation should remain "high on the continuing deregulation agenda" (ibid.: 19).

Finally it is worth noting that the HSC/E was urged to "assess all regulations against the 'small business litmus test'", while, more generally, the HSC/E should "continue to review all H&S regulations in the spirit of maximum deregulation, clarification, simplification etc." (ibid.: 19). More generally, the need to enforce legislation in a **de minimis** fashion, the prioritisation of cost-benefit analyses, and an assumed distinction between 'useful' regulation and bureaucratic 'red tape' pervade the document. While somewhat reminiscent of the 1985 and 1986 White Papers on deregulation, this report is more prescriptive, rather more bold. This boldness, for want of a better word, is reflected not least in the observation that "Business, especially small business, is concerned about potential criminal prosecution for trivial and technical offences. The HSC and Government should review such legislative requirements, including the defences available to employers under the Act, and the penalties which may be imposed, in order to ensure that there is a proper balance between the severity and persistence of the offence and the penalties" (ibid.: 20). While many health and safety activists would applaud the latter part of that quotation, the references to 'business concern' and to the ideologically significant phrase 'trivial and technical offences' (Pearce and Tombs, 1990) indicates that this is actually a recommendation designed to reverse what had begun to be a slight but nonetheless discernible trend towards the criminalisation of some health and safety offending, even if only the most egregious (Pearce and Tombs, 1994).

The significance of the Task Forces document is less in its concrete proposals, much more in the ideological tones that it reinforced. It provides a crucial - and highly constraining - ideological context in which regulators are to function. This explains the fact the Deregulation and Contracting Out Bill eventually laid before parliament contained just one clause pertaining to health and safety regulation. This appeared to mark the second time in less than a decade that Conservative Governments had backed away from any full-scale, frontal assault on safety legislation and its enforcement. But it is reasonable to assume that these 'retreats' were largely due to the widespread lack of legitimacy which such an assault would command, and a recognition that they were simply unnecessary; the ideological groundwork had been done, the HSC/E had implicitly accepted much of what the task forces recommended even before the recommendations had been made, and a new, dangerous, turn had been taken in the development of health and safety regulation in the UK. The ultimately truncated deregulation initiative had sent a clear signal to the HSC/E, and had inspired a compliance, a sense of defensiveness, a desire to pre-empt through 'autonomous' change that which central government might impose. As a commentary on the HSC's Annual Report 1992/93, which appeared at the very end of 1993, noted, "One recurring theme of the Report - which tends to give the impression of an HSC constantly looking over its shoulder - is the Commission's co-operation with the Government on potential privatisation, deregulation and internal cost-cutting' (Dawson, S., 1994: 14). Further, a clear signal had been sent to business communities. A private HSC survey of businesses found that "There is very widespread support for the framework of health and safety legislation established in 1974" (Guardian, 10/5/94); this extended, apparently, to the self-employed, a group on whose 'behalf' the Government had consistently lobbied (HSIB, 222: 2). Significantly, the HSC added that businesses had also made it clear that "the burdens arising from further large-scale changes would be unwelcome" (Guardian, 10/5/94). The concern of the latter is directed much more at new EU-inspired regulations than with existing UK health and safety law and its enforcement (Fairman, 1994: 16). Indeed, no widespread deregulation is needed by business, given the material and ideological frameworks in which existing law is now enforced. Business representatives seemed to take exactly the same stance in 1994 as they had when surveyed regarding deregulation in 1986 - namely, that it would invoke unnecessary public scrutiny and perhaps criticism and that, moreover, it was unnecessary.

Of course the significance of the work of the Task Forces is not confined solely to the ideological level. For the one clause contained in the Contracting Out and Deregulation Bill was probably the most dramatic of all - proposing that the Secretary of State for Employment be granted "potentially sweeping powers of repeal and revocation". This proposal apparently does not contradict Section 1(1) of the HSW Act. Thus the path to reform of the HSW Act is opened up; in other words, the "thin edge of the deregulation wedge" referred to above was put firmly into place.

'Unnecessary Burdens' and the EU: deregulation comes out into the open

Finally, it should be emphasised that there have been intimate connections between the Government's opposition to social legislation from the EU, its attacks on European 'red tape', and the its more general deregulatory assaults in the UK. Rhetorical interventions around these issues been mutually reinforcing.

Early in March 1993 Heseltine, Secretary of State for Trade and Industry, announced two inquiries into 'red tape'. One was to focus upon the formulation, implementation and enforcement of EU law. The aim was stated as ensuring that national legislation was "necessary and proportionate", minimising "burdens on business", and sustaining international competitiveness (HSIB, 207). The second inquiry involved the establishment of seven DTI task Forces which were to review, simplify and where possible abolish "over 7,000 regulations on business" (ibid.). The HSE immediately announced its intention to "co-operate fully" with the reviews (ibid.), and indeed within two months had announced its own plans to review over 400 health and safety regulations; within this latter review, particular attention was to be paid to the concerns of small businesses, and to the use of cost-benefit analyses to ensure that business in general did not face "unreasonable burdens" from such regulation (HSIB, 210).

By the end of the summer of 1993, the first DTI review - on the implementation of EU law - had reported. The report was mostly uncritical of the HSC/E's implementation of European Directives, though what was to become the vexed question of the extension of coverage to self-employed workers was raised. Rather curiously, given the brief of the review, some recommendations were made that went beyond the scope of the implementation of EU law. Moreover, the same report welcomed the HSE's decision to engage in a major deregulation review (HSIB, 213). Further, despite the nature of existing UK health and safety law, the HSE at this point committed itself to considering the "appropriateness of the coverage of the self-employed" by new EU regulations (ibid.). Here we have prime examples of the very fact of an external review being more important than its substantial findings - an almost symbolic effect which had the HSE rushing to undertake pre-emptive 'enquiries'.

This is not to cast the review of European Directives and HSE implementation of these as unimportant. Indeed, in what preceded, accompanied and followed this review, the tone it developed, and the responses it began to engender on the part of the HSC/E, all made it crucial.

The UK Government's ideological attack on EU Directives, and the subsequent review of HSE implementation of these, is significant not least because the EU has become the" main engine" of health and safety priorities and standards in the UK (HSC, 1991: 19), while the EU in general has proven to be a constraining force upon the UK's attempts at deregulation, particularly in terms of a systematic weakening of employment protection rights (Evans et al., 1992, Miller and Steele, 1993). Moreover, these attacks upon European legislative influence need to be understood alongside the often less public, but ongoing efforts to influence the general European agenda in very particular ways [ref]. The obstacle that the EU represents for the Government needs to be understood more generally in terms of the latter's aim of attracting inward investment as the site of low-wage, poorly protected, disposable and de-skilled labour in Europe (Evans et al., 1992). Both the Thatcher and Major Governments have thus seen the Social Charter / Chapter in general, but specific social dimensions in particular, as detrimental to potential UK 'Competitiveness'.

Of the so-called 'Six Pack' regulations introduced recently (HSIB, 1993) - that is, six pieces of European inspired health and safety legislation - the most important were those that came to be transposed in the UK as the Management of Health and Safety at Work (MHSW) Regulations (1992); and one particularly progressive element of the latter are new requirements for formalised risk assessments. However, the emergence of a new deregulatory climate has blunted this progressive potential, indicated by criticisms of these contained within a series of DTI publications (Business Deregulation Task Forces, 1994, Deregulation Unit of the DTI, 1993, 1994a, 1994b). In the context of demands that the HSE ensures "that UK law does not go further than required in the Directives .. and that regulations are only imposed where they are shown to be necessary and where no other means would achieve the desired results" (Business Deregulation Task Forces, 1994, :20), the 'form-filling' and 'unnecessary paperwork' apparently entailed in risk assessments has been subjected to a sustained critique. The Business Deregulation Task Forces concluded that risk assessment required by the Six-Pack should be limited to "significant activities", which have "clearly identified and potentially significant hazard associated with them", where there is "a real risk of injury", thus recognising that "'prevention' is impracticable and that 'risk minimisation' should be the objective" (Business Deregulation Task Forces, 1994: 14, 15, 19). The HSC/E was asked to provide guidance on the risk assessment requirements, something which it was apparently happy to do - a Risk Assessment Policy Unit had been established in 1992, and in its Annual Report for 1992/93, the HSE noted the deregulation initiative, and the importance of only requiring of employers action that is proportionate to risk (HSIB, 217: 13).

There are other aspects of the MHSW Regulations which appear to extend the protection offered by the HSW Act, even if the former are certainly not as progressive as they might have been (James, 1993, Walters, 1990). Yet some commentators - and most significantly the HSC/E in its own guidance on the Regulations - have claimed that regulations do little more than require that which is indicated by the HSW Act 1974. Indeed, in developing the Regulations, it has been commented that the HSE was "determined to avoid fundamental changes" to that 1974 Act (HSIB, 1993: 9); and this reflects a more general position of the HSC/E in negotiations at EU level, which is to maintain the "flexibility" of the UK approach to regulation (HSC, 1992: 6). Moreover, the HSC/E has made it clear that it is unlikely to enforce the MHSW Regulations in a way that would actually transcend the requirements of the existing HSW Act (Guardian, 12 May, 1993). In criticising what he calls a "narrow", "minimalist", and even "miserly" approach to the implementation of the Directive and subsequent intended enforcement of the MHSW regulations, on the part of the HSC/E (James, 1993: 25-27), Phil James cites a leaked internal document which shows the HSC/E's determination to "take no action" at points where the Directive "appears to go further than existing law" (James, 1993: 26). It is unsurprising that a survey found widespread non-compliance with the requirements of the 'Six-Pack' regulations, while just 1% of companies surveyed "had allocated funds for their implementation" (HSIB, 209: 3)

Notwithstanding all these points, HSE 'enforcement' of the Six Pack has invited consistent criticism from the Government in general, and from the deregulation task forces in particular. While the HSE are legally bound by these regulations, their 'pragmatic' position on non-enforcement of certain aspects is given high-level Government backing in these statements / recommendations. By the summer of 1993, the HSC was announcing that it "believes that it has generally done the minimum necessary to implement the Directives, and that any attempt to water down the requirements of the 'six-pack' may result in the UK being taken to the European Court for failing to adequately implement the Directives" (HSIB, 210).

There does remain one area in which the HSC/E has apparently gone further than required by the EU directives - namely in the extension of coverage of the 'Six-Pack' to the self-employed. This was done in order to keep the European inspired legislation consistent with other safety and health legislation in the UK. Moreover, given the increased numbers of self-employed persons, and the disproportionately high injury rates associated with this category (despite reporting levels estimated to reflect only 5% of the actual number of reportable injuries), this is a significant move by the HSC/E. However, it is one that has drawn constant criticism from the Government, and despite resisting this for some period of time, the HSC/E committed itself to reviewing the appropriateness of this coverage.

To emphasise, this general approach to EU inspired regulation is echoed in, if not encouraged by, the political climate in which UK regulators currently operate. The opposition by many of the Conservative party to closer political ties with the European Union has consistently been couched in terms of the statism, big government, inefficiency, and over-excessive regulation of the EU. While Thatcher is well-known for denouncing the Maastricht treaty as ushering in socialism by the back-door (Atkinson, 1990:20), and the Social Charter as 'Marxist' (Wedderburn, 1990:7), a fundamental suspicion of EU-inspired legislation has characterised all recent Conservative governments. Thus it was no idiosyncrasy to see Major reported as urging UK regulators to be "more relaxed" in enforcing the "excessive detail" of EU regulations. Major argued that the balance had tipped too far against the "small businessman". Echoing precisely the sentiments of the most strident deregulatory rhetoric of the first Thatcher government, and speaking at a conference examining the first year of Citizen's Charters, Major has stated: "It's time we stood back and looked again at the burden of regulation". As an example of this burden, regulations relating to fire safety were cited, as he argued for a "bonfire of controls" (The Times, December 4, 1992). Remarks such as these proved to be the rhetorical fore-runners to the deregulatory 'initiatives' launched by the Government, and the HSC/E itself, in the following year, and which have been assessed above.

Conclusion: whipping the fetishists?

In examining recent trends in injuries in manufacturing industries, it was found that the 'progress' towards lower injury rates which was interrupted in the early 1980s has not been re-established; further, in some areas, notably in respect of non-employees, there are clear signs of deterioration. The brief review of the data offered here leaves no room for complacency.

Moreover, regulation through inspection does seem to be breaking down as an enforcement strategy; indeed, the HSC/E had begun to recognise this fact as long ago as 1985 (HSC, 1985), this following the first deregulatory wave of the early eighties but prior to the latest deregulatory initiatives. Almost ten years after that admission, it is clear that the HSE and HMFI remain incapable of doing that which the UK system of tripartite, self-regulation requires of them. The changed situation is that these agencies are operating in an ideological climate which not only undermines their efforts, but questions their very rationale. This might help to explain why regulators continue to respond in ad hoc, piecemeal and generally supine ways to legislative changes and shifts in social and political priorities; but while they continue to do so, their present inadequacies are sure to be exacerbated.   Quite simply, through ideological and material initiative, certain issues have been organised off the political agenda - as Samson has argued, privatisation discourse operates on different levels, and in a variety of arenas, and involves the exercise of different 'faces' of power (Samson, 1994). Thus the claim of John Monks, cited at the beginning of this chapter, can only be sustained on the basis of an understanding of power which is one-dimensional, that is, focuses solely upon observable events and conflicts.

The fetishists have not been whipped; the deregulatory assault has not 'failed'. What has been documented in this chapter is a re-framing of the limits of acceptable discourse regarding safety regulation, so that the result of this recent set of deregulatory initiatives is a safety agenda which is once again shifted towards less worker protection; thus more progressive developments are less likely to proceed. Of particular interest is the way in which this has been achieved through sustained ideological assault coupled with more subtle, and sometimes even 'failed', material changes.

Finally, it remains to emphasise that while there has emerged a new hegemonic discourse which frames the limits of the 'acceptable', no hegemony can ever be entirely secure. Even during the hostile economic-political conjuncture characterised by Thatcherism in the early eighties pro-regulatory forces in the UK sustained their activity and won some gains. Unfortunately, these were ultimately won when long-term activity was able to capitalise upon a series of disasters; it is, quite literally, a gruesome thought to conclude that a further fightback is likely to require similar forms of public tragedy.

Notes

[1] While some more general data is presented, the main focus of this brief section will be on data pertaining to the manufacturing sector. There are two reasons for this: first, while manufacturing industries are, like all sectors, a site of gross under-reporting, injury data from this sector is estimated to be almost twice as reliable as that from construction, and over twice as reliable as injury data drawn from the service sector (HSC, 1993b: 2); second, manufacturing industries provide the key site of factory inspectorate activity, and it was this agency that provided the focus of an earlier paper (Tombs, 1990) with which some comparisons are made here.

[2] Of course, the picture is more complex if one moves from the level of considering the effects of disasters upon safety 'in general'. Since disasters affect perceptions then they can also serve to divert relatively fixed resources from one context to another. For example, following the Piper Alppha disaster in the UK, where 167 oil workers died, new legal duties were imposed upon offshore operators - yet enforcement of these duties has diverted some HSE staff from enforcinbg major hazards legislation associated with the chemicals industries. More generally, disasters can focus attention and resources upon high consequence but relatively low probability incidents, to the relative detriment of other safety issues, and certainly chronic problems associated with occupational ill-health.

REFERENCES

Atkinson, R. (1990) 'Government during the Thatcher Years', in S. Savage and L. Robins, eds., *Public Policy under Thatcher*, London: Macmillan, 8-22.
Barrett, B. and James, P. (1988) 'Safe Systems: Past, Present - and Future?', *Industrial Law Journal*, 17, (1), 26-40.
Business Deregulation Task Forces (1994) *Deregulation Task Forces Proposals for Reform*, London: Department of Trade and Industry.
Cmnd. 9571, *Lifting the Burden*, London: HMSO.
Cmnd. 9794, *Building Businesses ... not Barriers*, London: HMSO.
Dalton, A. (1991) *'Health and safety. An agenda for change'*, Workers' Educational Association Studies for Trades Unionists, 16, (64), JUne.
Dalton, A. (1992) 'Lessons from the United Kingdom: fightback on workplace hazards, 1979-1992', *International Journal of Health Services*, 22, 489-495.
Dawson, S. (1994) 'A Difficult Year: the HSC Annual Report 1992/93', *Health and Safety Information Bulletin*, 217, January, 11-15.
Dawson, S., Willman, P., Bamford, M. and Clinton, A. (1988) *Safety at Work: the limits of self-regulation*, Cambridge: Cambridge University Press.
Deregulation Unit of the Department of Trade and Industry (1993) *Regulation in the Balance. A Guide to Risk Assessment*, London: Department of Trade and Industry.
Deregulation Unit of the Department of Trade and Industry (1994a) *Deregulation. Cutting Red Tape*, London: Department of Trade and Industry.
Deregulation Unit of the Department of Trade and Industry (1994b) *Thinking About Regulating. A Guide to Good Regulation*, London: Department of Trade and Industry.
Employment Gazette, Occasional Supplement No.3, *Health and Safety Statistics 1990-91*, 100, September 1992.
Evans, S., Ewing, K. and Nolan, P. (1992) 'Industrial Relations and the British Economy in the 1990s: Mrs. Thatcher's Legacy', *Journal of Management Studies*, 29, (5), September, 571-589.
Fairman, R. (1994) 'Robens- 20 Years On', *Health and Safety Information Bulletin*, 221, May, 13-16.
Hadden, S.G. (1994) "Citizen Participation in Environmental Policy-making", in Jasanoff, S., ed., *Learning From Disaster. Risk Management After Bhopal*, Philadelphia: University of Pennsylvania Press, 91-112.
Harrison, K. (1992) "Manslaughter by Breach of Employment Contract", *Industrial Law Journal*, 21, (1), 31-43.
HSC (1985) *Plan of Work for 1985/86 and Beyond*, London: HMSO.
HSC (1991) *Plan of Work for 1991/92 and Beyond*, London: HMSO.
HSC (1992) *Health and Safety Commission Annual Report*, 1991/92, London: HMSO.
HSC (1993a) *Health and Safety Commission Annual Report* 1992/93, London: HSE Books.
HSC (1993b) *Health and Safety Commission Annual Report* 1992/93 Statistical Supplement, London: HSE Books.
HSC (1993c) *Plan of Work for 1993/94 and Beyond*, London: HMSO.
HSIB 205. *A Brief Guide to the MHSW Regulations*, January, 1993.
Hughes, A. (1992) 'Big Business, Small Business and the "Enterprise Culture"', in Michie, J. ed. (1992) *The Economic Legacy 1979-1992*, London: Academic Press, 296-311.
Institution of Professionals, Managers and Specialists (1993) *Health and Safety. Keep it Together. Deregulation, market testing and contracting out - the impact of Government policies on the Health and Safety Executive*, London: IPMS.
James, P. (1993) *The European Community. A Positive Force for UK Health and Safety Law?*, London: Institute of Employment Rights.
Keat, R. (1991) 'Introduction: Starship Britain or Universal Enterprise?', in Keat, R. and Abercrombie, N., eds., *Enterprise Culture*, London: Routledge, 1-17.
Massey, A. (1992) 'Managing Change: politicians and experts in the age of privatisation', *Government and Opposition*, 27, (4), 486-501.
Miller, K. and Steele, M. (1993) 'Employment Legislation: Thatcher and after', *Industrial Relations Journal*, 24, (3), 224-235.
Moore, R. (1991) *The Price of Safety: the market, workers' rights and the law*, London: Institute of Employment Rights.

Napier, B. (1993) *CCT, Market Testing and Employment Rights*, London: Institute of Employment Rights.

Nichols, T. and O'Connell-Davidson, J. (1993) 'Privatisation and Economism: an investigation amongst "producers" in 2 privatised public utilities in Britain', *Sociological Review*, 41, (4), 707-730.

Pearce, F. and Snider, L., eds. (1995) *Corporate Crime: Ethics, Law and the State*, Toronto: University of Toronto Press.

Pearce, F. and Tombs, S. (1990) 'Ideology, Hegemony and Empiricism. Compliance Theories of Regulation', *British Journal of Criminology*, 30, 4, 423-443.

Pearce, F. and Tombs, S. (1994) 'Class, Law and Hazards', submission to the *Permanent Peoples' Tribunal on Industrial and Environmental Hazards and Human Rights*, London, 28 November-2 December.

Richardson, B. (1993) *'Do Free-er Market Economies Create Crisis Ridden Societies?"*, paper presented at New Avenues in Risk and Crisis Management, University of Nevada-Las Vegas, August 12-13.

Samson, C. (1994) 'The Three Faces of Privatisation', *Sociology*, 28, (1), 79-97.

Savage, S. and Robins, L., eds. (1990) *Public Policy under Thatcher*, London: Macmillan.

Slapper, G. (1993) "Corporate Manslaughter: an examination of the determinants of prosecutorial policy", *Social and Legal Studies*, 2, 423-443.

Snider, L. (1991) "The Regulatory Dance: understanding reform processes in corporate crime", *International Journal of the Sociology of Law*, 19, 209-236.

Stevens, G. (1992) 'Workplace Injury: a view from HSE's trailer to the 1990 Labour Force Survey', *Employment Gazette*, December, 621-638.

Tombs, S. (1990) 'Industrial Injuries in British Manufacturing Industry', *Sociological Review*, 324-43.

Walters, D (1990) *Worker Participation in Health and Safety. A European Comparison*, London: Institute of Employment Rights.

Wedderburn, Lord (1990) *The Social Charter, European Company and Employment Rights. An outline agenda*, London: Institute of Employment Rights.

Wells, C. (1993) *Corporations and Criminal Responsibility*, Oxford: Clarendon Press.

DEREGULATION AND BSE

CHRISTOPHER GIFFORD
Consultant Mining Engineer
Hillcrest
Coed-y-paen
Pontypool
Gwent, NP4 0TH, UK

1. INTRODUCTION

The scholarship that has developed in the last 30 years in the analysis of disasters includes conclusions that question commonly held assumptions about the unpredictability and inevitability of disasters. Even natural disasters such as earthquakes have effects which depend on the preparedness or vulnerability of the populations involved.

Patterns have been described in man-made disasters. (Turner: 1978) Usually there are many causes which combine to make the disaster possible. Some of the causes, such as a precipitating human error, are predictable. When such errors are anticipated the effects can be minimised. Some causes lie dormant in an organisation and have no adverse effect until combined with other events which make the hazard unmanageable. Thus conditions in which some disasters can be avoided have been described. Conditions which make hazard management less successful have also been described with the conclusions that some systems are too complex and interactive to be managed successfully. (Perrow: 1984)

Cultural factors in the observed tolerability of risk and injury have also been studied. (Hood et al: 1992) It is acknowledged that the risk takers are seldom the risk bearers. The sharing of power between risk takers and risk bearers is not always in balance nor is information always equally available. Risk takers historically have underestimated or understated risk. There is less published hubris in the claims to be able to foresee causes of failure and thus quantify risk than there was two decades ago.

Statutory regulation is part of the government's role in risk management. It is the area where there has been least rigour in recent discussion. Two strands of development have been observed: the practical and the ideological. One of the first hazardous activities to be regulated by statute was coal mining. This is the industry in which practical procedures for successful risk management evolved out of the experience of many disasters and unprecedented health impairment. The procedures are easily summarised. There must be structured management in which responsibility and authority is assigned. Management competence must be measured. Selection, training,

E. Coles, D. Smith and S. Tombs (eds.), Risk Management and Society, 207–225.
© 2000 *Kluwer Academic Publishers. Printed in the Netherlands.*

supervision and thorough workplace inspection are essential. Risks must be anticipated, assessed and managed. Information must be available to employees. Employees are entitled to be represented and consulted and to have their representatives given access to the workplace and to documents relating to safety and health. Those procedures were remarkably successful in reducing deaths, injury and health damage from the ingestion of rock and coal dusts, falls of roof, failures of ventilation, fires, explosions, inrushes of water and gases from old workings, and contact with machinery and vehicles.

The risk management principles developed in the better managed and regulated coal mining industry have wider application. After the Robens report (Robens: 1972) they became the principles of the Health and Safety at Work, Etc., Act 1974 and they have since become the principles of European law enacted in several health and safety directives.

The ideological strand of development in the former government's role was discordant with much that has been learned in hazardous industries. It held that regulations were a burden on industry which made it uncompetitive and that deregulation was necessary even in the field of heath and safety. Any regulation which imposes a burden can still be revoked by order of a minister. (HMSO: 1994) Laws for the protection of employees and consumers are not exempt.

The first appearance of the new ideology in government came with the Thatcher administration of 1979. It became clear that employee representation was an early target and within two years the entitlement to safety representatives provided by the Health and Safety at Work, Etc., Act 1974 had been turned into a matter for the employer's discretion. In the second administration and after the defeat of the miners the government attempted to remove the provisions for employee representation from the Mines and Quarries Act 1954. Although in that respect it failed (because of the deviousness of the attempt and the protection afforded by European law) it did succeed in undermining the role of the unionised mine deputies and in revoking many other valuable sections of the Mines and Quarries Act. The deregulated free market ideology was supported by the Reagan administration and the free market was held to require the closure of most of the nations coal mines notwithstanding the £1.2 billion annual subsidy to the nuclear industry and the dearth of information on the costs of the gas fired power stations which, operated by regional electricity companies, were free of competition.

The dichotomy between the former government's commitments to the protection of consumers, employees, the public and the environment largely contained in European law and the activities of the Deregulation Unit of the Cabinet Office remained until the Conservative government was removed from office in May 1997. The former government had "mounted a challenge" in the European Court to the notion that the statutory regulation of hours of work was a health and safety matter. Not surprisingly, given the findings of Clapham rail collision inquiry and the fact that the government itself has erected signs on our motorways stating "Tiredness kills: take a break", it failed.

In the following sections of this paper there are examples of the effects in several industries of deregulation, non-regulation, weak enforcement of extant law, hostility to the notion of employee representation, cuts in the budgets of the Health and

Safety Executive and government research establishments, the privatisation of government science, the denigration of public service and the demoralisation of scientists and HMO Inspectors. That was the climate in which the former government, already diverted by ideology unconnected with hazard management from its duty to protect the citizen, responded to the appearance of Bovine Spongiform Encephalopathy.

COAL MINING

Although the statutory regulation of the conditions of employment in factories preceded by a few years that of coal mines it was in coal mining that the more intensive regulatory regime was developed. Early failures led to the pioneering of a unique model of regulation and hazard management in which nationalisation, trade union organisation and owner-management-employee relations were important ingredients.

The Annual Reports of the Chief Inspector of Mines and Quarries show that fatal accidents peaked in 1910 with 1775 deaths. In 1913, the year of the Senghenydd explosion which killed 439 men and boys, there were 1753 deaths. The industry then employed over one million persons. Fatal accidents exceeded 1000 per year from 1853 until the 1930's. In the 50 years 1900-1950 over 51,000 miners were killed.

Figure 1 illustrates the UK Coal Mining fatal accident rate per 100,000 manshifts from 1947 to 1994. In 1947, the year of nationalisation, 618 miners were killed. Output was almost 200 million tons and almost three-quarters of a million persons were employed. In 1987-8 the number killed in British Coal Corporation mines was 9. Output was 82 million tonnes and manpower was down to 38,000. (HSE: 1989) The risk of fatal accident had been reduced by a factor of 10. The improvement in productivity was equally impressive. World records had been established for both safety and productivity for deep coal mines. British mining equipment was exported world wide.

The number of deaths by pneumoconiosis over the period reviewed is probably greater than the number of deaths by traumatic injury.

In coal mining the story is thus one of unparalleled abuse followed by successful correction by good management, research and development, improved technology, good employer-employee relations, statutory regulation and enforcement of the law. The enforcement was by a government inspectorate with powers of direction and prohibition which was professionally qualified and experienced in management, and which was paid higher salaries than management

Figure 1 is indicative not only of a 40 years improving trend in mine safety but also of some reversal of that trend in recent years. They are the years in which changes in management ethos by the British Coal Corporation Chairman Mr Ian McGregor after the 1984-5 coal miners' strike led to the resignation of several senior officials and at least one Board member who recorded that he left "with alacrity".

It was not only in the BCC that changes took place after the miners' strike. The Health and Safety Executive and the Mines Inspectorate were put under pressure by the former government's engagement of American consultants to desist from enforcing the law, to grant unwarranted exemptions from the Mines and Quarries Act and to plan its repeal.

Figure 1

Fatal Accidents / 100,000 Manshifts

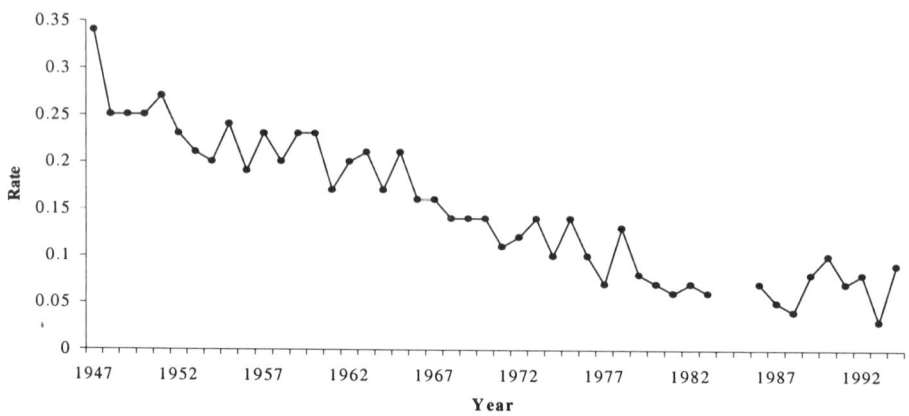

The Boyd Report No 2265.5 (Boyd: 1993) commissioned by the Department of Trade and Industry of the J T Boyd Co of Pittsburg USA and published in January 1993 proposed repeal of parts of the Mines and Quarries Act on support of mine workings, longer hours of work, increased use of non-flameproof equipment below ground, underground use of cutting torches and welding equipment, repair of damaged electric cables below ground, the "Deep Cut" system in which unnecessarily large areas of roof are created with no support and no roof bolt, the "Extended Cut - Big Bite" system in which even larger areas of roof would be without support and in which gas testing at the face would be impracticable. The report dismissed the objection that gas testing was impracticable, denigrated the officials' trade union NACODS, and misrepresented the British Mines Inspectorate as "an obstacle to new technology".

None of the recommendations was supported by any risk assessment as is required by the Management of Health and Safety at Work Regulations 1992. Accident rates in the countries where those methods of work are allowed are significantly higher than those for this country.

Some of the recommendations were adopted and accidents occurred. At Point of Ayr mine with some 40 square metres of unsupported roof two men were injured when the unsupported roof fell. The extensive fall of roof at Bilsthorpe Colliery became the subject of an inquiry by Professor Sir Bernard Crossland FRS. (Crossland, Gifford) Both systems of work were the subject of exemptions from the support regulations granted by a government inspector. Representatives of members of the Chemical Workers Union employed at the Middlebult mine in the Republic of South Africa told me that the "Extended Cut - Big Bite" system was used there when 53 miners were killed last in May 1993 in an explosion of gas and dust. (PC)

The Coal Mines Regulation Act 1908 (updated in 1975) regulated hours of work to 7« hours below ground plus winding time plus bathing and changing time. There were 19 exceptions to that rule for various occupations and contingencies, such as overtime to deal with emergencies. The then Secretary of State for Energy was given the power to repeal the 1908 Act by the Coal Industry Act 1991. The government claimed in Parliament that the Act would be replaced by an equivalent European provision but the European Directive on the organisation of working time is less than equivalent. This and a futile attempt to evade the 'no regression' provisions of the Directive would explain why the repeal was made in November 1993 three days ahead of the date on which the Directive took effect.

The repeal of the Act was in breach of Section 118a of the Treaty of Rome which makes it very clear that European legislation is not to be used as a levelling down process. The European Commission has been informed and its investigation of the alleged breach is expected. The possibility of referral to the European Court remains.

The 1908 Act was also protected as a Relevant Statutory Provision by virtue of Section 187 of the Mines and Quarries Act 1954 and was thus subject to the procedures of Section 1 of the Health and Safety at Work etc. Act 1974. It was for the Health and Safety Commission to propose its repeal when it was satisfied that standards of health and safety would be maintained or improved.

The Health and Safety Executive had for some time been working on proposals to erase some of the most effective parts of the Mines and Quarries Act 1954. Even the total repeal of Section 123 on worker involvement was proposed with no replacement provision in the October 1990 draft of the Management of Safety and Health at Mines Regulations. The quality of drafting was poor. Eleven indefensible drafts were produced and withdrawn before a 12th draft was enacted by the minister's order during the Parliamentary recess in the summer of 1993 resulting in The Management and Administration of Safety and Health at Mines Regulations 1993. The National Union of Mineworkers (NUM) and the National Association of Colliery Overmen, Deputies and Shotfirers (NACODS) invested tens of thousands of pounds in a second High Court Judicial Review. Although leave for review was granted the Court failed to hear the arguments and the minister's action was upheld. The judges could well have been influenced by the fact that the official opposition had attempted to censure HM Government and had failed to carry the motion.

The effects of the Management of Safety and Health at Mines Regulations was to weaken the law in almost every amended particular by lowering the level of duty or by complete erasure of a requirement. Thus, for example, company directors previously strictly liable jointly with others for any proven breach of the Act or regulations were given immunity from that liability. The duty of supervisors to inspect every workplace and to make tests for gas was delegated to others who lack authority to gain access or to instruct remedial action. The qualifications previously required of supervisors were lowered and a prohibition of the appointment of contractors as officials was removed. Even the powers of HM Inspectors were reduced.

The Conservative government's antipathy to the mining trade unions was far from satisfied by the defeat of the NUM in the 1984-5 strike. To dis-empower the trade unions the dismemberment of the industry and even of its legislation was required. By 1990 the closure programme had exceeded the prediction that Arthur Scargill made to justify strike action. In 1992 "middle England" protested so effectively at Michael Heseltine's announcement of another 30 pit closures that a stay of action was promised and a review was made. The justification of there being "no market" was disputed by experts far removed from the NUM who described the vast subsidy for nuclear power and the provisions made for the privatised generators and the Regional Electricity Companies to use more costly gas at the consumers' expense. The 'review' was unproductive. Profitable mines were closed and at the end of 1994 only 16 working mines remained to be privatised with only 7000 miners at work.

Schedule 6 of the Management of Safety and Health at Mines Regulations 1993 lists "Modifications of Regulations Relating to Deputies". Thirty five modifications of eight statutory instruments were made simply to erase the word "Deputy" from the legal vocabulary. It was the deputies' union NACODS which in August 1984 voted overwhelmingly to support the miners' defence of mining communities. That vote caused much apprehension in the Prime Minister's office and led to the offer of a review procedure. NACODS accepted the offer as a possible solution of the dispute and desisted from joining the strike. The review procedure was never properly applied after the strike and it did not save a single pit. It was the Government's disregard of that procedure that led to a High Court ruling that the closures had been "Unlawful and irrational" (Herbert: 1992)

NORTH SEA OIL AND GAS

In the North Sea the regulatory regime that had already been demonstrated as extraordinarily successful in coal mining could also have served as a model. The empowerment of employees is a proper part of a regulatory regime. This is not simply because employees can be whistle blowers. Good employers know the benefits of employee co-operation and involvement. The employees in any enterprise hold the skills and the intellectual property which the enterprise needs to succeed. Denial of that resource is not a characteristic of managers but of a minority of company directors.

For mainland industries in Britain the Safety Representatives and Safety Committees Regulations 1977 offered to employees the facilities for workplace inspection and access to health and safety information that had existed in mining for over 100 years and provided for the establishment of safety committees at the request of safety representatives. Significantly it was the Labour government of 1974-1979 that failed to apply the 1977 Safety Representatives and Safety Committees Regulations to the North Sea. The regulatory body remained the Department of Energy whose inspectors became the agents of the Health and Safety Executive.

Perhaps the Labour government intended to provide for employee representation by other legislation. But the owners of companies operating in the North Sea survived Labour's attempts at public ownership. Ownership and control remained with or reverted to the multi-national corporations and their contractors.

In the early 1980's it was obvious that the Thatcher government was ready to provide the de-regulated, de-unionised regime favoured by some of the oil corporations, not least Occidental Petroleum who were to become the owners of the Piper Alpha platform.

The Piper Alpha explosion and fire on 8 July 1988 caused the deaths of 167 persons and the worst offshore accident on record. Of the 225 persons on board the platform only 61 survived, some of them by jumping from the burning platform into the sea. (Cullen: 1990) The other losses from this disaster have been estimated at billions of pounds and they include not only the loss of the rig but the lost production from its satellites and the loss to the UK economy of tax revenue. The report of the public inquiry conducted by Lord Cullen focused on the regulatory regime that was separate and different from those provided for other industries by the Health and Safety Executive, noted that recent health and safety legislation, including the Control of Industrial Major Hazards (CIMAH) Regulations, had not been applied to the North Sea installations, and criticised the Department of Energy Inspectorate as "inadequately trained, guided and led" whose inspections had been "superficial to the point of being of little use..".

Lord Cullen's 106 recommendations, many of which have since been implemented, included that the Health and Safety Executive should be the single regulator body and that pro-active safety assessment should be a management routine overseen by a reformed inspectorate. Several former HM Inspectors of Mines and Quarries, specialised in mechanical and electrical engineering, were subsequently included in a reformed inspectorate. In 1989 the Offshore Installations (Safety Representatives and Safety Committees) Regulations were enacted. But in these regulations trade unions are not mentioned. The installation manager defines the constituencies for the election of representatives and conducts the ballot. Regulation 17 allows that representatives may seek advice and guidance from persons on the installation or elsewhere. Regulation 23 places a duty on the employer to facilitate the exercise by safety representatives of their functions, including the provision of communication facilities.

In March 1990 and before the publication of Lord Cullen's report the Department of Energy, in response to representations by the Manufacturing Science and Finance Union on the excessively long hours sometimes worked on offshore installations, issued Safety Notice 1/90 limiting non-stop working to 16 hours and recommending a normal limit of 12 hours work in any one day.

TRANSPORT DISASTERS

The investigation of the overturning of the Herald of Free Enterprise left no doubt that marine transport requires better regulation both at the national level and by international conventions. The hubris in the senior management of the owning company was breathtaking not least in the disregard of recommendations made by ferry captains. It is still possible to take photographs of ocean going ferries facing the Atlantic with their bow doors open. Few engineers have difficulty with the idea of providing devices, used on every domestic washing machine, which ensure that if a door is open the motor will not start.

The loss of the Estonia eight years later with 852 lives undermined Mr Brian Mawhinney's assertion made when he was Minister of Transport that there was nothing inherently unsafe in the design of roll-on, roll-off ferries. The failure of the bow visor in storm conditions caused failure of the vehicle deck door which rapidly led to the flooding of the vehicle deck. It is now accepted that the instability caused by the unrestrained movement of water on the vehicle deck is the principal design weakness and that remedies have to be found to maintain stability for at least the time required for evacuation. New standards have now been agreed by north European countries, including the UK, but six years have been allowed for implementation. The former government rejected the recommendation of a House of Commons Select Committee (with a Conservative majority) for a published star rating of ferries based on their stability and evacuation standards.

In the last 15 years over 150 bulk carriers have been lost at sea. Most recently the Braer and the Sea Empress in British waters caused massive pollution. If all of Lord Donaldson's recommendations made after the loss of the Braer had been implemented the Sea Empress could have been saved (Donaldson, 1994). The availability of a larger tug at Milford Haven would have made recovery possible with less damage and spillage of oil. There are other issues, not yet addressed in public debate, such as why, on an ebb tide with onshore winds, is it lawful for a vessel of 150 000 tonnes to enter the Haven with a ground clearance of one metre.

It was the merchant seamen's trade unions in their International Transport Federation that paid £400 000 for the surveys in the South China Sea which located the bow section of the 166 000 tonnes merchant vessel Derbyshire which sank in a storm with the loss of 44 crew. Only then was a further inquiry by government ordered. Lord Donaldson's inquiry did look into the possibility of a fracture at bulkhead 65 and the design of hatch covers that can withstand of head of six feet of water. Lord Donaldson, commenting of the much greater pressures that could be exerted on the hatch covers said "Personally, I find this quite astonishing." (Donaldson, 1995)

No public inquiry was held but five and a half years after the pleasure cruiser Marchioness sank in the river Thames with the loss of 51 lives an inquest jury returned a verdict of "Unlawful killing". The inquest heard that the helmsman of the dredger Bowbelle which was overtaking the cruiser could not see the river ahead because of the high bow and that the helmsman of the cruiser could not see to the rear. Terrence Blayney, who was posted as lookout on the bow of the 260 feet long dredger, one of the largest boats allowed on the river, was unable to communicate with his helmsman. Instead he tried shouting and whistling to the occupants of the Marchioness before the collision occurred. The inquest also heard that the crew of the dredger had consumed significant quantities of alcohol. The decisions to license the two vessels were questioned. There was clearly a failure of the regulatory bodies to look at design and operation as related matters. The report of the Chief Inspector of Marine Accidents had concluded that boats with such design faults "...ought not to have been allowed by the department in the first place. But to attach responsibility to any individual would be impossible, for the fault was simply part of a malaise which for many years affected not just the department but the entire maritime community." (Wells, 1995)

The fire at Kings Cross underground station in 1987 caused the death of 31 persons and injured 60 others. The Fennell Inquiry found that a discarded cigarette was the likely cause of the ignition of a build up of rubbish under the escalators. In parliamentary debate the criticism was made that wider questions of manning and funding had not been adequately addressed. The Inquiry heard that the Railway Inspectorate was 50% below strength and had 'misunderstood' its responsibilities.(Fennell, 1988) After allegations of an emphasis on cost cutting the chairman and chief executive of London Regional Transport (LRT) resigned. (LRT had been reformed by the Thatcher government after the abolition of the Greater London Council.)

The collision of two trains near Clapham Junction on 12 December 1988 was caused by faulty wiring in newly installed signalling equipment resulting in thirty six people being killed. Sir Anthony Hidden QC published the findings of his Inquiry which included reference to a "totally unacceptable" level of overtime which affected the concentration of the staff. British Rail's head of signalling resigned. Cecil Parkinson, Secretary of State for Transport, promised that the Railway Inspectorate would be staffed adequately and that new legislation would be introduced to widen its powers.

ASBESTOS

Acre Mill in Hebden Bridge, Yorkshire, illustrated the early failure to control the asbestos hazard and is one of the acknowledged failures of statutory regulation. Medical evidence was ignored. The law was inadequate and ineffective. Employees lacked representation and involvement in health protection. An under-resourced inspectorate, not professionally qualified, unable except by sanctions to claim the attention of management and lacking on-site powers failed even to confine the hazard to the factory.

In 1930 a Factory Inspectorate report by Merewether and Price confirmed that many asbestos workers were dying of asbestosis within 15 years of starting work. A link between lung cancer and work with asbestos was confirmed in 1947 by HM Medical Inspector of Factories. By 1960 it was confirmed that mesothelioma - a fatal cancer of the chest wall - was caused by asbestos. Workers whose X-rays showed early signs of asbestosis were not advised of their condition by doctors employed by the government's Employment Medical Advisory Service who later claimed to have been observing trends rather than attempting diagnosis.

On 29 March 1976 a report by the Ombudsman, Sir Alan Marre, criticised the Inspectorate for its lack of determination and decisiveness in its 30 years involvement with Acre Mill owned by Cape Asbestos. The investigation was requested by Mr Max Madden, Labour MP for Sowerby. Forty former employees had died of asbestosis or asbestos related cancer and 200 others, including the relatives of employees, had contracted asbestos related diseases. A judge in one of the common law cases which found the company negligent described the Factory Inspectorate as "supine".

Dr William Nicholson, a leading medical authority giving evidence in the United States, estimated that 13 million workers there had been exposed to asbestos between 1940 and 1979 to produce a then current death rate of 8500 persons a year. Looking beyond the United States, he estimated 400 000 asbestos related deaths during the 1980's. He described the situation as a "first magnitude occupational health disaster".

In August 1982 the Manville Corporation invoked US bankruptcy laws. The company, reported as "attempting to become a new company" had assets of one billion US dollars and estimated liabilities from 50,000 expected asbestos related claims of two billion US dollars.

In 1983 the UK Health and Safety Commission advised that even the tighter asbestos dust control levels set that year were not safe levels but were the upper limit of permitted exposure.

There was no statutory provision for worker involvement in hazard management in factories until 1977 when the Safety Representatives and Safety Committees Regulations were enacted more than 100 years after a similar provision for coal mines. They came into force in October 1978. By 1979 a Conservative government was in power and by 1980 the regulations had been undermined where they were most needed. Mrs Thatcher's new government removed the entitlement to appoint safety representatives by redefining a 'recognised trade union' in the Employment Act 1980. The Act removed from ACAS, the government Advisory and Conciliation Service, the power to make a binding recommendation for recognition of a trade union. Safety representatives could then exist only where the employer chose to recognise a trade union. This limitation was a breach of the Treaty of Rome and the European Framework Directive 89/391/EEC and remained so until new regulations in the form of The Health and Safety (Consultation with Employees) regulations took effect on 1 October 1996.

Deaths from asbestos cancers and other effects of ingesting asbestos are still rising. In the UK the total is expected to peak at about 9000 a year (Peto et al, 1995). This is a Bhopal disaster sustained for decades. For comparison the number of persons formerly employed in mining and quarrying in the UK whose death certificates included pneumoconiosis as a cause of death was 1108 in 1961. Since then the number has been falling.

The company now known as T & N which was formerly known as Turner and Newall (sometimes name changes are very important) applied the new UK asbestos legislation in the 1980's in its UK factories but evaded the provision of those standards abroad. Not only T & N but the insurance industry for these and similar misjudgements face unprecedented claims which Lloyds Names now contest. It has to be conceded that the contest by the Names has some justification. Did Lloyds know? Should the Names have known? Should they have been told? Were the Names a group of people generally disposed to support deregulation and self-regulation? Some are Conservative MP's who, if bankrupt, would have to resign. The implications of even one or two such by-elections for what was already effectively a minority government were considerable.

The statement 12 years ago that the risk was "tiny" was based on estimates made by Dr Julian Peto and Professor Sir Richard Doll. It was Professor Julian Peto who in March 1995 published with others in The Lancet (ibid) estimates of mortality indicating that deaths would peak at 9000 per year in the year 2025 instead of peaking at 3000 per year now as was forecast 12 years ago. The decade in question was the decade of deregulation in which the budgets for the Health and Safety Executive were reduced in relation to its tasks. In January 1996 the author wrote to the Health and Safety Executive for the attention of the Chairman of the Commission suggesting that enforcement of regulations rather than deregulation could still affect the shape of the asbestos mortality curve and that painful deaths on an atomic bomb scale do not have to be accepted. (Gifford, 1996)

This is a field of health and safety law making and enforcement in the UK where the hazards were known, where risk assessments have been made, where remedies are also well known, where the enforcement effort had been assessed and where monitoring was taking place. It is part of any effective monitoring process that deviation from an expected outcome be investigated and the causes of failure identified and removed. I believe that HSE Chemicals Branch and the branches concerned with health, policy, and resources and planning know the problem but have not removed the causes of failure. That will require resources that they do not have and, although some of them bravely faced the media with clear statement of what is required, they were constrained as were all civil servants vulnerable to privatisation and worse. Deregulation, privatisation and reduced enforcement were the policies of a government that was largely responsible for this failure and which, notwithstanding its massive defeat in a general election, has not yet been held sufficiently to account. The construction industry, from whose workers many of the victims of this failure will be drawn, was itself politically involved in deregulation and reduced enforcement by its financial support of the government party. The Scott Inquiry has already shown how difficult it can be, even for those with inquisitorial powers, to associate the makers of a policy with its consequences. The Nolan report recommendations may do something to break the links of influence between those who promote defective policy and those who legislate.

BOVINE SPONGIFORM ENCEPHALOPATHY (BSE)

The symptoms of BSE were observed in some cattle in 1985 and perhaps much earlier. In 1988 BSE was made a notifiable disease and since then some 170,000 cases have been reported and the cattle destroyed. The manner in which the diseased carcases were dealt with has not been fully described. The number of cases reported in the first six months of notification was over 2000. At its peak in 1992 round about 1000 cattle a week, mainly dairy cows, were reported. The current number is about 100 per week.

The Southwood committee in 1989 predicted that, under certain assumptions about the disease, the total number of BSE cases would not exceed 20,000. It also warned that there was a "remote risk" of the disease being transmitted to humans.

A ban on feeding ruminant animal protein to ruminants was introduced in 1988 on the recommendation of the Southwood Committee. A warning of a major problem by the Royal Commission on Environmental Pollution made in 1979 of the risk of transmitting disease bearing pathogens to stock and thence to humans had been ignored. Draft regulations on animal feed drafted by the Labour government in 1978 were abandoned by the Conservative government. In Ministry of Agriculture documents that was described as

> "...reflecting the wish of ministers that in the present economic climate the industry should itself determine how best to produce a high quality product.".

In the opinion of the World Health Organisation

> "The epidemic in the UK (the only country with a high incidence of the disease) appears to have been due mainly to the recycling of affected bovine material back to cattle before the July 1988 ruminant feed ban became effective".(WHO, 1996)

Thus some cattle infected with BSE as well as sheep infected with scrapie were fed into the animal food chain. Whether or not BSE is derived from scrapie there is little doubt that the feeding of diseased animals to animals was the cause of the rapid increase in the number of animals affected.

Because of the long incubation period of the disease in cattle, perhaps six years, it is likely that some animals incubating the disease must also have been sent to abattoirs for slaughter as human food. Dr Stephen Dealler, a microbiologist, estimated that 1.5 million BSE tainted cattle have been eaten in Britain. But the government's Chief Veterinary Officer, Keith Meldrum, interviewed by an ITV "World in Action" reporter in November 1995 four times refused even to discuss the possibility. He said that it would be "meaningless" to test for BSE at the abattoirs, that there was no evidence that BSE was a threat to man, and that the public was "perfectly protected". The programme makers concluded that "Most cows infected with BSE are eaten".

Since 1988 it has also been the government's intention that specified offal, even from apparently healthy cattle, should not be sold as food for human consumption. Because of evidence that the bans were not always enforced there remains doubt that human food has been disease free since the ban. Some cattle born since the ban on the use of diseased tissue in cattle feed continue to succumb to BSE and the explanation has not yet been found. Cattle feed delivered before the ban was not recalled and was found on farms as recently as May 1996 by European Community inspectors in spite of their earlier warnings. (Berlingieri, 1996) In August 1996 the government conceded that maternal transmission to calves was occurring.

BSE is known to have crossed 23 species barriers, by injection or by oral transmission. The species include zoo animals, laboratory animals including monkeys, and domestic pets. The government for several years insisted that transmission to humans was not possible and that even if it was possible it would be unlikely. These assertions rested largely on the "there is no evidence" argument. When the same argument is stated as "there is no evidence yet.." it becomes less convincing. Many current investigations will take several years to complete.

In October 1995 Dr Kenneth Calman, the government's Chief Medical Officer said

"There is no scientific evidence of a link between meat eating and CJD." Creutzfeldt Jakob Disease (CJD) is a spongiform encephalopathy found in humans, usually in elderly people.

On March 20 1996 the health minister, Stephen Dorrell, made a statement to the House of Commons to the effect that ten cases of a new strain of CJD had been found in young people and that the strain may be related BSE. He reported that the opinion came from the government's CJD surveillance unit in Edinburgh. (They were about to publish their findings in The Lancet)(Wills et al, 1996) At a joint press conference also attended by the minister, Dr John Pattison, the government's chief scientific adviser, said:

"The most likely explanation at present is that these cases are linked to exposure to BSE before the introduction of the specified offal ban in 1989." (The Times 21 3 96)

In a second Commons statement on the same day Douglas Hogg, Agriculture Minister, announced new rules. Carcases from cattle over 30 months old must be deboned in specially licensed plants supervised by the Meat Hygiene Service and the trimmings kept out of any food chain. The use of mammalian meat and bonemeal in feed for all farm animals was to be banned. He promised that existing controls would be "even more vigorously enforced".

Soon after the announcements by ministers in the House of Commons the European Commission voted unanimously for the imposition of an immediate world wide ban on the export of British beef and beef products. More than 20 countries, including the United States, had much earlier banned the importation of British beef but those facts were not widely known.

A team of senior officials was despatched from London to Brussels with a package of measures, including options for slaughtering cattle possibly infected with BSE, while ministers adopted a policy of boosting confidence in British beef at home and abroad. During the negotiations the government banned the sale of meat from all cattle over 30 months old and arranged for the slaughter, rendering and incineration of all bovine animals over 30 months old at the end of their useful lives. Thus 60,000 animals were destroyed between May and July 1996 and the total is expected to reach one million in the first year of the programme.(MAFF, 1996)   In addition other regulations with immediate effect prohibited the sale or supply or use of mammalian meat and bonemeal (MBM) as feed for farm animals including poultry, horses and farmed fish.

Alongside reports of earlier government proposals for the deregulation of food hygiene standards reports by European Commission inspectors and others of failed enforcement showed that one quarter of British abattoirs failed to comply with European standards and that there had been large loopholes in the "BSE Free" certification scheme for animals exported to Europe. In response to government claims that it had always acted on scientific advice the journal of the Institution of Professionals, Managers and Specialists, whose members include government scientists and veterinary surgeons, reported that "Scientists never gave the all clear". In July 1996 arrangements were still being made for stocks of feed containing bovine material to be recalled and destroyed.

The negotiations failed to achieve the lifting of the EU ban. A cluster of CJD cases was reported in Kent and a Glasgow girl age 15 died of the new strain of CJD. Polls showed that the public was losing confidence in the government and was blaming it rather than the EU for the crisis. But the government accused the EU ministers of a breach of faith and a new policy of non-cooperation with Europe was announced by the Prime Minister on 21 May 1996. Britain was to veto all decisions on all EU business requiring unanimity until the ban was lifted.

During the 30 days of the "beef war" a director of an animal waste rendering company in Kent whose plant was one of nine appointed by the government for the cull of cattle claimed that the spraying of BSE infected liquid waste as fertilizer on nearby land was in accordance with Ministry of Agriculture guidelines. The beef war ended with agreement on a UK plan to cull selectively cattle identified as being at particularly high risk of incubating BSE, and to provide an improved system for identifying all cattle and recording their movements. The world wide ban will be lifted progressively over an indefinite period of perhaps five years. Britain's contribution to a compensation scheme was agreed at œ2 billion.   Soon afterwards the British government published a 'Programme to Eradicate BSE in the United Kingdom' (MAFF, 1996) which included a proposed fivefold increase in the funding by the Department of Health of research into CJD.

In 1993 Paul Anand and Chris Forschner at Templeton College and the University of Oxford published a paper with the sympathetic title "Decision Making under Uncertainty: The Case of Government Public Affairs and Mad Cow Disease".(Anand & Forschner, 1994) They reviewed the decisions taken by government at each stage of the developing BSE crisis and, unlike the government, confronted with some prescience the possible economic collapse of the beef industry. Their conclusions were that the government did not acknowledge the uncertainties in its decision making, that it was too committed to the scrapie transmission model, that other hypothetical models were ignored, and that explicit analysis of costs and benefits, updating of beliefs on the basis of new information, concern for social preferences and an understanding the dynamics of media behaviour were absent or under used.

This examination is focused more on the institutional factors and especially the way in which the civil service, government scientists and other professional civil servants were constrained and influenced by the ideology of deregulation, privatisation, the intention to disband the major part of the civil service, the politicisation of the civil service, the creation of a culture of conformity and the intention permanently to reduce public sector expenditure. The Institute of Animal Health which includes the Neuropathogenesis Unit concerned with long term work on transmissible spongiform encephalopathies (TSE's) has suffered severe cuts in funding over the last two decades and has lost a considerable number of staff. The core budget has been pruned by around three percent per year but there has been some protection of the TSE work by short term funding some of which came from the private sector. (Almond, 1996)

With no claims to expertise in animal husbandry I asked my neighbour, a farmer, if one of her herd of cattle were to become sick and die, would she process the carcase and feed it to the rest of the herd. She was offended that I should even ask the question but she did not refute the information that the remains of sick animals had been processed into cattle feed. Her explanation in defence of the farmers who used that feed was that the bags were marked to indicate only that the contents were "Protein". Christopher Haskins, Chairman of Northern Foods, believes that fundamental changes in the British rendering and animal feed industries must have taken place in the early 1980's to have caused this catastrophe and that more needs to be known. (Haskins, 1996)

It was the rendering industry that was intended to benefit from the non-regulation of animal food processing in 1980. The connection with the increased incidence of salmonella poisoning in the same climate and with the same machinery of government becomes clear. In consumer protection the entitlement of the consumer to know what he is buying by accurate product labelling is not usually disputed. But this principle was breached for animal food perhaps because product labels with entries such as "diseased animal protein" and "animal excrement" would be disquieting.

Another farming friend commented on the government's policy of paying only 50% compensation for animals destroyed because of BSE. He said that an incentive to send an animal to market at the first sign of illness was created. Even with the full compensation now available the trading advantages for producers who remain disease free are an incentive to non-reporting and local burial. He advised than one should ask not only "Is the farm BSE free? but also "Is the farm JCB (excavator) free".

When a government minister was bound to silence and resignation for saying, truthfully, that salmonella was endemic in the poultry industry the civil servant who was then the government's Chief Medical Officer at the Department of Health, Sir Donald Acheson, was asked by the Prime Minister, Mrs Thatcher, to reassure the public in full page newspaper statements about eating eggs. I had then recently been described in government service as an "elderly officer" and I read with interest Sir Donald's statement that the estimate of 1000 people a year being affected by salmonella from eggs was probably low but that cooked eggs were safe to eat except by certain people, including the elderly, whose eggs "should be thoroughly cooked until the yolk is solid". I also was a professional civil servant and I admired the professionalism of the Chief

Medical Officer under political pressure. He clearly understood the difference between being true to his profession and being an obedient servant.

That was in 1988 when BSE was also appearing. Sir Donald's successor, Dr Kenneth Calman, was more supportive of the government in 1995 even using the "there is no scientific evidence" argument to maintain confidence in beef. Was he strong enough to resist political pressure? When he was tested in an interview with the journalist Megan Tresidder (The Guardian, 1995) he claimed that he did not see his job as that of a civil servant. But he made himself a hurdle over which he could fall. He could honestly say that over the past four and a half years "ministers have never not taken my advice on a health issue". Had he advised ministers on tobacco advertising? His reply was "....this is the one issue on which they have not.."

In contrast with Sir Donald Acheson's statement on salmonella, a full page newspaper statement in December 1995 by the Meat and Livestock Commission intended to assert the safety of British beef has been held to be misleading by the Advertising Standards Authority .(BBC, 1996)

A longer study of the machinery of government would examine the structure of the government departments, the priorities given to consumer protection as distinct from support for the food and agriculture industries, the way in which members of advisory committees are selected, and to what extent was conformity demanded in government science. The way in which Dr Harash Narang, a clinical virologist in the Public Health Service Laboratory, was excluded from government science after describing two cases of CJD in 1989 which closely resembled BSE with Dr Robert Perry in The Lancet in March 1990 was disturbing. Dr Narang identified the particles in brain tissue which he described as indicative of spongiform disease. His work was encouraged by an American Nobel prize winner, Carleton Gajdusek, who invited him to work in the United States. Dr Narang also devised a slaughterhouse test for diagnosing BSE in cattle, which he demonstrated at the Central Veterinary Laboratory at Weybridge but which the then minister John McGregor refused to use. Before being made redundant Dr Narang devised a quick postmortem test for CJD which would protect the recipients of donated organs. He was not the only person who questioned the former government's assertion: others who did so also suffered considerable abuse. (Martin, 1995) The World Health Organisation later recommended that there should be research on rapid diagnosis and agent characterisation of the human and animal spongiform encephalopathies.

The fact that the agriculture minister introduced 57 legal instruments after 1988 and pledged himself to further regulation is hardly surprising. But those late attempts were indicative of the free market deregulation and under-regulation of earlier years when the crisis was latent or developing. The Deregulation Unit remained in place in the Cabinet Office and ministers continued to swear their commitment to it until they ceased to be ministers after the 1997 general election. The Deregulation Unit was headed by the Deputy Prime Minister and it was represented in every government department by a minister.

By March 1997 16 young persons in Britain had died of the new form of CJD (Watts, 1977). The government's advisers hope that the number of people affected will remain small in relation to the estimate of up to 80,000 deaths in the 'worst case scenario' reported in a study by Peter Smith of the government's advisory committee. (Palmer, 1997)

CONCLUSIONS

There is extensive literature, an established discipline and much consensus in the studies of the prevention, mitigation and management of disasters with publications by bodies as prestigious as The Royal Society.

Risk management is not a new or wholly theoretical discipline. The disasters and other human rights abuses in coal mining were corrected by good management, employee involvement, risk assessment and control measures supported by statutory regulation. Those principles were generalised in the Health and Safety at Work, Etc., Act 1974 and they have since appeared in European legislation.

The disasters examined in this paper show that the Conservative government rejected the principles and resented and opposed the European legislation. Instead a laissez-faire, free market ideology competed with the government's treaty obligations.

The Deregulation Unit, headed by the then Deputy Prime Minister and represented at minister level in every department of government, was empowered not only to remove red tape but to repeal legislation essential to health, safety and the protection of the environment without Parliamentary debate. Employee representation was a an early target.

In the climate of deregulation and endless reviews of government science, under-regulation and non-enforcement of the law, cuts in research funding, privatisation of government science and the intention to dismantle the greater part of the civil service were not only indicators of the government's retreat from government and its reliance on the market but major causes of the BSE crisis and other disasters. A public inquiry is needed to name the persons who decided to ignore the warning of a Royal Commission, to find out how and when the animal food chain was polluted with disease agents, to find the reasons for and the effects of the under-regulation of the animal feed industry and to establish why draft regulations were discarded in 1980.

In the BSE crisis the deregulatory policies were a total failure even discounting the remaining threat to human health. Throughout, in the field of regulation and enforcement, too little was done too late. The consequences of the BSE outbreak so far have been mainly economic and political. The cost is already on a scale of billions of pounds. Compensation will be needed not only by those who own farms and the affected industries but by those who work in agriculture and those industries.

The credibility of the British government was damaged by its neglect of the precautionary principle, its unwarranted hostility to Europe, and its willingness to blame the customer and its trading partners.

The UK government's deregulatory policies tainted the European Commission's early responses to the British problem with the result that European regulation also failed. The losses incurred negate the claims for overall benefit from deregulation as a policy.

The lesson to be learned is that it is the proper role of government to ensure that the protection of the health and safety of the public and the protection of the environment by the enforcement of up to date regulations are given greater priority than any commercial interest or political dogma. The new government's Food Safety Agency, if it works independently of the Ministry of Agriculture and agriculture industry interests, may help to provide such protection.

REFERENCES
Almond, J W, (1996), BSE:Is British Science to Blame? *The Guardian* London
Anand, P and Forschner, C; (1994) "Decision Making under Uncertainty: The Case of Government Public
Affairs and Mad Cow Disease". In *"Changing Perceptions of Risk - Implications for Management"*
Conference of The Risk Society Bolton Business School UK
Annual reports of HM Chief Inspectors of Mines and Quarries, HMSO London UK
Annual Report of the Health and Safety Commission for 1987-8 HSE Books Sudbury CO10 6FS UK
BBC (1996), *Radio 4 News* 10 July
Berlingieri, C (1996) *Community Inspections on the Incidence of BSE in the United Kingdom,* Paper
prepared for the European Parliament by the EU Office of Veterinary and Phytosanitary Inspection and
Control.
Boyd J T (1993),: *Independent Analysis: 21 Pit Closure Review of Collieries*, Report No 2265.5, The
Department of Trade and Industry London UK
Crossland B, *Public Hearing Following the Extensive Fall of Roof at Bilsthorpe Colliery,* A report to the
Health and Safety Commission , London UK
Cullen (Lord) (1990), *The Public Inqirey into the Piper Alpha Disaster*, HMSO, London, UK
Donaldson, J T D (Lord) (1994), *Safer Ships, Cleaner Seas, The Prevention of pollution from Merchant
Shipping*, HMSO, London
Donaldson J T D (Lord) (1995), *Report into the causes of the sinking of MNV Derbyshire*, HMSO, London
Fennell D (1988), *Investigation into the Kings Cross Underground Fire*, HMSO, London, UK
Gifford, C Proof of Evidence presented to the *Public Hearing Following the Extensive Fall of Roof at
Bilsthorpe Colliery*, The Health and Safety Commission, London, UK
Gifford, C (1996) Comments on *Consultative Document CD96 Draft Consultation with Employees
Regulations and Guidance* , The Health and Safety Commission, London, UK
Haskins, C (1996), Beef Policy? It's all Been Madness, *The New Statesman*, London
Herbert, S, (1992) Law Report on the Queens Bench Divisional Court 21Ò12Ò92 Regina v British Coal, ex
parte NUM, NACODS, et al. *The Guardian,* London, UK
Hood, C et al (1992), Risk Management in *Risk Analysis Perception and Management* The Royal Society
London
Martin, P (1995), The Mad Cow Deceit, *Night and Day*, December 17
Ministry of Agriculture, Fisheries and Food⊗1996) *"Programme to Eradicate BSE in the United Kingdom"*
London
Palmer, J (1997), 'Thousands' Die in CJD Scenario, *The Guardian* , London, UK, 15 January
Perrow, C (1984), *Normal Accidents: Living with High Risk Technology*, Basic Books, New York
Peto J et al (1995), "Continuing Increase in Mesothelioma Mortality in Britain" *The Lancet*, Vol. 34 (3)
Private communications: the Chemical Workers Union of South Africa and the African National Congress.
Robens (Lord) (1972), *Safety and Health at Work*, HMSO, London, UK
The Deregulation and Contracting Out Act, (1994), HMSO, London, UK
The Management and Administration of Safety and Health at Mines Regulations, (1993) and Approved Code
of Practice, SI 1993 No 1897, HMSO, London, UK
Tresidder, M (1995), The Doctor will see us now, *The Guardian*, London, UK, 16 December
Turner, B A (1978), *Man-made Disasters*, Wykeham Press
Watts, S (1977), The Man who Cannot Tell a Lie, *The Guardian*, London, UK, 20 March
Wells, C; (1995), *Negotiating tragedy: Law and Disasters*, Sweet and Maxwell, London, UK
Wills, R G et al (1996), A new Variant of CJD in the United Kingdom , *The Lancet*, Vol 34(7)
World Health Organisation, (1996), *Consultation on Public Health Issues relating to Human and Animal
Transmissible Spongiform Encephalopathies*, Geneva

THE POLITICAL ECONOMY OF RISK:
PIPER ALPHA AND THE BRITISH OFFSHORE OIL INDUSTRY

MATTHIAS BECK
Glasgow Caledonian University
Dept. Risk & Financial Service
Britannia Building
Glasgow, G4 0BA, UK

CHARLES. WOOLFSON
Reader in Industrial Relations
University of Glasgow
Faculty of Social Sciences
Lilybank House
Glasgow, G12 8RT, UK

Introduction

In his article *The Titanic Disaster: An Enduring Example of Money Management vs Risk Management*, Roy Brander comes to a bold conclusion: the Titanic did not sink because of a number of technical flaws and human errors, but rather because of the prioritisation of money management (Brander, 1995). Brander's argument is as simple as it is convincing. In the mid 19th century, engineers aspired to build ships with a view towards maximising the physical protection of passengers from harm. Brunel's *Great Eastern* had an entire inner hull two feet inside the outer, inside that, the ship was divided into 15 transverse bulkheads and one lengthwise into 32 compartments. By the turn of the century, competition between more than ten shipping lines had grown fierce, with hundreds of passenger liners being built. As a consequence, the safety precautions that had marked earlier ships like the Great Eastern were chipped away. The new designs included fewer and smaller bulkheads, whilst unsafe lifeboats were provided in inadequate numbers. None of that seemed to upset the regulators, and the industry had ways of making sure that it would not. The *Titanic's* only life-boat drill was conducted on deck with two boats and a hand picked crew. When asked during the Titanic enquiry why he cleared the Titanic for sailing following this drill, the inspector, Maurice Henry Clark gave the following response:

E. Coles, D. Smith and S. Tombs (eds.), Risk Management and Society, 227–262.
© 2000 *Kluwer Academic Publishers. Printed in the Netherlands.*

Did you think your system was satisfactory before the Titanic disaster?
No, sir.
Then why did you do it?
Because it was the custom.
Do you follow a custom because it is bad?
Well, I am a civil servant sir, and custom guides us a good bit.
(Bender, 1995: 1)

Bad custom, apathy, the prioritisation of monetary concerns, and regulatory collusion with industry still guide many decisions relevant to the physical prevention of harm to employees and the public in general.

Unfortunately, the language of risk management or, to use a more pretentious term, risk science has done little to alter this. Indeed, in the past decade a number of different strands of risk management have come to legitimate top-down authoritarian approaches to workplace management, and in some cases deregulatory initiatives. Neo-classical risk analysis, for instance, perceives a certain level of 'societal risk' as unavoidable. With the real sources of risk and harm safely put to the side, it is then able to assign a key role to the neo-classical economist as risk manager. The economist as risk manager has two principal tasks. Firstly, she or he, has to, based on economic skills, prevent individuals from pursuing or demanding the elimination of small risks, which they chronically overestimate. Secondly, and more importantly, she or he has to advise governmental agencies not to squander excessive resources on risk prevention measures demanded by a diverse set of 'special interest groups' such as trade unions, the elderly, the disabled.

The gurus of neo-classical risk management, Richard Zeckhauser and Kip Viscusi, give an eloquent summary of this agenda:

Individuals overestimate small risks, are averse to imprecisely understood risks, and give excessive weight to errors of commission over errors of omission. The challenge for the government is to strike and appropriate balance in its risk regulation efforts and to avoid institutionalising common irrational responses to risk. Excessive expenditures on risk reduction, often undertaken by or required by government, not only squander resources but also may increase risks to us all; they can divert expenditures that could have been used to enhance our standard of living and, directly or indirectly, our health (Zeckhauser and Viscusi, 1996: 144-5).

To call the ideological presuppositions of this analysis suspect would be an understatement. Clearly, there is nothing wrong with taking an anti-regulatory stance based on the, albeit largely unproven assumption that the overestimation of small risks has lead to over-regulation in many areas. What is worrisome, however, is the deliberate failure of neo-classical analysis to recognise that most risks, and especially those subject to a long history of regulation, are man-made, intentional and controllable. To merely take a given risk level, such as that of physical injury to a construction worker on a job, as given and then weigh costs and benefits of harm prevention measures, intentionally obscures the underlying structures of harm creation and the possibilities of harm prevention. In a Marxian sense, the neo-classical analysis of risk represents an anti-humanist mystificatory construct aimed at preserving and obscuring existing power relations. We will return to this argument at a later stage of this chapter.

Mystification, however, is not the preserve of neoclassical risk analysis. Whilst willing to accept individual risk perceptions as valid basis for policy making, postmodernist analysis, not unlike its neo-classical counterpart, *apriori* rejects the possibility of or need for successful concerted harm reduction efforts. This is perhaps nowhere more apparent as in the recent literature on *autopoiesis*, which has found ever expanding areas of application. Using the framework of *autopoiesis*, John Paterson and Richard Teubner, for instance, have examined the pre-Piper Alpha regulatory regime in the North Sea. With a minor twist, their conclusions confirm the basic tenets of *autopoeisis* theory. Regulatory failure occurs because of an insufficient adaption of the existing prescriptive framework to the needs of industry, which in turn causes industry to self-steer into a low level control system of health and safety matters. Regulatory reform will in all likelihood fail, because it does not allow for a flexible self-steering of industry. Without going too deep into the self-enclosed catechism of *autopoeisis*-speak we quote a brief passage from Paterson and Teubner's work:

Regulatory success depends then on the ability of the regulatory system to recontextualise in its turn the recontextualisation in the regulated system. In other words, the regulators must direct their attention to the codes and programmes of the systems they seek to regulate. This observation will still of course be on the basis of their own distinctions and so the theory [*autopoeisis*] offers no hope of direct intervention. But as problematic as this sounds, it is important to realise that failing to problematise this situation appropriately [sic] is only likely to make matters even worse.

These systems, then, continued to evolve in rather path-dependant ways, steering according to their own difference-minimising programmes and largely indifferent to the self-steering of others. It comes as no surprise then to discover that in due course, following the worst offshore accident in 1988, a further public inquiry found little to praise in the regime and plenty to condemn (Paterson and Teubner, 1998: 26)

Clearly, the scrutiny of any regulatory regime which fails to prevent 167 mass fatalities is going to find 'little to praise and plenty to condemn'. What is absent in this analysis, however, is any concrete analysis of the dynamics of regulatory failure and renovation rooted in an historical political economy. In its place we are offered a pre-rehearsed meta-discourse in which there is neither structure not agency. Whilst it is a matter of dispute whether postmodernist analysis merely exchanges the God of market forces for that of *autopoeisis*, Teubner's analysis confronts us with a new double disingenuity. On the one hand we are faced with the analysts' explicit Pontius Pilatus abdication from policy making, whilst on the other hand, we are made to believe that we will be unable to analyse existing contingencies of risk properly, were we not to accept the *autopoeisis* catechism.

In the area of occupational risk prevention, abdication from class politics is nothing new. Indeed, Kit Carson identified the de-contextualisation of health and safety as one of the outstanding characteristics of state intervention as well as of much academic research. Said Carson:

Factory legislation helped to 'defuse' or 'declass' the employer/employee relationship at one of its most critical and socially obtrusive points, the price being extracted from industrial wage-labour in terms of occupational injury and disease. More broadly it served to 'mask' one of the potentially most unpalatable aspects of class relations in an industrialising society. It did so by purporting to pluck issues such as questions of occupational health and safety out of the fraught and, at that time, potentially disastrous arena of industrial conflict and making them a matter of 'classless' state regulation.

As a result of the above process an ideological separation between occupational health and safety as one category, and industrial relations as an another, took place (Carson, 1985: 146).

Unfortunately, modern approaches to risk management are still characterised by attempts to separate out legitimate spheres of influence and discourse. This often means that processes of risk creation are ignored and the analysis focuses on the management of the status quo. Even in the examination of potential managerial approaches, meanwhile, typically only a limited set of options is examined. Thus, whilst management practices and systems possess great vogue in the literature, remarkably little has been written about the benefits of safety management and auditing from below, or even more so concerning the potential benefits from the involvement of the representatives of organised labour.

What our brief, eclectic and perhaps somewhat unfair, review of different views of risk management suggests is that certain strands of mainstream risk analysis, contribute little to an adequate understanding of the policy analysis of harm prevention. This is our working hypothesis. The aim of our chapter, meanwhile, is to present an alternative view of risk management which places risk in its social and political context. For lack of a better term, we call this alternative view a political economy approach to risk, if only to signal our attempt to embed applied risk research into a broader context of institutional, economic and political forces. As yet, this political economy approach is more of a research programme than it is a set agenda. In this paper we explore its intellectual potential through an analysis of one of the areas where different approaches to risk management have led to great deal of controversy and confusion, namely that of regulation and risk prevention in the British offshore oil industry. Our paper focuses on the dynamic interaction of three forces: firstly, and most importantly, the economic context of offshore oil production; secondly regulatory inputs of the state and thirdly, industry responses. We proceed as follows. In the first section we examine the early phase of oil exploration in the North Sea which was characterised by a regulatory void. The second section examines the post-Piper Alpha reconstruction of the offshore regulatory which was marked by an industry containment strategy aimed at exploiting notions of self-regulation and goal-setting in order to avoid costly regulatory interference. We note that the industry's adoption of 'modern' risk management methods has resulted in new patterns of workforce exclusion. Section three describes that latest phase in the offshore safety debate which is marked by industry attempts to dictate a new safety agenda outwith any real workforce consultation. The concluding section highlights the relationship between industry profits, workforce exploitation and offshore industry which appears to define the political economy of offshore safety irrespective of government and industry claims to the contrary.

Oil Exploration in the Regulatory Void

NON-REGULATION AND THE 'ECONOMY OF SPEED'

The geopolitical background of UK offshore oil production has been discussed in a number of works. Here we will give only a brief summary of the economic context of oil exploration on the British continental shelf. By the late 1950s, US control over client regimes in Latin America, North Africa and above all, in the Middle East, no longer looked certain. In 1960 OPEC was formed as a producers' cartel and the continuing Arab-Israeli conflict placed question marks over the continuity of supplies. These developments triggered a radical shift in the corporate policies of multinationals. Henceforth, oil exploration centred on politically stable areas, with the aim of establishing sufficient supplies from outside OPEC.

It was at this point that oil exploration began in Canada, Alaska, Australia and the North Sea. The oil multinationals realised that oil production in these locations would be more expensive. But its principal value was now seen as a bargaining counter. Exploration in non-OPEC provinces could enable oil companies to break the Middle East price monopoly. In this context, the discovery of North Sea oil in 1969 became part of an evolving strategy aimed at gaining independence from OPEC producers. With a leap in oil prices in the early 1970s oil extraction had become economically feasible in the North Sea, even though the costs involved in exploiting remained high.

At the time, the scale of investment required, calculated at one fifth of all industrial investment for a decade, was beyond the resources of either the British Government or private investors. North Sea oil exploration therefore required a strategic alliance with US capital. This alliance was conditioned on the importation of a US style production regime. Britain's oil was to be extracted at the fastest rate possible, with limited State control and under conditions of close commercial partnership between American oil companies and banks. Both Labour and Conservative governments enthusiastically endorsed what Carson was to call 'the political economy of speed' (1982: 84).

As a consequence of this philosophy, safety concerns took a backseat to economic considerations. In his path-breaking analysis, Carson (1982) analysed the slow reactive response of UK regulatory agencies to safety problems in the offshore industry. In 1965 a jack-up rig, the Sea Gem collapsed with the loss of 13 lives. Yet it was a full six years before the Mineral Workings (Offshore Installations) Act (MWA) of 1971 was put on the statute book. The MWA empowered the Secretary of State for Energy to require that installations be certified as fit for purpose. Further regulations were even slower to emerge. It was not until 1973 that the requirement for the reporting of offshore casualties were imposed. A further three years elapsed before regulations for occupational safety, health and welfare were created and it took until 1978 before standards for fire-fighting equipment were established.

The MWA was highly prescriptive. Although modern risk management principles were developed in the context of the physical prevention of harm in such areas the petro-chemical industry, the MWA relied heavily on the imposition of minimum standards for equipment used. This gave the offshore industry a seemingly legitimate cause to evade regulatory impositions. Indeed, over the years, MWA-based regulations were never rigorously enforced.

Carson described this regulatory regime as the 'institutionalised tolerance of non-compliance'. In this regime, non-compliance evolved from collusion between the regulator and the regulated industries, rather than from resource or technical limitations of the regulator. Underlying this complicity was a process which some analysts describe as a pattern of 'regulatory capture'. Regulatory capture describes a process whereby a regulatory agency comes to wholly identify the public good with the interests of the industry. In the UK offshore oil industry, such 'regulatory capture' was near to complete.

All through the 1970s, the Department of Energy repeatedly lined up with the industry to prevent the encroachment of other agencies onto its territory. This was so most notably at the end of the 1970s, when a Labour government inquiry into offshore safety under J. M. Burgoyne was conducted. Its report recommended the continuation of the current arrangements despite the worsening safety record of the industry, and submitted evidence in favour of the exclusion of involvement of other agencies (Burgoyne Report, 1980).

As a consequence of outdated regulations and the regulator's complicity with industry, safety regulation in the North Sea was seriously out of line with onshore practices by the 1980s. Onshore, following the Robens Report into Safety and Health at Work, safety regulation had moved away from prescriptive rules towards a goal-setting regime (Robens, 1972). Goal setting implied a shift in enforcement philosophy from externally-policed regulation towards industry 'self-regulation'. Onshore, this process was overseen by a new unified independent agency, the Health and Safety Commission/Executive (HSC/E). This agency was run along tripartite corporatist lines, involving representatives from industry, the unions as well as government agencies.

The Robens committee had envisaged that, in time, the offshore industry would come under umbrella of the accompanying Health and Safety at Work Act (HSWA) 1974. This was not to be. Aware of the potential of the HSWA to impact on existing offshore practices, the Department of Energy jealously guarded its 'special relationship' with the oil industry. This relationship had been cemented by frequent interchanges of personnel and the recruitment of senior officials and even government ministers by the industry. Regulatory integrity was undermined by a number of forces which included processes which US writers have observed in a number of industries and have come to describe as 'deferred bribery' (Spiller, 1990).

In the 1980s, supporters of the Department of Energy argued that it alone had the necessary specialist expertise to respond to the rapidly changing needs of this technologically frontier industry. The industry itself was only too happy to endorse this view. The consequences amounted to the creation of regulatory void. While the HSWA was formally extended offshore in 1977, the operation of an 'agency agreement' between the Department of Energy and the HSE prevented the direct supervision of offshore safety by the HSE. The development of offshore regulations, meanwhile, remained with the Department of Energy. This simultaneously ensured the maintenance of a sympathetic regulatory regime for the industry, and strengthened the industry's ability to resist the offshore application of key regulations, such as those onshore regulations dealing with hazardous substances as well as those permitting organised trade unions to appoint workplace safety representatives.

The combined effect of these factors allowed the oil majors to create a virtual 'zone of exclusion' offshore in which they could conduct their activities largely unmolested by regulatory interference. In this zone of exclusion, offshore safety regulation was placed in a paradoxical context. On the one hand the key ministry, the Department of Energy, was charged with ensuring the rapid development of offshore oil. On the other hand, it was also responsible for ensuring the safety and health of those who worked in the industry.

THE REALITY OF ON-SITE SAFETY MANAGEMENT

If the management of offshore risk was flawed on the macro-level of regulation, it was even more so on the level of on-site management. Out on the platforms, attitudes to safety and occupational welfare of employees were casual, as was the general view on existing UK regulations on safety. In these early phases of North Sea development the priority was to get the oil out of the sea bed. Money was no object, and the drilling companies' and operators' focused on quick production. To this end a site-level management style was adopted which attached a great deal of authority and independence to line and platform managers. Auditing and monitoring mechanisms such as existed were implemented only to the degree that they did not infringe on speedy production.

In this context, workforce participation in safety matters and even more so, as collective bargaining partners was unwelcome. An internal 1976 analysis by a union noted that oil companies, virtually without exception, employed a number of strategies aimed at obstructing the expansion of unions offshore. These included:

The insistence on full ballots, not only for collective bargaining rights but also for simple representational rights; company initiated anti-union propaganda being spread in the run up to the ballot; prolonged delays in holding ballots, and delays in affording rights where the ballot has been successful; the setting up of staff consultative machinery to undermine the activities of bona fide trade unions, more favourable conditions of service to non-unionised areas and asking prospective employees their attitudes to trade unions (ASTMS, 1976).

Even in the early days of offshore activities, UK trade unions voiced dissatisfaction with management-dominated consultative committees, and particularly their ineffectual role as concerns safety matters. This is not surprising given the role assigned by the oil majors to these committees. An unpublished doctoral thesis by Thom cites an industry handbook which defined 'consultation' as 'a process for communication between staff and management to enable the views of staff to be expressed, discussed and taken into account before management makes a decision on a matter' (Thom, 1989: 101). The handbook moreover stated that the consultative committee is not 'a forum for negotiating terms and conditions' (Thom, 1989: 102). While Thom's study suggests some minor variations in the scope of these committees, there is evidence that offshore management commonly perceived them as a 'safety valve' (Thom, 1989: 102). Employee input in safety matters was generally sought within this committee structure and potential gains in terms of risk management from improved communication with the workforce were largely unrealised.

This casual attitude towards communication with workforce was illustrated by examples of management responses to workforce demands for change. In one instance a consultant, hired by industry to investigate workforce dissatisfaction, depicted the employees' desire for collective representation as a from of neurotic response. In reality it was largely driven by workforce concerns for greater occupational safety. The report, written by Robert de Board of Henley Management College, found that workers on the platform were suffering from 'acute anxiety'. The workers were 'looking for the feminine mothering side of human nature which is being deliberately excluded in the macho management style'. The consultant concluded that the wish for union representation was 'a cry for help, "come and look after us"' (cited from Dalton, 1998).

The use of management consultants was particularly important for companies such as Mobil, who were strongly opposed to unionisation on their platforms. Mobil delayed a request for a visit by union officials to the Beryl Alpha platform for over three months, ostensibly on the grounds that there had already been one such earlier visit and that two visits per year were a 'reasonable' number. Industrial relations problems that summer had led Mobil to bring in Henley Management College consultants. Mobil denied that this was part of a 'union avoidance' strategy. The Mobil employee relations manager noted:

'Sensitive discussions were taking place with our employees to resolve the difficulties identified earlier. As we informed you, a union visit during these discussions would endanger proper focus on the actual problem at hand, and jeopardise our immediate objective to resolve the issues through direct consultation with our employees'. (Letter from the Employee Relations Managers, Mobil Northsea Ltd to Campbell Reid, Inter Union Offshore Oil Committee, 26/7/82).

There is evidence that the employers' casual attitudes towards the needs and fears of the workforce seriously undermined their own awareness of safety and related managerial deficiencies. In the authoritarian management culture offshore, the informed knowledge of the platform worker could not filter upwards to inform the risk perception and awareness of management.

Perhaps even more importantly, the authoritarian management style adopted by the industry in the first decade of oil exploration and production eroded the possibility of consensual co-operation between management and an organised workforce especially in the are of health and safety management. Offshore two fundamentally different sets of expectations had clashed. On the national union level there had been the expectation of an extension of collective bargaining to all sectors of the economy. The oil multinationals, based on US experience, meanwhile had expected that unionisation could be radically circumscribed, if not avoided altogether. This belief led the multinationals to oppose workforce involvement in safety matters out of fear that such involvement may create a bridgehead for unionisation. Based on their own experience with an authoritarian management unwilling to listen, offshore workers, had little ground to trust in management's ability to create and maintain a safe and well-managed working environment.

These problems were compounded by the specific nature of offshore production. The North Sea production regime established in the 1970s included a dependent layer of specialised sub-contractors. Only a quarter to a third of the offshore workforce were direct employees of the oil companies. The majority of offshore employees were employed by contractors who provided services to their client oil companies. This production system further compounded the difficulties of risk management which existed in this 'frontier' industry of authoritarian managers and distrustful workers. In more ways than one, the Piper Alpha disaster was an incident waiting to happen.

Piper Alpha and the Smokescreen of Self-Regulation

REGULATORY INITIATIVE AND INDUSTRY RESISTANCE

Survivors' transcripts submitted to Lord Cullen's public inquiry into the Piper Alpha disaster reveal the total breakdown of emergency procedures during that cataclysmic event. Communications were knocked out, sprinkler deluge systems failed to operate and support vessels could not perform rescue functions adequately. Management on pipeline linked platforms failed to shut down production and continued to feed the fires on Piper. Those responsible for emergency action proved totally unprepared for a major emergency of this sort. Those workers who survived only did so because they ignored the existing safety procedures. We have documented the narrative of these dramatic events and their consequences elsewhere (Woolfson, Foster and Beck, 1996).

In his 800 page inquiry into the Piper Alpha disaster, Lord Cullen concentrated on the causes of the disaster and the measures that could be taken to prevent a recurrence of such incidents (Cullen, 1990). In this context he was eager to examine the issue of 'workforce involvement' in the safety process. Cullen suggested that, where the unions could demonstrate that they had achieved substantial recognition and membership on a given installation there might be a case for union appointed safety representatives. Cullen noted that this 'could be of some benefit ... mainly through the credibility and resistance to pressures which trade union backing would provide' (1990: 21.84). Moreover, he conceded that the issue of victimisation needed to be addressed and recommended that legal protection be made available to offshore safety representatives.

Cullen's criticisms of the risk management procedures implemented by the operators of Piper Alpha, Armand Hammer's Occidental Petroleum, were scathing. Cullen noted Occidentials failure to operate a safe system of work despite a number of previous incidents on the platform. Offshore workers testifying at the inquiry pointed to the cutbacks in maintenance following the oil price downturn in the mid-80s which had led to extensive maintenance work and production taking place simultaneously. Both micro-level management failures and macro-level development in the industry were identified as causes of the Piper Alpha disaster. All appeared to have contributed to the disaster.

Faced with this evidence, Lord Cullen's report severely criticised the previous regulatory regime administered by the Department of Energy. Cullen noted that only five inspectors had been responsible for policing the entire North Sea. As a rule, an installation would be visited perhaps once every two years. More specifically, Cullen described the inspection of Piper Alpha in the weeks before the disaster as described as 'superficial to the point of being of little use as a test of safety' (1990: 15.48). The Department of Energy's approach to offshore regulation, meanwhile was described as being marked by 'overconservatism, insularity and a lack of ability to look at the regime and themselves in a critical way' (1990: 22.20). Little had been learned from the more modern onshore approach to hazard characteristic of the HSWA or from the more forward-looking regime in Norway (1990: 22.20).

For Cullen, and a number of experts, the time had come to look at 'modern' approaches to risk management. At the core of these modern approaches stood the notion that safety risks had to be assessed within a comprehensive framework. The concept of 'Formal Safety Assessment' (FSA), advocated in the Cullen report, involved 'the identification and assessment of hazards over the whole life cycle of a project' through all its stages of development to final decommissioning and abandonment. Included in the concept of FSA were a number of analytical techniques of risk assessment. Formal Safety Assessment was to lead to the development of a 'Safety Case' for each installation. The Safety Case was meant to provide a systematic documented review of all hazards potentially existing on an installation, and the safety management systems put in place to deal with them.

The creation of the Safety Case became the centrepiece of Lord Cullen's proposals for the new offshore risk management regime. This marked the dual realisation that offshore safety management had been out of line with modern onshore practices, and that offshore operators now had to be brought up to desirable standards by the regulator. The Safety Case concept had evolved onshore, particularly as a consequence of the Seveso disaster and a subsequent European Directive. Within the UK the explosion at the Flixborough petrochemical plant had resulted in a reappraisal of major hazard regulations, and led to the implementation of a number of comprehensive risk assessment based regulations.

Central to the Safety Case was the notion of a 'temporary safe refuge' (TSR). Each installation was now required to possess a TSR, that is an area of specified durability which, in the event of a major incident, could shelter employees for sufficient time to effect a safe evacuation. Quantitative Risk Assessment (QRA) provided the specifications for this system, with the remit 'to assess the risks, to identify and assess potential safety improvements, and to ensure that the TSR meets the standard set' (1990: 1761). We discuss QRA in more detail below. Immediately, and predictably, conflicts arose as to the status of the new FSA and Safety Case regulation. In this context, Lord Cullen did not concur with the industry's view that FSA and the Safety Case were sufficient to ensure adequate safety management. In Cullen's view, FSA and the Safety Case required the complement of a system of largely prescriptive regulations 'setting intermediate goals [which] would give the regime a solidity it might otherwise lack'. These included regulation regarding construction, fire and explosion protection, evacuation, escape and rescue (1990: 17.67).

Cullen's focus on FSA and the Safety Case raised questions about the appropriate body for the evaluation of the operator's Safety Cases. The choice was between a continuation of safety regulation under the Department of Energy (DEn) and the HSE, which had experience in administering a goal-setting regime onshore. The oil operators at the Piper Alpha inquiry claimed to be 'agnostic' in this matter; a position which stood in marked contrast to their previous hostility to the HSE. Based on the view that the offshore management of safety under the DEn lagged 'a number of years behind the approach onshore' (1990: 22.21), Cullen formally recommended a transfer of responsibility for offshore safety from the DEn to the HSE (1990: Vol. 2, Recommendation 25). This included the creation of a new division of the HSE responsible for offshore safety, namely the Offshore Safety Division. The Offshore Division came into existence in the spring of 1991, with the remit implementing, a total of 106 recommendations made in Lord Cullen's report, which the government had accepted in full.

Perhaps predictably, the UK Offshore Operators Association (UKOOA) was anxious to show that, in the aftermath of the disaster, the industry was 'pro-actively' reforming itself. Thus the industry claimed that it had in its advocacy of Formal Safety Assessment (FSA), already anticipated the main thrust of Cullen's recommendations. Behind the facade of compliance with the future regulatory regime, however, lay a different agenda. This agenda was driven by two goals: the avoidance of costly specific regulations through the exploitation of the notion of flexibility and goal-setting, and the avoidance of mandatory union recognition as against workforce involvement in safety matters.

The Cullen inquiry had sent several delegations to Norway to examine the workings of the Norwegian Petroleum Directorate (NPD). The NPD had a much more prominent regulatory profile in the industry than the DEn. The essence of the Norwegian system has been described as a system of 'internal control'. Internal control makes the operators responsible for organising the safety of their installations within a general goal-setting framework laid down by the NPD. This system avoids a mass of externally-policed detailed prescriptive regulations.

For the operators organised in UKOOA a Norwegian-type system was attractive, as they believed it allowed them latitude in retaining control over most aspects of installation safety. A 'goal-setting' regime, based on the Safety Case, UKOOA spokespersons argued, went back to the 'first principles' of safety management in which there was little place for detailed regulations. UKOOA's advocacy of Formal Safety Assessment and the Safety Case at the Cullen inquiry was driven by the perceived flexibility such an approach would offer to the operators. An industry paper entitled 'Offshore Safety—The Way Forward' noted that FSA had 'many advantages' in that was is 'flexible' and could take account of the different types of installations in the UK offshore environment. According to the industry, FSA had the advantage in that it did 'not dictate to the operator how safety should be achieved' (Taylor, 1991: 6).

UKOOA's 'strategy' in dealing with media on the publication of the report was outlined in the same briefing document:

We should use the media opportunities presented to us to state UKOOA's aims and objectives and where possible show how they have been advocated by Lord Cullen. The advantages of FSA can be used to answer any detailed question. Using this strategy we should be able to avoid being dragged into detailed argument about specific proposals. If Lord Cullen makes a specific recommendation, which is counter to the actions already under way in the industry, we should embrace it objectively, and agree that it seems sensible and promise to look at it (UKOOA, 1990: 1).

Among the 'useful phrases' which UKOOA had ready for media 'sound-bites' were statements such as: 'Safety is our Number 1 Priority' and, 'A Safe Platform is a Profitable One'. The language of modern risk management had become industry's principal weapon in its fight against public recriminations and, more importantly costly regulatory impositions.

UKOOA claimed that in the two years since Piper Alpha, the operators had spent £750 million on safety improvements, roughly £1 million a day. Approximately £230 million of this expenditure was on the fitting or relocation of topside emergency shutdown valves on risers at the interface with the platforms. These alterations were required by the *Emergency Pipe-Line Valve Regulations*, passed in July 1989 in the immediate wake of Piper Alpha. A further £230 million was spent on subsea isolation systems, and a further £300 million on other safety measures. In any event, up to 80 per cent of this total expenditure could be offset against petroleum revenue tax.

Although difficult to quantify, probably the greatest bulk of this expenditure was occasioned by the postponed maintenance work which was left over from the oil price downturn of 1986. Measures to improve offshore safety were almost impossible to separate out in cost terms from essential maintenance. Nevertheless, UKOOA's claims of huge safety expenditure went largely uncontested.

Industry claims of its own pioneering role in adopting modern safety principles, however marked only the first phase of its containment strategy. Once Piper Alpha had disappeared from the headlines, a second phase followed which was marked by open hostility to the new requirements imposed by the new regulatory authority, the Offshore Safety Division (OSD) of the HSE. Cullen had recommended that every installation have a temporary safe refuge or safe haven (1990: 17.38). The first draft of the Safety Case regulations was published in December 1991 and in March 1992, as a consultative document (HSE, 1992). In response to representations from UKOOA, the HSE indicated that the TSR requirements might be modified where a satisfactory alternative could be demonstrated. It was conceded that TSR requirements might be relaxed with respect to certain normally un-manned installations or mobile drilling rigs. For the operators this concession did not go far enough. At a major conference, industry spokesman Harold Hughes (then Director General of UKOOA) issued his challenge to the new regulatory authority:

Although the HSE have already signalled that exemptions will be available for such cases, it is a fact that over half of current UKCS installations are likely under this proposed legislation to be the subject of applications for exemptions... It does not seem to us, to put it mildly, to be good law that demands, *ab initio*, the exemption of over half the installations to which that law is supposed to apply (Hughes, 1992: 6).

The cost of each TSR was expected to amount to about half the total costs of the installation Safety Case. HSE estimated the outlay on TSRs alone as between £13 billion and £17 billion for the industry as a whole (HSE, 1992b). Viewed from this angle, UKOOA's sensitivity to what it regarded as the 'prescriptivity' of TSR requirements was perhaps understandable.

Between the publication of the consultative document on the draft Safety Case regulations and the accompanying guide to the Safety Case in March 1992, the HSE met with UKOOA on a regular basis. HSE also met with the International Association of Drilling Contractors (IADC) and British Rig Owners Association (BROA) representing the drilling rig owners. During the course of the consultation over TSRs, the concept of a refuge was successively redefined by the substitution of the term 'temporary refuge' for temporary safe refuge. The latter definition stressed the functional rather than structural dimension of survivability (Pape, 1992). The HSE guide to the Safety Case regulations eventually described these requirements in the following terms:

> '...measures to protect the workforce should include arrangements for temporary refuge from fire, explosion and associated hazards during the period for which they may need to remain on an installation following an uncontrolled incident, and for enabling their evacuation, escape and rescue'. (HSE, 1992a: vii).

The industry argued that simplistic notions of a refuge in terms of a 'protected box' or physical entity, normally the accommodation block, needed to be replaced by a more 'sophisticated' concept of a processual 'flow'. Personnel could be removed from the hazard source in a series of protected access or escape routes to evacuation points. In redefining the notion of a refuge it was even suggested that in certain circumstances a lifeboat could fulfil the necessary function. In terms of safety engineering, it was argued, a broader view of what a refuge was had much to commend. It offered flexibility in adapting to different installation requirements. However, as with the concept of 'goal-setting' itself, such flexibility also created space within which the operators could freely redefine safety parameters and re-interpret regulatory objectives.

In its negotiations with the new regulator UKOOA effectively sought the re-create of a zone of compliance discretion. Recourse to the notion of flexibility became the leitmotiv of the industry's newly adopted strategy of containment. Evidence for a systematic misinterpretation of goal-setting, based on a desire to contain the impact of future regulations, can be found in a number of industry publications. We refer to a paper by Dr Harold Hughes to an HSE-sponsored conference on offshore safety. Here UKOOA's response to the draft Safety Case regulations was three-pronged. First UKOOA argued that there were a large number of mobile installations unlike Piper Alpha (a northern waters fixed installation). Implicitly UKOOA asserted that the draft regulations assumed there was a uniformity. Second, UKOOA criticised the requirement that each installation have a temporary safe refuge as being overly 'prescriptive'. Lastly, UKOOA opposed the proposal that the new regulatory authority, the Offshore Safety Division (OSD), would formally 'accept' the Safety Case of each installation; a requirement which went well beyond the existing onshore Control of Industrial Major Accident Hazard (CIMAH) regulations. Hughes pointed out that the cost and time-scale for the preparation of Safety Cases would be a considerable burden on the industry. The crux of UKOOA's objection on this point was described by Hughes as follows:

> "The concept of acceptance places a great deal of power and discretion in the hands of the HSE without any of the safeguards often brought into such legislation to cover instances where real issues of difference arise between the regulatory body and industry". (Hughes, 1992: 4).

Reluctant to accept the regulator's guidance in the creation of a new risk management regime, the second phase of UKOOA's containment strategy was marked by attempts to discredit the new regulator. This included purported 'concerns' that the Department of Employment, to which the HSE and OSD at that time ultimately reported, was 'non-technical' and, therefore, not well equipped to judge 'quite arcane technical issues', as well as repeated references to the industry's own superior expertise (Hughes, 1992: 5). The foundations of this purported expertise were rather less firm than the industry sought to claim and indeed actively disregarded and excluded alternative sources of expertise located in the offshore workforce itself.

RISK ASSESSMENT AS A NEW TYPE OF WORKFORCE EXCLUSION

Indeed, the new framework of risk assessment endorsed by Cullen itself was unconducive to workforce involvement. In his report Cullen endorsed the techniques of Quantitative Risk Assessment (QRA) which provide a more sophisticated cost-benefit analysis based on statistical probabilities. He saw this as a useful way of enabling the limits of what is 'reasonably practicable' in terms of risk management offshore to be accurately assessed (1990: Ch. 17.61). As a methodology, QRA itself had been 'a matter of some controversy'. Even the HSE at the Cullen inquiry was 'only cautiously enthusiastic' (1990: Ch. 17.53). Cullen felt, nevertheless, that QRA was an important tool which had an educative role for the operators, making them more rigorously define and monitor their procedures for risk control (1990: Ch. 17.49). Placing the final onus of risk management on the operator as duty holder, Robens emphasised that the specialised techniques of QRA were to be understood by the operators themselves, who were to apply them to issues such as the endurance capability of the TSRs and more generally, to the Safety Cases themselves.

The general thrust of the Cullen report was that acceptance standards for QRA should be proposed by the operator. To this purpose the operators were to apply the ALARP principle, which seeks to specify the boundary of 'tolerable risk' at a level that is 'as low as reasonably practicable'. ALARP involves a cost-benefit analysis compatible with a notion of reasonable practicability, albeit one that is statistically arrived at. However, what is considered by management as reasonably practicable and what workers may consider to be tolerable risk, on the basis of their on-the-job experience, may not necessarily coincide. The seemingly 'scientific' nature of the ALARP calculations and QRA has made these judgements difficult to challenge by the workforce. Today, judgements of acceptable risk hence are almost exclusively managerially determined; a fact that is obscured by the guise of probabilistic theory. With the adoption of QRA and ALARP, the power to decide what constitutes acceptable risk has shifted upwards, and in the final event rests with management. Moore, a critic of QRA, has commented:

There exists no available economic or statistical techniques which can readily provide 'quick fixes' as far as improvements in health and safety at work are concerned. Safety is about effective workplace risk control and public accountability, not pliable mathematical exercises in statistics or economics (Moore, 1991: 13).

What the QRA approach fails to recognise is that in developing a safe working environment it is not simply specialised quantitative 'expertise' exclusively concentrated in the hands (or heads) of management, but also what is described as 'low level safety intelligence' which counts (1991: 11). Hazards are often identified and controlled most effectively by those most immediately involved in the work-tasks, through a process of constant monitoring or 'risk valuation from below' (1991: 11). An example of such risk valuation from below is provided by an offshore incident on the Amoco Montrose installation. A pipefitter on the Amoco Montrose was given a 'hot-work permit' and instructed to cut into a length of pipe. The pipe contained potentially lethal explosive gases, but the worker did not proceed with the job using oxy-acetylene cutting gear and thereby forestalled what could have been a major accident. In this instance, on-the-job monitoring of safety saved the day. It did so only because the worker was prepared to exercise his initiative and adopt what he felt to be a safer work practice. The individual worker contested an already managerially approved task assignment on the basis of his 'own tacit knowledge'. That contest indicates how unequally valued are the relative risk valuations emanating from the bottom as opposed to the higher levels of the authority structure.

There is an additional dimension to risk valuation from below which is illustrated by the pipefitter incident. The individual concerned was an experienced craftsman with a knowledge of the job built up over many years.   The offshore employers, however, by recruiting large numbers of 'green labour', are seeking to create maximum labour flexibility through new forms of training so that workers would be 'multi-skilled' (OIAC, 1992).   This attack on craft boundaries enables operators to reduce the number of workers required.   But it also brings with it a reduction in the 'quality assurance' and therefore safe working practices, that are an in-built part of the tradesman's traditional training.   Multi-skilling is attractive to employers because it means that each worker can perform a wider range of functionally related tasks and thus progress the job more quickly.   The pressure not to question management orders is thus even greater, particularly where there is ambiguity about, or blurring of, task demarcations, especially amongst inexperienced employees who provide the 'flexible' workforce.   While the employer gets on with the pressing business of meeting client deadlines, workforce safety concerns too often are simply secondary considerations.   The pipefitter case was one that derived from the pre-Cullen context, that is, before the new goal-setting Safety Cases and attendant risk assessment procedures had been put in place in the industry.

Meanwhile, the newly applied QRA techniques have been found wanting in a number of contexts.   In the voluntary Safety Case (for Cormorant Alpha) submitted by Shell to the HSE, a risk measure called the Individual Risk Per Annum or IRPA was utilised.   Shell's own literature described IRPA:

The frequency per annum of potential loss of life of an individual from all work-related hazards taking into account the fact that he is not always on the installation or at work and will, if given sufficient warning, try to escape from the hazardous event (Shell Expro, 1993).

This measure combines the risk from travelling offshore, being offshore and working offshore with the probability of escaping from any incidents.   IRPA, says Shell, 'is one of the most important risk measures from a quantitative risk assessment' (Shell Expro, 1993).   Given the importance to Shell of IRPA measures, it could reasonably have been expected that great care would be taken in their calculation and use.   Similarly, it could be expected that HSE as regulator would subject these calculations to the closest scrutiny.

Yet this most critical measure revealed certain basic mistakes.   First, Shell calculated the IRPA using the time spent offshore of Shell staff, for example, at an average value of 1.5 x 10^{-3}.   This ignored the fact that about 80 per cent of those on board the installation are not Shell staff.   They are contractors' personnel who work a two weeks on, two weeks off shift pattern, making a working year of twenty-six weeks offshore, as opposed to twenty-one weeks for Shell staff.   The effect is to underestimate the 'average' IRPA by about 29 per cent for an installation with a 150 roster of 'Personnel on Board'.   Shell even presented the contract catering staff IRPA based on staff tours of duty.   Second, Shell noted that the risks from travel and actual employment on board were relatively small, at around 3 x 10^{-4} as an IRPA.   But these figures too will depend on risk exposure, and contractors' staff make about 25 per cent more flights and generally work longer hours in higher risk occupations.

With eleven offshore workers, mainly contractor employees on a helicopter flight to Shell's Cormorant Alpha only recently killed, the offshore union OILC commented:

"It appears to be astonishingly inept and insensitive to make such a mistake in respect of the very installation that was to sustain a tragic loss of life in March 1992 when contractors' men in a shuttle flight were exposed to a risk not borne by Shell staff". (OILC, 1993: 3).

This basic mistake, Shell would correct, but it is revealing for the glimpse of the mistaken assumptions that underlay seemingly rigorous risk calculations. It illustrates the corporate mental block which ignores workers who make up the majority of the offshore population. In the context of risk management hierarchical organisations tend to concentrate 'expertise' on the core of their organisation. As a consequence risk assessment and management has a tendency to become a part of managerial control. Moore's comments are pertinent to the offshore industry:

"For it is a fact that companies and managers remain reluctant to divest themselves of organisational or administrative power, leaving workers thereby poorly informed. The result of which is that, despite legislation — the power to decide or determine what constitutes a risk — and how risky a process might be — shifts upwards, always finally resting with management, technical experts and administrators. Therefore this acceptable risk approach is in truth about managers holding power and workers facing risks" (Moore, 1991: 11).

For 'risk valuation from below' to be effective, the workforce needs to be informed. Much more than this, they need to be involved in the process of risk assessment as fully legitimate active participants able to articulate their collective concerns without constraint.

GAUGING SUCCESS

In 1995 the HSE produced a series of commissioned reports which collectively were described as the Interim Evaluation of the Safety Case. This was the first attempt to look at the new goal-setting regime 'objectively'. The report relied heavily on QRA procedures, and indeed, tied much of its claim of improved safety to such assessment (1995a: 63). The industry itself, while endorsing the 'central role of QRA' in formalising and systematising company approaches to risk assessment and control, also expressed 'doubts' and 'scepticism about the extent of reliance on QRA and about the value of the effort required to carry out such studies' (1995a: 28). Hughes of UKOOA conceded:

"Without the solid foundation of a good data base, QRA can become a self-delusion, a mere refinement on non-knowledge". (Hughes 1994a: 9)

The report itself recognised the obstacles to workforce communications created by the new QRA regime. In this context the interim report noted the observation of many managers that 'the technical nature of QRA was seen as a potential obstacle to workforce involvement in the Safety Case process' (1995a: 28). A workforce survey conducted by Whyte and Tombs (1995) provides documentary evidence, admittedly from a small sample, of a range of attitudes indicating inadequate participation and dissatisfaction on the part of the workforce, with the nature of consultation over the Safety Case.

The newly adopted technocratic approach to safety management actively hence appears to have created a situation which runs counter to Lord Cullen's first goal of increasing workforce participation (Cullen, 1990: Vol. 2, Recommendation 27). This situation appears to have been worsened by informational deficits on the part of the workforce and safety representatives. The HSE interim report states that one fifth of safety representatives 'appear to be unaware that management has a statutory duty to consult' in the preparation of the Safety Case (1995a: 43). One third of safety representatives meanwhile 'felt inadequately informed about the Safety Case' and a 'slightly higher proportion believed that their constituents were not adequately informed' (1995a: 33).

In addition the Interim Evaluation of the Safety Case notes that there has confusion 'as to the respective role of companies and HSE in the Safety Case process, with almost one in four of the sample believing that Safety Cases were HSE instructions to companies to carry out safety improvements and one in five believing that the HSE was responsible for producing Safety Cases' (1995a: 43). In this context, the HSE's analysis reports that 'over a quarter of the sample said they required additional information: summary information and information tailored to particular worksites' (1995a: 45).

One part of the HSE report, a workforce survey noted that 'formal training specifically relating to the Safety Case regulation or process, is rare among the general workforce' (1995a: 47). 'Over 25 per cent of the workforce indicated that they wanted additional information on the Safety Case regulations, and in particular, how the regulations would affect their work task in a day-to-day sense.'

Taken together these indicators strongly suggest that persistent communication problems offshore have emerged from the way the new safety system was set up and is currently being implemented on a day-to-day basis. A significant portion of the workforce feel unable to use the Safety Case as a safety barometer when assessing the safety conditions of their actual workplace. Thus, even senior management admits to only mixed success in involving the workforce in the Safety Case (1995a: 46).

By the mid 1990s then, the implementation of the new risk assessment based Safety Case has led to the erection of significant barriers to auditing from below. Safety Case procedures place the major burden of responsibility on line management and do so in a highly technocratic way. Safety Cases, produced largely by consultants, are extremely technical documents involving a range of both engineering and risk analysis skills such as QRA. Only a very small segment of the workforce (including management) understand how and why they specify certain procedures — only that these *are* the procedures. This is compounded by the fact that compliance is assessed in a technocratic way. The HSE audits the processes set down in the Safety Case. This means that inspectors are no longer required to come offshore to investigate whether 'their' regulations are being observed.

Meanwhile the existing system of safety representatives, lacks a trade union based system of support. This means that direct workforce involvement would remain individualised, weak and ambivalent, often lacking the technical knowledge or support necessary to question management interpretation of Safety Case requirements.

Industry Resistance and 'Leadership' in the Third Phase

NEW PATTERNS OF RESISTANCE

UKOOA's policy has focused on limiting the impact Lord Cullen's recommendations during the post-Piper Alpha reconstruction of the offshore regulatory regime. Industry resistance has intensified as the third phase of post-Cullen safety legislation was launched. In the first phase the Offshore Safety Act had been passed. In the second, the Safety Case regulations were implemented. In the third phase, existing offshore health and safety legislation was to be revised to create a less prescriptive framework style which would complement the newly established Safety Case regime. This included the overhaul of offshore-specific regulations on evacuation, escape and rescue, fire and explosion protection, on design and construction, and on management and administration.

The new goal-setting approach to regulation was typified by the Offshore Installations (Prevention of Fire, Explosion, and Emergency Response [PFEER] Regulations and Approved Code of Practice (ACOP) (HSE, 1993). PFEER Regulations were implemented in the latest phase of regulatory reconstruction (HSE 1995b). They were intended to simplify, rationalise and replace existing prescriptive regulation governing the prevention of fire and explosions, and emergency response on offshore installations. Related to these regulations, Lord Cullen had emphasised the need for a degree of regulatory 'solidity' in the Safety Case goal-setting regime. Said Cullen:

> "The regime should not rely solely on the Safety Case ... the regulation requiring the Safety Case should be complemented by other regulations dealing with specific features.... These regulations would complement the Safety Case by setting intermediate goals and would give the regime a solidity which it might otherwise lack" (Cullen, 1990: 17.63).

The PFEER regulations played and important role as the kind of 'intermediate' regulations Lord Cullen had intended to complement the more general goal-setting regulations of the Safety Case. Indeed OSD officials asserted that 'PFEER does contain more detailed and more specific primary duties with which duty holders need to comply' (Patterson, 1995). In responding to such intermediate regulations, the operators established a false opposition between goal-setting and prescription. In a speech to the Offshore Engineering Society in Aberdeen, Harold Hughes argued that UKOOA 'wants to ensure that all elements of prescription are done away with'.

This view extended to the industry's position on the legal status of other regulatory instruments, such as Approved Codes of Practice (ACOPs). ACOPs prescribe the particular means of compliance with the legal requirements of the regulations which, if not followed, oblige the operator to demonstrate that an equally effective system has been devised. The normal burden of proof is then reversed and it is up to the 'duty holder' (the employer) to satisfy the default requirement in any legal proceedings. UKOOA suggested that ACOPs be reshaped into more general non-mandatory 'Guidance Notes'. This was based on the view that Approved Codes of Practice were 'too prescriptive', mainly because of their potential 'legal implications'. A UKOOA paper stated to this issue that 'by supporting goal-setting Regulations with ACOPs rather than non-mandatory Guidance, the flexibility provided by goal-setting Regulations is greatly reduced' (Hughes, 1994a: 49). This was the reiteration of a line of argument, seeking to minimise regulatory oversight, which the operators had put forward since the Burgoyne inquiry (Burgoyne, 1980: 134, para E). Industry authored guidance, rather than regulator-based codes of practice backed by law have increasingly become the norm in the new offshore regime.

Similar conflicts about the more or less binding nature of new safety requirements have arisen over the matter of standby vessels. Again, economic considerations have appeared to dictate UKOOA's policies. The statement of one serving standby vessel ship's officer is indicative of this approach:

> "It appears that the whole industry is trying to paper over the cracks with the cheapest possible options, and hope that another large-scale disaster does not happen". (Letter to *Telegraph, The Journal of NUMAST*, November 1995).

Lord Cullen had urged that the standard of the existing standby vessel fleet be improved 'with dispatch' (1990: 20.41). The code covering such vessels had first been proposed in 1974, its 'binding force ... based on a voluntary agreement whereby members of UKOOA undertook to abide by the standards set out in the code' (1990: 20.37). The Department of Transport felt that the voluntary agreement had been honoured. Lord Cullen, however, had expressed concern over the time it had taken to update the voluntary code for standby vessels. The third edition of the code had been proposed in 1986, but was not published in draft form until 1989. It had still not been agreed at the date of his report's completion at the end of 1990. In July 1991 it had finally come into effect. The history of the industry suggests that the voluntarism of its guidance standards offers, when convenient, almost limitless scope for delay and or course, strategic regulatory avoidance.

SETTING THE TONE OF THE DEBATE

The operators' strategy was aimed at limiting the new risk management requirements to a degree that made their implementation palatable for industry. However, it left industry with the image of a recalcitrant mover, who is only willing accept those elements of a new risk management regime which do not substantially infringe on its authority. This image was not without dangers. Regulatory initiatives, be they related to risk management, or to other areas, emerge not only from the sphere of domestic post-Piper Alpha policy making. Rather they evolve from a number of arenas, which increasingly include that of European policy making.

Whilst the Social Charter in itself creates no enforceable rights, the resulting Social Action Programme and the Directives on worker protection which flow from it have the potential to do so. Under the Single European Act (1986) which introduced Article 118A into the Treaty of Rome, the European Commission was given the authority to adopt health and safety Directives laying down 'minimum requirements' on the basis of simple majority voting rather than member state unanimity. The European Framework Directive (89/391/EEC) of 1989, which sought to encourage improvements in the safety and health of workers, and its 'daughter' Directives, therefore impact on British domestic legislation (DTI, 1993).

One European initiative which has directly impacted on the offshore operators is the European Working Time Directive, which the operators have found particularly objectionable. In a frank 1991 newspaper article, Chris Ryan, the external affairs director of UKOOA, spelled out the trepidation with which the oil operators' association viewed the 'increasing deluge of European legislation ... threatening to swamp UK industry, including firms involved in the North Sea offshore oil and gas industry' (*Aberdeen Press and Journal*, 25/2/91). According to industry sources, the proposed 'Directive Concerning Certain Aspects of the Organisation of Working Time', the so-called 'Working Time Directive' limiting the maximum working week to 48 hours, presented a major threat to its viability (Working Time Directive, 93/104/EC). The Directive was adopted on 23 November 1993, its provisions to be implemented by member states within three years. The UK secured an additional seven year opt out period during which employees can work longer than 48 hours a week on a voluntary basis.

In 1996 the then Conservative government, with the operators' support, sought to challenge the legal basis of this Directive in the European Court of Justice. This action was motivated by an alleged lack of evidence of a link between working hours and accidents.

The former government moreover argued that the proposal related to working conditions and employment matters and, therefore, required unanimous approval (Bercusson, 1994). The European Court decided against the UK government.

UKOOA launched a campaign to oppose limitations on offshore working hours when it appeared that the European Commission might attempt to reclassify oil and gas production as a 'continuous industrial process' to which the directive would apply. Application of the directive would have made the operators subject to the same regulations setting the maximum working hours per shift, or over a week, as governed onshore activities. This would have been particularly unwelcome, in the context of the cost reduction exercise the industry had launched in 1993 in collaboration with the DTI. This initiative, called Cost Reduction Initiative for the New Era (CRINE), had resulted for several major operators, in the alteration of shift patterns to three-week tours offshore, from the previous two-on, two-off pattern, thus increasing the number of actual hours worked (CRINE, 1993).

In mid-1994, the operators learned that the Exclusion which had been granted from the Working Time Directive for 'Other Work at Sea' was to be reviewed and that a special review group had been established. Renewed representations were made by UKOOA's Employment Practices Committee to the Commissioner for Social Affairs, Padraig Flynn. The industry argued not only that its safety record was improving, but also that it was safer than a number of hazardous onshore industries. Industry spokespersons argued that the termination of the offshore industry's exclusion from the Directive would result in more shifts offshore, requiring more changeovers and more helicopter flights per individual; a change which would eventually pose increasing risk to personnel. The chairman of Esso warned the UK Government that if the Directive was implemented 'energy-intensive industries ... will simply migrate to other, less foolish parts of the world' (*The Scotsman*, 21/2/1998). Thus an argument, essentially about risk exposure (the number of helicopter flights undertaken by personnel) was used in an attempt to seek to avoid progressive European safety legislation.

The are indications that the offshore industry has at least formally lost this battle, however without the dire consequences predicted by the operators. The likelihood at the time of writing is that the Working Time Directive will apply to 'excluded sectors' such as offshore oil. Despite this, the new rules are unlikely to impact in a major way on existing work rotation patterns, as the total time-reference period has been extended to one year for 'other work at sea'. This extension limits total working time over a year to 2,300 hours, which is above the normal working time of most offshore employees of 2,184 hours.

What the Working Time Directive controversy has shown to the industry, is that it is not immune from outside policy interventions as long as it can be considered a deficient in matters of employment quality. This has prompted industry officials to take the lead in new initiatives, which are aimed at portraying the offshore industry itself as a leader in the management of its work environment.

In September 1997 the industry launched a new initiative, 'Step Change', in tacit recognition of the fact that workforce involvement in safety at platform level has continuing weaknesses. This initiative brings together three major sectors of the industry, the operators as represented by UKOOA, the exploration side as represented by the North Sea Chapter of the IADC, and the contracting companies grouped together in the OCA. Step Change has three aims. First, to 'deliver a 50% improvement in the whole industry's safety performance over the next three years'. Second, through 'safety performance contracts' Step Change 'will demonstrate visibly our personal concern for safety as an equal to business performance'. Third, the industry will 'work together to improve sharing of safety information and good practice across the whole industry, through the act of involvement of employees, service companies, operator, trade unions, regulators and representative bodies'.

For the first time the trade unions are invited to play a direct role in the management of safety offshore. Potential union involvement is part of a 'change the culture' of the industry which aspires to 'empower' employees. It is accompanied by attempts to raise safety awareness through safety packs, common induction programmes, posters, the wearing of green hardhats for 'newstarts' and a safety video featuring a 'Terminator lookalike'. The efficacy of these interventions is questionable. As Nichols has argued earlier, safety campaigns which are contradicted by 'day-to-day experience at the point of production' are likely to make 'workers sceptical of the propaganda' (Nichols, 1997: 54).

Both managers and workers have voiced concerns over the goals advocated in the Step Change programme. One issue of concern is the attempt to deliver a '50% improvement' in the safety performance of the industry over the next three years. The manifesto of the Step Change initiative concedes that the industry's annual improvement in safety performance as measured by the all injury rate frequency, has slowed in the past two years' and that fatalities and injuries remain a matter of concern (UKOOA, IADC, OCA, 1997: 5). Meanwhile, the initiators of Step Change, have hit upon a much more ingenious way of achieving their 50% improvement target. The more 'traditional' statistics, now deemed to be 'lagging indicators', are to be supplemented 'with some more proactive measures of safety activity, often referred to as "leading indicators"'. These include visits by senior management to work sites 'to carry out safety audits and to lead workforce discussions' as well as safety training 'as measured by programme days and spend' are to be deemed leading indicators. With inverted logic Step Change proposes that 'measuring increased "input" activity will be just as important as measuring "output"' (UKOOA, IADC, OCA, 1997:6). Success is pre-programmed as the safety circle is squared.

That the Step Change programme faces growing expressions of workforce scepticism is predictable. We have collected a number of critical statements from which we cite brief excerpts. Here is the voice of one highly committed safety representative:

"You say you want workforce involvement in 'Step Change', but you have already set up the whole basis of the scheme. There is simply no mechanism for achieving involvement at a fundamental level. What you really want is workforce compliance,

compliance with decisions already arrived at. What you want are people who will nod in agreement with these edicts you hand down to them. You want Nodders". (Stephenson, 1998a: 8).

As with so many such top-down initiatives, exhortation by senior management has run into the sands of middle management obduracy rather than 'cascading' down through the hierarchy. Offshore most supervisors have retained an intractable focus on meeting production targets. This is documented in an a survey of Offshore Installation Managers (OIM)s conducted by Aberdeen University. The survey noted that installation managers 'rate corporate culture, senior management commitment, a participative style of leadership and effective communication as very important to the role of safety management. However many OIMs believe that their own organisations perform poorly on these issues'. This view of a poor performance with regard to participative leadership was combined with a 'high level of consensus about the need "to get back to basics" on safety'. At the root, the OIMs felt 'overburdened with new safety initiatives, procedures and legislation which they have real difficulties in translating into changes in working practices at the workforce level'. This does not bode well for the industry's aspirations for a proactive approach towards safety management. Indeed, the report comments on 'the need for improvements in corporate communications ... evidenced by the significant numbers of OIMs who had never heard of the Step Change initiative' (OIM Safety Survey, 1998).

The failure of Step Change to ignite the imagination of the offshore workforce is evidenced by the low level of participation in the initiative even by elected safety representatives (*Shell Platform Safety Minutes*, 19/4/1998). Recognition of this problem has led to the introduction of a 'workforce involvement trial' on the Brent Charlie installation (Shell Expro, Brent Charlie trial, 10/6/1998). In this trial, the dedicated presence of workforce representation is 'seen as an essential element in making the Step Change a success'. Now, members of the platform crew are being seconded onshore for six months to take part in Step Change team meetings. A number of additional cross-industry workshops have been targeted at approximately 200 workers. When the workforce on the Brent Charlie were invited to put their names forward as volunteers, platform management selected the lucky individual not from the ranks of elected safety representatives, but from six names drawn out of a hat. Conduct of this kind, of course, is only likely to further undermine the credibility of the programme.

The minutes of Shell's internal quarterly safety representatives meetings, further signifies the scepticism of the workforce towards the new initiative. We quote these minutes at some length because they reveal around very different concerns as compared to those imposed by management:

'The consensus was that a marked apathy has developed within the workforce with regard to 'Step Change'. The programme is generally viewed as, 'yet another management initiative on health and safety with no direct impact on the day-to-day life of the employee'.

A major part of the current level of 'apathy' and the general unsettled feelings ... derives from the insecurity and the conflicting messages which are being received. This in particular related to changes to contract without any formal notification being received before its implementation. Also assurances of no imminent redundancies immediately followed within a week by this being applied to two long-term ... employees.

A lack of feedback and information on actions or progress on the Step Change initiatives, within the stated policy of active employee involvement, made identification with the initiative difficult to maintain.

Indeed after the initial 'star-spangled' launch, the programme seems to have disappeared into limbo apart from the introduction of the 'Green Hat' campaign. (It was noted by the reps attending that this programme had already been in practice for some time).

It was agreed that the concern over conflicting information with regards to the workforce stability and its knock on effect in creating apathy with regard to Step Change be raised at the main meeting under 'Representative's Items'.

The representatives while identifying the level of concern also believed that the offshore personnel recognised the problems were having to be tackled in a period of decline of production on older units, this aggravated further by rising unit costs and falling returns. Improved safety was a goal they all agreed with and would like to achieve. It was however difficult to relate to this while worries over their personal future was so much in the forefront of their minds. Both the long experience and goodwill of employees were being regularly jeopardised. Until a constructive approach in terms of the honest and consistent handling of this valuable and irreplaceable asset replace(s) what was often perceived as an insensitive and shortsighted style of management (nothing will change). A 'Step Change' in the real and perceived value placed on the on the workplace also has to be nurtured if we wish to achieve the 50% improvement in performance'. (Quarterly Safety Representative Meeting, 7/5/1998).

Step change, immodestly, envisages a cultural 'revolution', akin to Mao's 'Great Leap Forward'. Indeed, Heinz Rothermund, managing director of Shell, actually uses such metaphors to describe the 'birth' of the programme:

"We knew that we were doing something revolutionary. The idea of setting a breakthrough target in health and safety as a means of stimulating a breakthrough response was new and contrasted with the continuous improvement targets we had always settled for". (Rothermund, Memo, 1/6/1998).

It is perhaps this very rhetoric which has discouraged safety representatives and workers from buying into the new programme. Yet as far as the regulator is concerned step change seems to have already produced some desirable effects for the industry.

The meeting at which the above discussion of Step Change among safety reps took place is described in the minutes as a 'pre-meeting'. The 'main meeting' which followed on later, included safety representatives, management and an HSE representative. The main meeting minutes do not record HSE reaction to the concerns of the workforce. The HSE official is reported as indicating that the 'Safety Case Guidance' is currently being reviewed 'with the prime objectives of simplification and introducing more of a "goal setting approach"' (Blowout, 1998b: 15-6). The HSE official further advised that he felt 'the current Safety Management System to be satisfactory and that, based on the assessed level of risk, he shall be reducing his scheduled offshore visits 'from six-monthly to an annual programme, with an "open agenda" format rather than a "specific inspection" itinerary'.

This and other evidence indicates that the industry's new policy of 'taking the lead' in safety initiatives is paying off. In the largely self constructed dialogue between regulator and the industry intentions and projections appear to bear greater weight than facts. In part, this may be a reflection of the new-found disciplinary prominence of risk management which attaches great importance to programmatic statements and initiatives, while paying far less attention to concrete measurable outcomes and even more so their real world determinants. It is on the level of outcomes, however, that we must centre the core of our critique of offshore risk management.

Summing Up-The Political Economy of Risk

Our previous analysis has described the offshore oil industry, as an industry sector in which the balance of power is shifted away from the workforce in favour of oil multinationals. This uneven balance of power has allowed the oil operators to manipulate the discourse on safety to their advantage and ultimately to prevent a significant improvement of working conditions offshore. Following the Piper Alpha disaster, the language of risk management which accompanied the Piper Alpha enquiry, has served to re-legitimise the systematic exclusion of the workforce from safety management and safety auditing. These processes have been accompanied by a continuous attempt by industry to downplay the real safety record of the industry.

During the past decade, the offshore operators have suggested that injury rates of those employed in oil and gas extraction are generally in the low range, falling, amongst others, below Postal Services and Telecommunications (rank order 10) and Construction (rank order 15). This view was presented in a paper by Hughes on behalf of UKOOA to the Offshore Northern Seas Conference 1994 (Hughes 1994b), which is similar in content to others presented to oil industry forums (Hughes and Taylor, 1995). According to Hughes, in 1992/93 'Extraction of Mineral Oil and Natural Gas' could be placed at a rank order of the 22nd most dangerous industry. By 1993/94, its position, using all-injury data, had further improved to 34th.   [See Table 1. UK Injury Rates by Industry (All Reported Injuries),1992/93]

This denial of the real risks of offshore work, has provided a poor basis for realistic approaches to offshore risk management. Rather, it has helped a distorted view of the relative dangers of offshore work to prevail amongst those for whom a realistic assessment would have been most important, namely senior and line managers. Despite the picture painted in industry publications, the reality of offshore risk exposure, was and is much more bleak. Once incident data are stripped of distortions arising from the inclusion of onshore employees and minor injuries which are chronically under-reported offshore, the offshore industry turns out to be the third most dangerous industry. Under realistic assumptions, taking into account fatalities and major injuries only, the extraction of oil and gas ranks only behind (1) Open Cast Coal Workings and Coal Mines and (2) Forestry. This rank order of third most dangerous industry moreover is stable for a number of years from 1990 onwards. [See Table 2. Ranking of Industries by Injury Rates using Combined Rates of Fatalities and Serious Injuries 1990/91, 1991/92, 1992/93 and 3-Year Average: Offshore Rates Disaggregated].

This picture of the offshore industry as one of the most dangerous industries is confirmed when the 'hardest' figure of all, fatalities alone, is examined. The fatality rate for the oil and gas extraction industry, averaged over the eight year period 1986/87 to 1993/94 has been reported by the HSC's publications (HSC,1994). It confirms a ranking of third equal most dangerous industry. [See Table 3. Average Fatality Rate, UK Offshore Sector, 1986/87 to 1993/94 (excluding Piper Alpha)].

TABLE 1. Ranking of industries by injury rates per 100,000 employees using combine Incident rates (fatalities and serious injuries) 1990/91, 1991/92, 1992/93and three-year average: offshore rates disaggregated

		1990/1	1991/2	1992/3	Average 1990/3
1.	Open cast coal mining	-	729.2	952.4	840.8
2.	Coal Mines	651.7	709.3	646.8	669.3
3.	**Offshore extraction of mineral oil and natural gas**	370.0	318.5	329.4	**339.3**
4.	Forestry	291.7	307.7	324.7	308.0
5.	Coke ovens	200.0	428.6	285.7	304.8
6.	Construction	289.3	247.3	247.3	261.3
7.	Railways	245.9	224.6	224.8	231.8
8.	Metal manufacturing	231.2	222.5	224.3	226.0
9.	Repair of consumer goods and products	206.7	193.2	221.4	207.1
10.	Food, drink, tobacco manufacturing	227.5	204.6	218.5	216.9
11.	Production of man-made fibres	269.8	112.9	218.2	200.3
12.	Supporting services to transport	160.0	172.6	193.2	175.3
13.	Manufacture of metal goods not elsewhere specified	208.1	221.1	191.1	206.8
14.	Manufacture of non-metallic mineral products	209.5	214.5	190.9	205.0
15.	Timber and wooden furniture industries	200.5	190.1	187.9	192.8
16.	Processing of rubber and plastics	182.4	178.0	171.5	177.3
17.	Non-energy mineral extraction	199.2	179.7	171.3	183.4
18.	Water supply industries	145.6	126.4	135.4	135.8
19.	Manufacture of other transport equipment	130.8	138.0	126.9	131.9
20.	Manufacture of motor vehicles and parts	142.4	114.7	125.7	127.6
21.	Production and distribution of energy	135.6	113.2	109.3	119.4
22.	Postal services and telecommunications	118.7	89.3	105.2	104.4

Source: HSC Annual Report, various years; HSE/OSD Offshore accident and incident statistics report, various years.

Whilst it is relatively straightforward to document the industry's misconception of the real risk exposure of offshore workers it is much harder to contextualise this exposure.

However, the long term safety record of the industry gives some indications as to the underlying causes of the industry's relatively poor safety performance. The combined number of fatalities and serious injuries rate has fluctuated substantially during the past three decades.

TABLE 2. Injury rates by industry (All reported Injuries, including fatalities. List in order of rate per 100,000 employees – 1992/93

1.	Coke ovens	5571.4
2.	Open cast coal workings	4142.9
3.	COAL MINES	**3957.8**
4.	RAILWAYS	**3151.6**
5.	Non-energy mineral extraction	2786.3
6.	Food, drink and tobacco manufacturing	2767.0
7.	Forestry	2026.9
8	METAL MANUFACTURING	**1993.6**
9.	Manufacture of non-metallic mineral products	1831.3
10.	Postal Services and telecommunications	1709.3
11.	WATER SUPPLY INDUSTRY	**1693.8**
12.	Processing of rubber and plastics	1655.8
13.	Repair of consumer goods and vehicles	1646.2
14.	Manufacture of metal goods not elsewhere specified	1611.8
15.	CONSTRUCTION	**1602.8**
16.	Manufacture of motor vehicles and parts	1538.3
17.	Production and distribution of energy (gas etc.)	1533.8
18.	Supporting services to transport	1526.0
19.	Production of man-made fibres	1500.0
20.	Timber and wooden furniture industries	1277.4
21.	Manufacture of other transport equipment	1263.4
22.	EXTRACTION OF MINERAL OILS AND NATUREAL GAS	**1196.0**

Source: HSC, Annual Report 1992/93

For the first six years from 1968 onwards, the number of combined incidents fell below 30. This was commensurate with small offshore employment figures of less than or about 1,000 employees. From 1973, the offshore oil industry experienced a steep incident rise, which paralleled the expansion of construction work on the British continental shelf.

The number of incidents during the construction phase peaked in 1976 with almost 80 combined incidents and declined thereafter up until 1984 (see Woolfson et al., 1997: 396, Figure 8.7a). This process paralleled a shift of the industry from exploration and construction which was accompanied by a phase of unparalleled prosperity of the industry.

TABLE 3. Average Fatality Rate 1986/87 to 1993/94 (*This excludes the 167 fatalities in the Piper Alpha disaster. The average with these deaths would have been 64 per 100,000 employees

	Average rate per 100,000 employees
Extraction of minerals/ores other than coal, oil and gas	25
Forestry	21
Extraction of mineral oil and natural gas	15*
Coal Extraction	15
Railways	10

Source: HSC Health and Safety Statistics 1994/95:8

Between 1980 and 1985 the oil the oil price remained US $ 25 per barrel, allowing the industry to reap revenues in excess of £ 15 billion in a number of years (see Woolfson et al., 1997: 406, Figure 8.11).

The rapid and massive fall in the oil price from 1985 onwards was accompanied by a massive increase of accident rates which included two years of during which there were in excess of 100 combined incidents (excluding Piper Alpha). Despite industry claims of improvements the number of combined incidents remained near or above 80 up until 1992 (with the exception of the 1987 reporting year). Increased oil revenues from 1992 onwards were accompanied by a renewed fall in the number of combined incidents. The ensuing brief period of improvements terminated in 1997, when a renewed increase in incidents occurred paralleling a renewed fall in the oil price and oil revenues.

The basic pattern of three cycles of increased vulnerability is confirmed by our depiction of combined incident rates from 1972 onwards. Incident rate computations face some limitations in the offshore context due to the absence of reliable workforce figures. Indeed, the HSE and industry calculate accident rates on the basis of workforce estimates derived from tax returns multiplied by a factor of 1.6. Our data was computed on the basis of estimates derived by various surveys conducted by the Scottish Office and Grampian Region. Yet our figures also must be taken as rough approximations, different reporting conventions affect workforce counts. Using this latter data we find three phases of upsurge of combined incident rates one occurring around 1973/74, the next in 1985/86, followed by smaller less secondary upsurge around 1990 and the most recent upsurge manifest in the 1997 increase in the combined incident rate.

Taken together these data indicate a link between the economics of oil production and incident rates. Cost-cutting drives leading to a significant deterioration of safety levels been identified as a recurrent cyclical phenomenon in the offshore oil industry. The economist Pike, for instance, has addressed the dramatic impact of the 1985/86 downturn on the level of activity on the structure of the oil industry. This economic downturn had dramatic effects on the phasing of offshore safety. After several years of postponing essential maintenance activities, a renewed investment cycle, resulted in an upturn in activity offshore. The combination of renewed maintenance activities with increased production provided the immediate context to the Piper Alpha disaster.

The wake of the disaster was characterised by demands for a completion of previously postponed maintenance and costly new investment in safety technology. These demands took place against the background of a stagnant sales take, and contributed to a profit squeeze. In the context of the employer dominated offshore industrial relations system, this profit squeeze translated into a wave of wage decreases (see Woolfson et al., 1997: 408, Figure 8.13). It is plausible that the periods of increased offshore incidents are directly linked to periods of an intensification of the labour process, which are marked by relative hourly wage declines and a concomitant increase in hours worked (ibid.: 409, Figure 8.14).

Comparing the years from 1979 to 1984 to the years from 1985 to 1990, we can identify a move from a low accident, relatively high wage regime to one of high accidents and low wages. This bifurcation of clusters corresponds to the discontinuity in combined incident data earlier observed in the form of a new higher incident plateau following the 1985/86 downturn. Further evidence for a link between economic pressures and an increase in incidents is found in the analysis of the years from 1990 onwards. While initially an increase in relative wages brings with it a 'backwards' shift towards the low incident level area, renewed pressures wages are accompanied by a second phase of increased incidents, of which data for the reporting year 1998 represents the most recent evidence. These developments are a reflection, both of the increased economic pressures the industry has experienced in the current low price regime as well as the industry's own attempts to renew and intensify previously initiated cost-cutting efforts. Most recently these have led to the most substantial down-manning of personnel in the industry since the oil price down-turn of 1985/86.

Risk in the offshore industry, when measured in terms of the real damage to the workforce, then, is very much a man-made or better, industry-made phenomena. It is a correlate of the economic status of the industry as much as it is a by-product of the industry's own perception of managerial necessities. These perceptions, despite the masquerade of safety, behaviour, risk management and communication speak, centre on the maintenance of profit margins. Risk prevention and especially the prevention of risk to employees play at best a side role in this theatre. When economic prosperity appears to permit it, fewer corners are cut, more resources are spent on safety and fewer incidents occur. During periods of economic downturn and profit pressures the opposite takes place, with labour intensification and limitations on safety-designated resources quickly leading to a notable deterioration of safety indicators. All of this, needless to say, is subject to the cover-up of either outright denial or documentable untruths, and more recently, the slick public relations speak of safety initiatives and their industry created indicators.

This, unfortunately is not a caricature of the industry. Offshore, the seeming sophistication of culture, behaviour or communication based discourses on safety and risk loses meaning in the face of the crude and unsophisticated reality of the political economy of oil; a political economy in which the balance of power is firmly shifted in favour of industry allowing it to consecutively de-prioritise or re-prioritise safety. Risk to employees does not, as safety culture advocates would make us believe originate from the absence of a safety-centred workplace culture, or as behaviouralists would argue, inadequate communication and procedural arrangements, rather it originates from the board-room in which profit margins are assessed and priorities determined.

This leaves with both an unconstructive puzzle and a constructive question. The puzzle is a follows: If, in an industry, risk is blatantly related to underlying economic forces, why does the masquerade of culture, behaviour, communication, or even neo-classical or *autopoiesis* risk maintain such vogue. The question is, if risk levels are determined by factors evolving in the broader political economy, such as the balance between capital and labour, the level of labour intensification and most importantly return on investment, what role does risk management have to play? As to the puzzle, there is no clear answer. What appears to be an unproductive discourse from a broad socio-political perspective, may provide mutual returns for the providers and consumer of such a discourse. The symbiotic relationship between safety consultants promising to provide safety improvements without reducing profit margins and without involving organised labour, and an industry unwilling to publicly acknowledge that there is antagonism between safety and profits or workforce disempowerment and safety, is only one variant of a broader consensus between industry, consultant-academics and the regulator. Yet, ultimately these relationships are too intricate and too long lived to establish even a clear causality between interests and output. Indeed, risk management itself, insofar as it offers the illusion of control, where evasion from responsibility is sought, may have become the ultimate 'moral hazard'.

The question as to the implications of this political economy view on the possibility of risk management is even more complex. We would argue that there is a possibility of risk management, but that this is conditional on the realisation of the real causal factors of risk creation and a concomitant shift in the forum from a primarily managerial discourse to a socio-political discourse focused on the role of regulation and the creation of individual and collective rights in the prevention of risk. Risk, we believe, must be understood above all as an industry generated phenomena whose extent is contingent on regulatory inputs, regulatory oversight and the balance of power between risk generating capital and risk bearing labour. As such, risk assessment and ultimately the assignment of 'tolerable' risk must be firmly relocated in the political arenas of state regulation on the macro-level and collective bargaining on the micro-level.

REFERENCES

ASTMS (1976) *Industrial Relations in the Oil Companies: an ASTMS View* (unpublished document).
Bercusson, B., (1994) *Working Time in Britain: towards a European Model, Part 1: The European Directive*, London: Institute of Employment Rights.
Blowout, (1998) 'Rough 47/3B Friday 13–Lucky for some?', Issue 54, April/May, pp. 15–16.
Brander, R., (1995) 'The Titanic Disaster: An Enduring Example of Money Management vs. Risk Management' <http://www.cuuug.ab.ca:8001/~branderr/risk_essay/titanic.html>
Burgoyne Report (1980) *Offshore Safety*, Cmnd 7866, London: HMSO.
Carson, W.G. (1985) 'Hostages to History: Some Aspects of the Occupational Health and Safety Debate in Historical Perspective,' in W.B.
Creighton and N. Cunningham (eds) The Industrial Relations of Health and Safety, Sydney: Croom Helm.
Carson, W.G., (1982) *The Other Price of Britain's Oil*, Oxford: Martin Robinson.
CRINE Secretariat (1993) *Cost Reduction Initiative for the New Era Report*, St Paul's Press, November.
Cullen, The Hon. Lord, (1990) *The Public Inquiry into the Piper Alpha Disaster*, Vols 1 and 2, London: HMSO: Ch.21.84.
Dalton, A., (1988), Labour Research
DTI (Department of Trade and Industry) (1993) *Review of the Implementation and Enforcement of EC Law in the UK*, London: HMSO.
Esso Chairman in *The Scotsman* (1998) 21 February.
HSC (1994) *Annual Report 1993/94*, London: HMSO.
HSE (1992) *Draft Offshore Installation (Safety Case) Regulations*, London; HMSO.
HSE (1993) *Draft Offshore Installations (Prevention of Fire and Explosion, and Emergency Response) Regulations and Approved Code of Practice, Consultative Document*, London: HSE.
HSE (1995a) *An Interim Evaluation of the Offshore Installation (Safety Case) Regulations 1992*, London: HMSO.
HSE (1995b) *Prevention of Fire and Explosion, and Emergency Response on Offshore Installations, Offshore Installations (Prevention of Fire and Explosion, and Emergency Response) Regulations 1995 and Appended Code of Practice*, London: HMSO.
Hughes, H., (1992), 'Towards a goal-setting regime–plans and issues–the operators' view' in HSE (1992b) *Health and Safety in the Offshore Oil and Gas Industries*, Proceedings of Conference, 6–7 April, Aberdeen 1992.
Hughes, H., (1994a) 'The operators' response' Paper to conference on *Offshore Safety Case Management*, reprinted in *Safety Management*, Vol 10, No 6 , p.49.
Hughes, H., (1994b) 'Developments of Safety, Health and Environmental Standards and Regulation Systems in North-West Europe', Paper to Offshore Northern Seas Conference, Stavanger, Norway.
Hughes, H., and B. Taylor (1995) 'Offshore Safety–an Update of Progress', Leith International Conference, Aberdeen, 24 October.
Letter from the Employee Relations Managers, Mobil Northsea Ltd to Campbell Reid, Inter Union Offshore Oil Committee, 26/7/82
Letter to Telegraph, The Journal of NUMAST (1995) November.
Minutes Quarterly Safety Representatives HSE Meeting (1998), Ardoe House Hotel, 7 May.
Moore, R. (1991) The Price of Safety: The Market, Worker's Right and the Law, London: Institute of Employment Rights.
Nichols, T., (1997) *The Sociology of Industrial Injury*, Mansell: London.
OIAC (1992) Guidance on Multi-skilling in the Petroleum Industry, London: HMSO.
OILC (1993) 'Comments on the Draft Offshore Installations (PFEER) Regulations and Code of Practice', submitted to HSE/OSD, 3 December.
OIM Safety Survey (1998) University of Aberdeen, May.
Pape, R., (1992) 'Risk assessment in UK Offshore Installation Safety Cases',
Paper to *Risk Assessment: International Conference*, 1992.

Paterson. J., and Teubner, G., (1998) 'Changing Maps: Empirical Legal Autopoiesis', forthcoming in Social & Legal Studies.

Patterson, Robert (1995) 'Offshore Safety–the New Regime: Part 2 Implementation', *Leith International Conference*, Aberdeen, 24 October.

Robens, The Hon. Lord (1972) *Safety and Health at Work,* Report of the Committee 1970–72, Cmnd 5034, London: HMSO.

Ryan C. in Aberdeen *Press and Journal* (1991), 25 February.

Shell Expro (1993) 'ALARP in practice: an Industry View', paper presented to the Offshore Safety Case Conference, 1-2 April 1993.

Shell Expro (1998*) Step Change–Workforce Involvement, Brent Charlie Trial*, 10th June.

Shell Platform Safety Minutes North Cormorant (1998) 19 April.

Speech given by Stephenson D. at Shell Expro HSE Conference, 20th March 1998, reprinted in *Blowout*, Issue 54, April/May 1998.

Spiller, P., (1990) 'Politicians, Interest Groups and Regulators: A Multiple-Principles Agency Theory, or "Let them be Bribed"', *Law and Economics,* Vol 33, pp. 65-101.

Taylor, B., (1991) 'The UK Offshore Operators Response to Piper Alpha and Lord Cullen's Report', Paper to Society of Petroleum Engineers, pp. 349–356 in *First International Conference on Health, Safety and Environment*, The Hague, 10–14 November.

Thom, A., (1989) *Managing labour under extreme risk: collective bargaining in the North Sea oil industry*, unpublished PhD thesis RGIT, Aberdeen.

UKOOA (1990) Internal Briefing Documents on the Release of the Cullen Report.

UKOOA, IADC, OCA (1997) *A Step Change in Safety: A Cross-Industry Commitment to Improve Our Safety* Performance, September.

Whyte, D., and Tombs, S., (1995) 'Offshore Safety Management in the "New Era": Perceptions and Experiences of Workers', pp. 35-53 in ICHEME Symposium Series, No 138.

Woolfson, C., Foster, J. and Beck M. (1997) *Paying for the Piper: Capital and Labour in Britain's Offshore Oil Industry,* London: Mansell.

WTD (1998) *Working Time Directive* (93/104/EC).

Zeckhauser,R., and Viscusi, W. K., (1996) 'The Risk Management Dilemma', Annals of the American Academy of Political & Social Science, Vol 54, pp. 144-56.

LEARNING THE LESSONS OF PIPER ALPHA?

Offshore workers' perceptions of changing levels of risk

DAVID WHYTE,
Department of Sociology
Manchester Metropolitan University
Geoffrey Manton Building
Rosamund Street West
Manchester, M15 6LL

Introduction

Since the death of 167 offshore workers on the Piper Alpha installation in July 1988, and the subsequent establishment of a new regulatory regime in the offshore oil and gas industry, much has been made of a renewed commitment to safety shown by the operating oil companies and service companies in the industry. The operating companies cite the o5 billion targeted for improvements in safety since Piper Alpha [for example, Brandie, 1994/5], and the apparent success of the post Piper Alpha safety committee system [for example, BBC TV, 1996] as proof that a former complacency towards standards of safety offshore has changed. In addition, the United Kingdom Offshore Operators Association (UKOOA, the oil companies' collective organisation) have begun regularly to refer to official Health and Safety Executive (HSE) data, which appear to show a gradual decline in reported injury rates [for example: UKOOA, 1994b and BBC Radio Scotland, 1996].

However, amidst the corporate, governmental and academic interest that has been generated around the post-Piper safety regime, relatively little attention has been paid to the opinions and experiences of workers [Whyte, Tombs and Smith, 1996]. This chapter seeks to redress that omission. It uses qualitative data from interviews with offshore workers to challenge the predominant view of safety improvements within the industry [1].

The chapter begins by briefly tracing some of the economic and cultural developments in the industry during the 70s and 80s, and highlights some elements of the circumstances which led to the Piper Alpha disaster and the subsequent Cullen report, before outlining the new economic challenges that face the industry. It then turns to examine workers' recent experiences of offshore safety, focusing in particular upon the perceived effectiveness of new safety committees regulations, and the involvement of the workforce in the preparation of safety cases. These experiences are further detailed in relation to the nature and predominance of pressures to maximise production offshore and

E. Coles, D. Smith and S. Tombs (eds.), Risk Management and Society, 263–283.
© 2000 *Kluwer Academic Publishers. Printed in the Netherlands.*

in the context of the cost-cutting project led by operating companies in the sector. In conclusion, there is a brief assessment of the prospects for offshore safety.

The Context for the New Offshore Safety Regime

Commentators are in broad agreement that domestic policy in the early years of development of the sector was largely driven by government desperation to deal quickly with the balance of payments crisis of the mid 70s [see, for example: McKay and McKay, 1975, Corti and Frazer, 1983 and Noreng, 1985].

One effect of this approach was the relative downgrading of the need to establish a comprehensive and independent system of safety regulation. In addition, as the industry was allowed to evolve relatively free from regulatory interference, an intimidatory and authoritarian style of management developed on the platforms [Carson, 1982].

Traditionally, the bulk of the workforce in the industry has been sub-contracted, with the percentage of contract workers (as opposed to those directly employed by the operating oil companies) varying between 75 and 85 per cent. The 'advantages' of such a labour force are clear. The operating companies "bear none of the hiring/firing costs of the peripheral workforce and so they can adjust supply to demand without such costs which are passed on to the 'satellite' oil-related companies" [Gasteen and Sewell, 1995: 244]. This section of the workforce has been largely non-unionised, and, where pockets of a unionised workforce have existed, they had either no bargaining rights, or the recognition agreements that were achieved weren't worth the paper they were written on [OILC, 1991].

Thus, the industry emerged from the early days of development with an unorganised workforce which, in the main, had no security of tenure and, due to this status, had few rights of employment. In this environment, there was little resistance to a dominant management. Where resistance did occur, managers had the option of disposing of workers at will. The infamous NRB system [2] - an industry-wide means of dispensing with "troublesome" workers - was used against those who dared to challenge the supremacy of management decisions, on many occasions over the issue of safe working practices.

The Piper Alpha Disaster and its Aftermath

These long-term features of offshore production are crucial to the understanding of the underlying causes of the Piper Alpha disaster. On the night of the 6th July 1988, a series of explosions tore through the platform, and in the resulting inferno 167 died, 61 survived and twenty bodies were never recovered. One example of management reluctance to communicate on safety matters on the Piper Alpha is highlighted by the documented reports of gas leaks in the weeks and months leading up to the disaster that were not acted upon [Tombs, 1991].

A more immediate set of explanatory factors can also be found in the effects on the industry of the collapse of the OPEC cartel quota system in the mid-eighties. This saw the price of oil per barrel plummet from $30 in November 1985 to just $10 in April 1986, and was followed by a general reduction in oil company budgets by between 30 and 40 per cent across the board. At the same time, pressures from the boardrooms of the oil majors to increase production levels intensified [Harvie, 1994].

The Piper Alpha installation was not a rogue nor exceptional operation. Indeed, it was subject to the same pressures from senior management as almost all other installations in the UK sector. Thus, for example, the maintenance programme for Piper Alpha was cut in successive years after the oil price crash of 1985 [Channel 4: 1994].

The Piper Alpha disaster can only be understood in terms of a combination of related factors which militated against an effective safety regime: an economic context which created the twin pressures of cost reduction and increased production; and a relatively powerless workforce that was marginalised from the process of safety management [Tombs, 1991], and was under pressure to adopt "work practices where safety considerations have a relatively low priority" [Lavalette and Wright, 1991: 65].

Within 3 months of the disaster, Energy Secretary, Cecil Parkinson, under intense pressure, announced that a statutory system of safety committees was to be established offshore. He did not, however, see fit to grant the trade unions the same role in the committees as in onshore workplaces [Guardian, 8 October 1988].

The Cullen Inquiry

In the meantime, Lord Cullen had been appointed to oversee a Public Inquiry. His conclusion was a damning indictment of Piper Alpha's operator Occidental, which he stated was guilty of a "string" of errors and lapses, amounting to a "superficial attitude to likely risks" and "gross negligence" [Cullen, 1990].

Cullen was equally scathing as regards the inadequacy of the regulatory regime. He pointed out that the Department of Energy were in no position to effectively regulate safety at the time of the disaster, and recommended that responsibility for regulating safety in the industry should transfer from the Department of Energy to a "discreet division" of the Health and Safety Executive [Ibid.].

In addition, he proposed that the application and assessment of safety standards was to be driven by the principle of self regulation. The new regulatory system was to be based around 'goal-setting' as opposed to 'prescriptive' regulations. Thus,
"...principle regulations in regard to offshore safety should take the form of requiring that stated objectives are to be met....rather than prescribing that detailed measures are to be taken." [Ibid.]

The new goal-setting regime was to be achieved largely through the use of the 'safety case' approach. Cullen modeled his system of safety cases on the existing system for regulating hazardous industries onshore. He recommended that, as in the onshore industries, each installation should prepare a safety case document which should demonstrate that the operator or the owner of the plant has identified the major accident hazards and that measures have been taken to reduce these risks to 'as low as is reasonably practicable.' This system was to be monitored by the newly established Offshore Safety Division of the Health and Safety Executive (HSE).

The Cullen report was clear in its view that workforce involvement in safety was crucial, and asserted that
"It is essential that the whole workforce is committed to and involved in safety operations" [ibid.: 276-277; see also 281-289].

It was also intended that the principle of workforce involvement should be a key factor in the compilation of safety cases. The Offshore Installations (Safety Case) Regulations 1992 directed that safety representatives should be consulted on the preparation of each installation's safety case.

Cullen recommended that safety representatives should be protected from victimisation by management and ensured that section 58 of the Employment Protection Act was extended offshore. He did not, however, specify a formal trade union role in safety committees [ibid].

While it has only been possible to indicate in outline here, the Piper Alpha disaster and the Cullen Report might have proven a watershed for offshore safety. It certainly did act as a moment of exposure for the industry [Woolfson, 1995a], highlighting the supremacy of management decisions and the securing of profits taking priority over the safety of the workforce.

The Emergence of CRINE

More recently, a significant development in the industry has been the emergence of the CRINE (Cost Reduction Initiative for the New Era) initiative as a joint enterprise between government and the oil industry, through the operators' organisation UKOOA. CRINE has been promoted as an strategy for survival which was necessary in order to prevent the industry from going out of business [CRINE: Cost Reduction in the New Era, founding statement]. Indeed, the formal establishment of CRINE towards the end of 1992 was preceded by a series of public statements from senior managers in the operating companies and government ministers warning of the need for the reduced costs to stop the imminent decline of the North Sea [for example: The Independent, 23 October 1992 and Lloyds List, 4 November 1992].

Thus, CRINE aims to establish 'new partnerships' in the industry and closer relationships between oil companies and their suppliers. The aim of this strategy is to achieve a 30% reduction in capital costs and a 50% saving in operational costs within 2 to 3 years [Risley, 1995].

The government have worked closely with UKOOA to construct a deregulation agenda for the industry [UKOOA, 1994a]. The CRINE initiative has been enthusiastically encouraged by President of the Board of Trade, Michael Heseltine, and a number of government ministers, as another supporter of and contributor to the deregulation agenda [for example: Department of Trade and Industry (DTI) press notices, 2 March 1993 and 24 November 1994].

Industry and Energy minister, Tim Eggar, has been particularly keen to support CRINE. He has enjoyed a close personal association with the oil industry for many years. In the past he has worked as a banker specialising in financing the oil industry, and served as a non-executive director of Charterhouse Petroleum. As a backbencher, he was involved in a campaign to restructure offshore taxation in the early eighties [Lloyds List 15 July 1992]. Eggar promotes CRINE as the common ground where the interests of government and industry meet. In a speech at the inaugural CRINE conference he pledged a new partnership between the DTI and the oil companies: "Partnership means making your concerns our concerns. It means the DTI searching relentlessly to find ways to help the environment in which you [oil companies] operate" [Eggar, 1993: 5], and that "I.want to confirm that the DTI will be a full partner in pursuing the changes needed." [Ibid.: 1]

Thus, the DTI has been an equal partner with the industry in the CRINE project. Conferences are jointly organised by UKOOA and the DTI, and in 1995, the Department donated o100,000 for the work of the CRINE Office to match the operating companies funding [European Offshore Petroleum Newsletter, 30 November 1995].

Cost-cutting is now viewed as part and parcel of life in the industry. The inherent dangers of implementing expenditure cuts on this scale have been swept away by the doctrine of CRINE as a necessary measure which has saved the viability of North Sea operations.

The establishment of CRINE and related cuts in operational costs have presented the industry with economic challenges which are, in magnitude, broadly comparable to the pre-Piper economic squeeze. These circumstances, alongside the demands of the new offshore safety regime, have also created new challenges for the effective management of offshore safety.

In the remaining sections of this chapter, the extent to which these challenges have been met will be assessed. The various dimensions of offshore safety in the post-Cullen era are discussed, through the voices of those least often heard in such assessments, namely offshore workers themselves.

Workforce Experience of the Safety Representative System

In this section, the views of offshore workers are used to assess the efficacy of the safety representatives system. General perceptions of the system amongst the workforce are examined, as are the experiences of those who have served as safety representatives. The section then discusses some problems associated with the organisational support available to safety representatives.

Although most agreed that the introduction of safety representative and committee regulations represented a distinct improvement on previous informal and non-standardised systems, the majority of respondents also believed that the current system of safety committees is unable to adequately deal with workforce input to safety.

Those who were uncritical of the safety representative system were, without exception, working at supervisor level or above. The tendency amongst management was to point out that the system is working perfectly well:

"With most safety fears, it's their own fault. It's stupidity on their part. They've all got a constituency safety rep and I bet some of them don't even know who their safety rep is. The OIM's (Offshore Installation Manager) door is always open." [W32]

The general view is, however, that serious problems exist which prevent the system working effectively.

Two broad profiles of safety representatives were constructed by respondents. Firstly, there were those who were known as "compliant" or "tame" safety representatives. They would largely follow the management agenda, raising the odd minor matter of concern that entailed no threat to the rate of production. It was reported that often the members of this group were attracted to the position of safety representative because of the extra payment they were given, or because they saw it as a good career move to secure the stability of their job. Within this group there were also those who had realised, after they had been elected, that they could not operate as strong safety representatives for fear of dismissal. Secondly, respondents pointed out that there were some "strong" safety representatives who were willing to raise workers' concerns freely and openly. The main difficulty encountered by this group was that they risked their position of employment because of this openness. The general view was that this type of safety representative is much less prevalent in the industry than the first group.

Those respondents who had served as safety representatives were amongst the most critical regarding the efficacy of the system. Respondents who had experience either of serving as a safety representative themselves or of using their safety representative to raise concerns, pointed out the obstacles that often existed to prevent safety representatives acting effectively.

The perception amongst those who had experience as a safety representative was that due to most monthly safety committee meetings being chaired by the OIM, the agenda could be manipulated by management. It was reported that there were certain issues that could not be raised by representatives in safety committee meetings for fear of reprisal. Those issues would tend to be the more significant problems which may have an impact on the process of production. One safety representative said: "Safety meetings are not safety meetings, they're safety lectures." [W8]. Another reported that the OIM on his platform would bar certain topics from being raised, such as helicopter safety: "There's things your [safety] rep. can't say at a safety meeting.........They'll ask you if you've got points to raise, but you only raise some things. You can't raise political hotspots" [W12]. Often questions from safety representatives were ignored during the meetings, and on one platform, the OIM would wait until the end of the meeting before allowing safety representatives to raise points. This part of the meeting would not be minuted.

In some cases, it was reported that the safety representative was under pressure not to report a problem because the supervisor had been alerted, but not acted. Thus, the safety representative was sometimes under pressure to cover for them: "The supervisors are in the same boat as you are. They don't want to upset the applecart either." [W33]

Four respondents who were known to management as trade unionists reported that they had been deliberately prevented from standing for election to the position of safety representative because it was thought that they would cause management too many problems. One was told that he was about to be moved to another platform, and that there was little point in him standing. Two were told to change crew and found that the new constituency already had an elected a safety representative in post, which effectively removed them from their position. Another was told by the OIM that he was waiting for nomination forms to be sent out from headquarters. As he waited for his nomination form to arrive, an election was organised and an alternative candidate elected.

One respondent reported that he had been NRBd for raising a safety related issue as a safety representative, and 4 others knew of safety representatives who had been disposed of in similar incidents. Although Cullen recommended in the strongest terms that safety representatives be protected from management intimidation, as one respondent pointed out, the form of protection enshrined in the Cullen report only comes into effect after the event, and is, therefore, no compensation for the lack of trade union support for safety representatives. One worker summed up the view of the majority: "The safety rep system can't work in the North Sea because of the age old problems with the system of hire and fire." [W35]

Lack of Trade Union Support

The difficulty of operating effectively without any organisational back-up was stressed by a number of respondents. The continued absence of collective bargaining agreements with trade unions offshore and general hostility shown by operators to trade unions [Foster and Woolfson, 1992] means that even if safety representatives are organised collectively in a trade union, it is difficult to operate openly as a trade unionist. Although opinion was split, a large majority of workers agreed that trade union support for safety representatives would increase their ability to carry out the role. Most of those who disagreed with this view were in lower management positions. Trade union support was seen as important both because it gave the safety representative strength in numbers, and also because of the advice, training and information services that a trade union can offer. One worker who had not been a safety representative recalled that "the only good safety reps I remember are the one's that's in the union." [W9]

Respondents reported that, in general, both operators and contracting companies are still opposed to the notion of trade union representation of the workforce. One described how: "You can walk in for a job and the first question is, are you a member of a trade union? Then they ask you to sign a form that says you're not in a union." [W9]

Although some of those interviewed said that they were trade union members and that this did not cause any problems, those respondents also said that they would not announce the fact in front of management.

Almost half of those interviewed talked about the existence of a 'blacklist' of known trade unionists and 'troublemakers' that was still used to prevent certain individuals gaining employment offshore. A large majority of respondents were able to point to specific cases of people who had been barred from the industry because of previous trade union activity.

There are a few trade union recognition agreements scattered around the industry, but in the main they only protect a small minority of the workforce and do not cover health and safety matters [Foster and Woolfson, 1992]. Recent ballots held by the OILC union on the Brent Charlie and Delta platforms resulted in the workforce voicing its overwhelming support for recognition [Blowout, March 1996]. This request has been rejected by senior managers in Wood Group Engineering, the largest employer the Brent field. Their reaction is an indication of the continued hostility to trade union organisation which exists offshore.

HSE Study of Safety Representatives

A group of researchers at Aberdeen University prepared an assessment of the offshore safety committees system for the HSE in 1993. [Spaven et. al., 1993] The data was taken from a sample of 686 safety representatives in the UK offshore sector and a series of structured interviews and case studies with managers and workers in operating an contracting companies. Although the report appears to be broadly positive about the system, careful reading of it reveals that their findings identify problems with the safety representatives system that have much in common with the evidence presented above.

On the trade union question, the Spaven et. al. conclude that "a majority of this sample of the workforce has a positive attitude towards some form of increased trade union involvement in the safety representatives system." [Spaven et. al., 1993: 119]

They also uncovered a good deal of evidence for the continued existence of a 'climate of fear' offshore, finding that 47.7% of safety representatives "had feared some sort of victimisation". In addition, a clear majority of 55.8% believed that safety representatives are subject to victimisation by management. A further 16.6% admitted to avoiding raising an issue because of fear that they might be victimised [Ibid.: 111].

Spaven et. al. also provide some detailed evidence of victimisation. One case concerned the sacking of a group of safety representatives the company had regarded as "a bunch of troublemakers" [Ibid.: 113]. On another platform, one operating company safety officer reported the practice of dismissing undesirable safety representatives by moving them to a platform "where the contractor's managers knew that downmanning was about to occur."[Ibid.: 115]

Thus, there are undoubtedly parallels to be drawn between the findings of the Aberdeen University group and the data presented here. This combined evidence leaves little doubt that offshore safety representatives on many platforms risk intimidation and bullying, and ultimately the loss of their job, if they are high profile and willing to confront management.

The current situation has severely affected the faith that the workforce have in the safety committee system. Approximately half of those in this survey reported that they did not bother to raise safety concerns with their safety representative because "nothing will get done" [W11] or because "nobody listens to what they say". [W21]

The problems encountered by safety representatives and their constituents raise serious issues for the efficacy of the safety committee system, and for the management of safety generally. Even if what is presented here is, at best, only a perception, and not an accurate reflection of the current situation, this perception is actively preventing workers from getting involved in safety decisions.

The widespread fear of victimisation indicates that safety representatives do not feel adequately protected by the extension of section 58 of the Employment Protection Act to the offshore sector. One form of protection available in other industries, that of an organised trade union is resolutely denied to offshore safety representatives. Given the wealth of evidence that links improved health and safety performance with an active trade union presence in the workplace [for example: James, 1993 and Walters, 1995] , it is puzzling that the operators and contracting companies have retained their intransigence on this issue.

The Safety Case Consultation Exercise

The central role for workers urged by Cullen in the compilation of safety cases, and the significance placed upon the consultation exercise by the HSE and UKOOA [Petroleum Review, December 1990], makes this exercise of marked importance for the success of the regulatory regime. The following section examines the degree to which the expertise of the offshore workforce has been utilised in the compilation of safety cases generally, and also uses safety representatives experiences of the safety case consultation process in order to bring the issue of workforce involvement into sharper focus.

There were two reported ways in which workers were allowed an input into safety case documents. These were: by the operator calling a safety case meeting and inviting selected workers or selected safety representatives to participate as representatives of the workforce, or by the operator selecting individuals to contribute because of their expertise in a particular area.

Less than a fifth of respondents who had been working offshore at the time of the safety case exercise had been given the opportunity to get involved in the preparation of safety cases. Some of those interviewed reported that their safety representatives and other work mates had been given the opportunity to contribute to or comment on the safety case, although others had never heard of safety cases.

Two respondents had been asked to participate and spend a number of days onshore because of their technical expertise. Both of those respondents were offshore supervisors. One supervisor said that the company he worked for had hand-picked a few 'specialists': "They selected a few people that they knew would give them what they wanted." [W35]

The majority of respondents reported that they had not even been invited to make a contribution to the safety case, even in this manner, although most could recall receiving some type of brief presentation informing them of the content of the safety case. Three had participated in short sessions of half a day or a few hours as safety representatives, and three workers were invited to a mass meeting with onshore managers where 'suggestions' were called for.

Where some form of contribution from workers to the preparation of safety cases was invited, suggestions were usually rejected by management as inappropriate or obstructive. Generally, respondents regarded these meeting as being a "paper exercise", in that the meetings did not allow them a forum for raising matters in relation to the safety case. One respondent recalled that at one of these meetings, a company executive held for an hour long meeting in the cinema with any workers on the platform that wanted to attend and discuss the safety case. The first half hour was taken up with the company executive explaining the process of preparing a safety case. The second half hour was allocated to questions and points from the floor. None of those present had been given a copy of the safety case or had received any official communication about the safety case until that day. Perhaps unsurprisingly, no questions were asked and no points were made.

In some cases, management were accused of selecting 'compliant' safety representatives for safety case meetings: "On the **** **** [installation], they consulted two safety reps, but the reps didn't take up any issues or add anything to the safety case, because they were compliant safety reps." [W3].

In other cases, management were perceived to have conducted a "fake" consultation exercise simply to "look good" and had not seriously considered comments or questions from workers. One respondent recalled how during one meeting, the majority of workers had been concerned about the location of life boats, and wanted them resited. They were told that this was not possible because the safety case had been finalised and was at the stage of being approved.

A particular problem was experienced by safety representatives when they were invited to a safety case meeting. It was reported that the usual format for such a meeting was to gather together all of the safety representatives on a platform and supply each with a copy of the safety case. They would then be given a period of time (usually between an hour and half a day) to read selected sections of the safety case, and asked for comments at the end of this period of time. It was made clear that this would be their only chance to have an input to the safety case.

Several workers pointed out, however, that they would have been unable to properly to contribute to the safety case had they been invited to, recognising that a quality submission for a safety case can only be made by workers if they are well organised. A number of respondents said that only the trade unions could facilitate proper worker involvement the preparation of safety cases:

> "Only an organisation can provide a professional input" [W 5]. "I'm all for trade unions because I realise how bad it is out there, and because it's probably the only answer ... I can see that the unions getting onto the platforms is the only way for getting any input" [W 17].

Respondents were keen to stress that "the real experts", shop-floor workers, were the best people to consult:

> "The companies don't accept that the guy doing the job is probably the best qualified safety officer they could have. He will be very attentive to the safety case because it's his fucking life" [W4].

Although usually made somewhat differently, such a point is commonplace amongst academics and the HSC/E [Foster and Woolfson, 1992, HSC, 1994, Tombs, 1991].

There is a clear view amongst offshore workers, then, that their experience and expertise is not being utilised by the operating companies as it could be. As one worker pointed out: "The attitude out there is, you're a low life. You don't have the experience to know the risks, so they don't tell you anything. You are not here to think." [W33]

This type of consultation exercise has understandably generated scepticism of the authenticity of operators' commitment to workforce involvement in safety. Workers experience of this process reflects the earlier issues discussed in relation to the effectiveness of the safety committee system.

The HSE's own evaluation of the safety case regulations found that a third of safety representatives said they were not adequately informed about the safety case and that a slightly higher number felt their constituents were not adequately informed [HSE, 1995]. Whilst the evidence presented here indicates that the HSE report may have understated the degree to which offshore workers have been excluded from the consultation process, it certainly identified problems of a similar character.

In light of Cullen's eagerness for workforce involvement to be a central feature of the safety case process, it is dismal that so few respondents reported serious involvement in the formulation of safety cases. While Cullen specifically stated that the safety case itself should include details of "operations which are aimed at involving the workforce in safety" [Cullen: 392], the vast majority of the workforce have been denied an input to consultation, and consequently, the industry has missed a unique opportunity to utilise the expertise of the workforce.

Cost-cutting and the CRINE Initiative

There are clear signs from the interviews, then, that in crucial respects little has changed offshore (or, that if it has changed, this has been for the worse). That is to say, there is little faith amongst workers in the formal mechanisms put in place after Cullen through which they might have their safety concerns represented. Yet these concerns can be further concretised, through reference to workers' perceptions of the effects of a new CRINE "initiative" amongst operating companies.

CRINE-related statements consistently assert an absolute complementarity between the goals of cost-reduction, efficiency, quality, and safety and environmental protection. The original CRINE statement, 'Cost Reduction Initiative for the New Era', notes that, "The correct implementation of CRINE recommendations is fully synergic with the overall safety process followed by the industry in the post-Cullen era .." This view has been endorsed by government ministers [Lloyds List 30 November 1994] and the HSE [Todd, 1996].

Evidence of the complementarity of CRINE and improving safety performance, however, was not supported by the testimonies of workers. Almost all respondents believed that 'cost-cutting' was now a significant threat to safety.

The elements of the cost cutting project having the most serious impact upon safety were identified as: a steady reduction in the workforce and the related operators led demand for a 'multiskilled' workforce, changes in shift patterns, cuts in maintenance budgets, a general reduction in safety related training, a lack of personal safety equipment and a re-emergence of the practice of employing 'cheap labour'. Each of these elements will now be outlined.

Reduction in the workforce

Central to the cost cutting strategy has been a sweeping reduction in the workforce on offshore platforms [Press and Journal, 25 November 1994 and 23 February 1996].

One example was reported of a platform on which a group of three respondents worked that had recently switched operator due to the previous operators pessimism about the economic viability of the field. The new operator had cut the workforce by a third over a 4 month period in order to sustain a profitable operation.

The HSE's evaluation of the safety case regulations points out that "....the changes perceived as having a negative effect on safety are in the area of reductions of manning and associated increases in workload" [HSE, 1995: 32] Most workers reported instances where a reduction in POB (persons on board) had implications for safety.

In some cases, respondents believed that risk was increased because of a lack of cover for safety critical procedures: "They've cut the crew hours by making the HLO (helicopter landing officer) work the minute he gets off the chopper. The rest of the crew wait on the chopper while he goes down two flights of stairs below deck, gets his survival suit off and his overalls off, comes back up to man the guns and lets the other guy get on the chopper. Meanwhile, everyone waits. It's a fucking comedy routine." [W25]

There were a few reports which indicated another dimension to the impact of this reduction in the workforce; that workers are reluctant to 'rock the boat' , particularly towards the end of a contract, because they are aware of the impending selection and disposal of workers. One example of this problem concerned workers on one platform who said nothing when told to carry out minor maintenance work without a permit [3] being raised: "I saw guys who were totally intimidated by authority and were brow beaten into it. A lot of the lads were keen to impress and needed the future work." [W20]

Fear of dismissal was invariably cited as the reason to "keep your mouth shut for two weeks" [W14] instead of reporting hazards to safety or taking a pro-active approach to safety on the platform. When the jobs are fewer and the competition for jobs greater, workers (especially the workers who are seen as "moaners" or a "trouble makers") become increasingly more disposable: "Now people are scared to do anything or else they'll end up with the undesirable tag" [W16]

Multiskilling

Respondents were also concerned at the pressure being created by the increasing requirement for workers to become skilled in more than one type of job. This was identified as a general trend on the majority of offshore installations and is referred to in the industry as 'multiskilling.' Examples of this included a maintenance crew on one platform being told to erect scaffolding, and on another platform, crane mechanics with no previous experience of operating cranes now working as crane drivers. Clearly, as the workforce is reduced across the industry, there are going to be 'skills gaps' left by the process of downmanning. These gaps are now being plugged by 'multiskilling.'

A major concern of the workforce is that 'multiskilling' adversely affects the quality of the job: "It's just becoming a joke. You've got that many jobs, that you can't concentrate on the one." [W21]

The erection of scaffolding is a major area currently being tackled by 'multiskilling.' One worker described the process of putting up a scaffold as a multiskilled worker:

> "We have to put our own scaffolding up. The gaffer will come and check it over. You've normally got a guy there who knows what he is doing, but he will probably just leave you to it.....They do have competency testing for the quality of the job, but when it actually comes to it, they overlook quality. You might have 1 person doing the job of 3. Safety just goes out of the window then. They always tell you to take your time. But no job is not worth taking the time to do. The truth is you never get enough time......It's the one real scaffolder that checks the job after it's done. But he's under pressure as well, and he knows the importance of getting it right first time." [W29]

The principle complaint is that workers are not being trained properly:

> "Scaffolders are being sent on rigging courses and riggers are going on scaffolding courses. But you'll never learn it in 3 days or 10 days. They want a lot more out of a lot less men. They're trying to create 'riggolders.' You can't multiskill someone in 3 days. I've been a scaffolder all of my working life, you can't learn that in three days." [W25]

It is no coincidence that the HSE Offshore Safety Division recently highlighted safety standards for scaffolding as a priority in the offshore industry [Offshore Oil International, November 1995]. The publication of the booklet 'Offshore Access Safety Guidance' in October 1995 was in direct response to 180 accidents in the 7 years since Piper Alpha which have killed one worker and seriously injured 12. [Evening Express 11 October, 1995]

Cuts in Maintenance Budgets

Cuts in planned maintenance programmes were reported on a number of platforms:

> "It's ridiculous. It's being cut down to the bare bones...Its the oldest [platforms]
> that they are running into the ground, and that's the ones that need more....[W2].

On one platform the night shift had been stopped for all crews except production and drilling. There are no maintenance crews on night shift. The maintenance mechanic who related this said that alarms are going off much more frequently, and that they are wakened from their beds regularly to carry out essential work when they are off-duty.

Another effect of such cutbacks is that a backlog of work can accrue where there are reduced maintenance crews. One respondent recalled how one maintenance team would falsify permits to work and timesheets in order to cover for jobs that they did not have the resources to finish in the time allocated. He discovered a valve lying on the deck beside a pipeline waiting to be fitted. According to the paperwork the valve had been installed. Evidence of this problem was recently detailed on BBC TV's Frontline Scotland programme. Len Thompson, an offshore worker from Shell's Cormorant Alpha platform spoke of a backlog of maintenance in 1993 which created a "thousands of hours backlog in vital equipment; in particular, fire and gas, IR detectors, smoke detectors, fire pumps, sprinkler systems." The programme also made the point that similar backlogs of work exist on this platform today [BBC TV, 1996].

In general, respondents were clear about the increase in risk at the workplace due to cuts in maintenance budgets. One commented:

> "It's like working on a time bomb. The operators are aware that it's a time bomb,
> but it just comes down to cost cutting. They're playing a gamble. The gamble is
> money to them, but to us it's people's lives". [W21]

Reduction in Safety Related Training

The duration of the industry standard RGIT offshore safety and survival course and the associated refresher course has been cut from 4 days to 2 days and the refresher course has also been shortened [see for example, Scotland on Sunday, 2 Oct. 1994; Lloyds List, 7 Feb. 1995]. In addition, there is an increasing requirement of workers to pay for the safety and survival course where previously the operator or contractor had done so. Several respondents said that the majority of offshore workers are now paying for their own RGIT certificate. Increasingly, offshore workers are being required to attend courses such as permit to work and firefighting training in their own time at a basic rate of pay or less. In some cases, these courses have to be paid for by workers as a requirement of their terms of employment.

Reduction of support vessel cover for installations

In many cases, support vessels are now being required to guard two platforms where previously the requirement was that one vessel should be allocated to each platform [The

Scotsman, 29 September 1994]. This is a direct result of the extension of goal-setting under the Prevention of Fire and Explosion, and Emergency Response Regulations [PFEER] 1995, and is deeply unpopular amongst offshore workers.

Cuts in rescue cover have become high priority for the operators' strategy of cost reduction. Both BP and Shell were in the public eye in May 1996. The former had reduced standby vessel cover on the Magnus and Ocean Guardian installations [Press and Journal, 3 May 1996], and the latter had withdrawn the emergency helicopter allocated to the Brent field, cutting rescue response time from 20 minutes to 2 hours [Press and Journal, 13 May 1996].

The operating companies point out that many of the new standby vessels are of superior design and are therefore more able to cope with 'incidents', and that the workforce have little to worry about [Brian Taylor, Technical Director, UKOOA, interview with author, 21 May 1996]. This view is strongly contested by offshore workers. The unanimous opinion of offshore workers is that having a standby vessel in close attendance is of crucial importance to safeguarding their workplace "Every rig should have a standby boat as a basic safety requirement."[W17]. One support vessel worker pointed out the impracticality of guarding more than one platform: "How can you cover two rigs? If you've got to cover two rigs, it's impossible with the time factor. The oil companies say time is money. But time is lives if somebody goes in the water."[W18]

Change in Shift Patterns

The industry wide move towards the use of 3 weeks onshore and 3 weeks offshore working shifts where previously workers were on 2 week shifts has now been adopted by at least 3 major operating companies. Again this trend was unanimously viewed by respondents as a threat to safe working conditions: "3 weeks working and you're going to get hurt, and people are getting hurt. It's quite surprising how many people get injured on the final shift."[W10

Some respondents identified another problem related to changes in working hours. He described the shift patterns of one workmate who was often asked to work shifts of up to 17 hours. It was reported by many that shifts of a similar length are becoming common place as a result of reduction in the size of the workforce.

UKOOA have waged a vigorous campaign against the EC 'Working Time Directive' which seeks to impose statutory limits on shift lengths [Woolfson, 1995a]. In light of the evidence presented here, it is perhaps not surprising that the operating companies are keen to retain the ability to overcome work backlogs and reduced crews by this means, despite the clear safety implications of such shift patterns.

A Lack of Personal Safety Equipment

An increasing reluctance of operators and contractors to issue personal safety clothing (such as boiler suits, hard hats, safety boots and safety goggles) was reported. Some employment agencies were reported to be demanding that employees supply their own personal safety gear. One interviewee who had experience of working on a Norwegian platform recounted that all necessary safety clothing was readily available on demand. At the end of the trip, the British workers would keep their old worn out safety gear and take it home in their kit bag. It was also common for British workers to steal tools out of the storerooms because they knew that if their next trip was on a platform in the UK sector, there would be a shortage of tools:

> "You were totally degraded. The Norwegians were saying what the fuck are these scumbags doing. We were just made to feel like animals. And if you treat people like animals, then they behave like animals." [W5] Another reported that some workers have started to take their own medical supplies offshore, such as elastoplasts, because they are generally unavailable."

Re-emergence of the Use of Cheap Labour

There is evidence of a re-emergence of 'cheap labour', typically from Eastern and Southern European countries and South-east Asia, at significantly lower wages than British workers. This phenomenon was described by one worker with more than 20 years offshore experience as "a return to the bad old days" [W17]. In the examples reported by respondents, these workers are typically employed as semi-skilled labourers, and as stewards to carry out domestic cleaning jobs. It was reported that in many cases, imported workers are earning less than half the normal rate for the job. Those workers are generally employed for longer hours, are asked to spend up to 2 months on the platform, and often may be required to take more risks than the rest of the workforce. One respondent recalled having seen a group of Malaysian painters using odd bits and pieces of rope as a safety harness. The Safety and Health Practitioner has also documented this occurrence. In a recent edition, it was reported that Portuguese and Lithuanian workers on a drilling rig "had a poor grasp of English and had difficulty understanding instructions during safety drills." [October 1995, page 6] Some of these workers earned less than a fifth of their British counterparts .

If one overwhelming conclusion can be drawn from this section, it is that the workforce does not share the view that CRINE and the cost cutting agenda are compatible with the provision of safe working conditions, and that, if anything, as CRINE and the more general cost-cutting trends continue to spread, standards of safety may be further eroded.

If there really is a 'new era' in offshore safety following Piper Alpha, it may actually be one which is less, rather than more safe; certainly many workers perceive it as such. The following section provides further signs that the offshore industry has not learned from some of the lessons of Piper Alpha.

The Pressure to Maximise Production

The view of many workers is that the programme of expenditure cuts has been matched with an intensification of the pressure from senior management to maximise the production of oil. According to most, one result of this strategy is that corners are being cut. The following section discusses this claim in more detail and examines the impact of this productivism on the workforce.

A minority reported that the pressure to carry out their job had not increased in the past few years. The following comments in this section therefore represent a majority and not a unanimous view. One worker described the atmosphere on a large platform during the hook up (construction) phase:

> "The pressure to get work done was intense. The pressure was felt at every level of the hierarchy and made for a chaotic situation.......There was so much going on that they couldn't keep track of it. There seemed to be a lot of gaffers running around not knowing what was going on." Corners were cut on safety: he was asked to do jobs on a ladder that should have been done on a scaffold because it needed two free hands. "They tried to apply macho pressure, "c'mon you wanker, what's wrong with you," "I saw other guys who were totally intimidated and were brow beaten into it. A lot of the lads were keen to impress and needed the future work." [W20]

Most respondents identified the root of these pressures in the orders their supervisors and the OIMs were given by production managers onshore: "The guy above you is under pressure from the one above him, and he makes sure you know about it."[W34]

A number of respondents reported that conflicting goals of safety and productivism led to an inconsistent approach by management:

> "At the toolbox talks [crew safety meetings], the company management may be there, and we will be told, take your time, there's no rush. But when the job starts, it's 'where's the crane? What are you doing, move your arse.' Behind closed doors they say, 'beef the guys up a bit.'" [W33]

One worker expressed concern at witnessing drillers coming in after working 10/11 hours without a break when the pressure is on. Others noted that there has been a recent increase in pressure to work overtime:

> "They want you to work for 12 hours without a break. They just want you to go like fuck without breaks, act crazy for an hour and you'll get paid for 7. They had to get a job done yesterday, and the deal was so good because they had a team of mechanics standing there on o3000 a day." (rig superintendents make these decisions according to the priority of each job.)

Respondents described the current use of bonus systems in the industry, particularly in the drilling sector, and outlined the potential threat to safety when drillers are working for substantial performance bonuses. Some pointed out that the use of bonus schemes by the contractors was a direct result of the continued promulgation of risk/reward type contracts by the operators. For drilling companies, downtime is a key variable in the contract, and the amount of downtime accumulated may considerably affect the value of the contract. This contractual risk is passed on to the workforce, and thus downtime is related to the remuneration that drillers get.

Two drilling contract workers described the type of situation they commonly faced:

"....some will shove safety down your throat, but when you have accidents, they still keep on at you to keep it quick...Safety is today's big thing. The supervisors are always on about it....But whoever you work for, time is the main issue. Safety gets blown out of the water because you need to get a job done. It takes second place if you want to get something done. You don't think about it, you take the short cut."

One of the drillers cited the example of company policy of workers having to use tag lines for lifting everything over a certain weight. But when the pressure is on, everyone is expected to manually handle anything it is possible to lift by hand, regardless of the weight.

Two other drillers employed by a different company presented an almost identical account:

"The targets for drilling are responsible for the pressure to work too quickly."[W33] and reported that major jobs are given time targets; for example, 90 days to drill each well. Incentive bonuses are also given on their rig for depths of casing. For four sections on each well drilled, the top bonuses are o380/o400.

Elizabeth Pate-Cornell has described the process of the how production pressures can directly influence safety on offshore oil platforms in the North Sea:

"In some oil companies, the philosophy seems to be "production first" and the time horizon seems limited to the short term....In an organisation that rewards maximum production....and faces a demanding world market, the culture is marked by formal and informal rewards for pushing the system to the limit of it's capacity. Production increases sometimes occur with little understanding of how close one is or might be to the danger zone" [Pate-Cornell, 1993: 227]

Pate-Cornell's description of the industry is entirely consistent with the conclusions of workers presented here. The experience of the workforce highlights how safety may be compromised where conflicts between the twin managerial goals of maintaining safe working conditions and maximising production arise, and that such conflicts may be reinforced by the widespread use of bonus schemes. There are, then, clear warnings for the industry to consider, particularly in terms of the impact of risk/reward contracts on the way in which offshore operations, particularly in the drilling sector, are prioritised.

Conclusion: the Prospects for Offshore Safety

Current economic conditions and the continued existence of organisational weaknesses in the industry appear to be alarmingly similar to the pre-Piper Alpha cultural climate. The industry is experiencing a significant intensification of the pressure to produce oil, alongside the most severe programme of expenditure cuts witnessed since 1985.

There is much evidence that the general context of productivism within which safety had been disregarded prior to the establishment of the new safety regime remains. Whilst the twin pressures of cost-cutting and production demands are of primary concern to the workforce, their continued exclusion from the process of safety management has reduced the potential for workers to resist unsafe working practices.

If one general conclusion can be drawn from all the strands of evidence taken from the interviews, it is that management have continued to dominate the interplay between managerial control and worker resistance.

CRINE in general, and downsizing in particular, have exacerbated the widespread fear to raise safety issues. An ever increasing proportion of offshore workers appear to be drawn from contract labour (OILC, 1994: 4). The NRB still exists, and the workforce appears as disposable and insecure as ever.

Despite the experience of other industries that an organised trade union input to health and safety improves safety performance, the almost Victorian opposition to trade unions continues to marginalise workers contributions to the management of safety.

Industry claims that safety performance is improving do not stand up to scrutiny. Recent, and it seems to me incontrovertible, evidence from researchers at Glasgow and St Andrews Universities reveals a rather different picture to that presented by the HSE and UKOOA. Woolfson and Beck [1995b] reexamine the bases upon which the HSE have calculated their own figures for fatalities and serious injuries in order to challenge the industry's conclusion that postal workers suffer more injuries at work than offshore workers [BBC TV, 1996]. First, in order to minimise problems associated with accident reporting in the industry, problems which are in fact, general across industries [Stevens, 1992], they exclude reportable minor injuries. They then weight the workforce based on Grampian Regional Council data. Their conclusion is that, with the exception of the coal mining, the offshore industry is the most dangerous sector in the UK.

The extension of Cullen's goal-setting regime as a form of self-regulation in the industry is worrying in this climate. As Tombs has argued, a combination of a weakened trade union movement, recession and the reassertion of a managers right to manage results in a disintegration of self-regulation. [Tombs, 1995] The cumulative effects of deregulation which occurs as a result of this process and the present government's explicit policy of deregulation is likely to militate against safe working conditions.

There is no doubt that large sums of money invested by the operating companies have provided a range of hardware improvements from emergency shutdown systems to temporary safe refuges. However, the words of Dave Lambert, a survivor of Piper Alpha, are perhaps appropriate as a reminder of the limits of applying technological solutions to more general organisational problems:

> "I was just going to bed and there was a bang. I thought it was metal to metal dropping. They came in and there was a fire - 'Get out'. I thought this couldn't happen - there were that many safety devices. That's what they told you. I know now that is not the case." [Guardian 8 July 1988]

If these issues remain unresolved, there will be little to reduce the likelihood of another Piper Alpha disaster occurring, or indeed to improve the appalling rate of accidents offshore.

ACKNOWLEDGEMENTS

Thanks to Steve Tombs for his comments on an earlier version of this chapter.

NOTES

[1] The data used in this paper is based upon a total of 83 semi-structured interviews, all conducted between February 1995 and January 1996.

38 interviews were conducted with offshore workers. A total of 57 people were included in the sample (24 respondents were interviewed individually and 33 in small groups of 2, 3 or 4).

The sample of workers interviewed was gathered from two sources. Firstly, the group of trade union members were recruited in one of the trade union social clubs in Aberdeen, and, secondly, the larger group of (mainly) non-trade union members were recruited from three public houses normally frequented before and after their trip offshore. Interviewees were selected on the basis that the total sample reflected the variety and depth of experience held by those working in the industry.

The length of time taken to complete interviews averaged just over an hour. The shortest interviews lasted about half an hour and the longest was recorded at 4 hours 5 minutes.

[2] 'Not Required Back' is the expression still used in the industry to describe the practice of on the spot dismissal. The expression has its origins in the early years of the industry when, it is said, the worker's card would be stamped with the letters 'NRB' in order to prevent him/her gaining employment in the industry again.

[3] The Permit to Work system is the administrative system that allows all on-site work to be controlled and coordinated. It is thought that a break down in the Permit to Work system was a crucial factor in the Piper Alpha disaster.

REFERENCES

BBC Radio Scotland (1996) 'Colin Bell Show', 5 February.

BBC TV (1996) Frontline Scotland. Paying the Price of Piper, 16 May.

Brandie, E (1994/5) 'Achieving the Balance Between Safety and Cost Reduction' in Offshore International, Winter.

Carson, WG (1982) The Other Price of Britain's Oil: Rutgers University Press.

Channel 4 (1994) Wasted Windfall, October.

Corti G and Frazer F (1983) The Nation's Oil: A story of control, London: Graham and Trotman.

CRINE Report (nd) Founding Statement for the Cost Reduction Initiative For The New Era.

Eggar, T (1993) 'Notes for Speech to the Cost Reduction Conference,' 2 December.

Pate-Cornell, E (1993) 'Learning From the Piper Alpha Accident: A postmortem analysis of technical and organisational factors', Risk Analysis, Vol. 13 No 2.

Foster, J And Woolfson, C (1992) Trade Unionism And Health And Safety Rights In Britain's Offshore Oil Industry, London: International Centre For Trade Union Rights.

Gasteen, A and Sewel, J (1995) 'The Aberdeen Offshore Oil Industry: Core and Periphery', in Rubery, J, Employer Strategy and the Labour Market, Oxford: Oxford University Press.

Harvie, C (1994) Fools Gold: The story of North Sea Oil, London: Hamish Hamilton.

HSE (1995) An Interim Evaluation of the Offshore Installations (Safety Case) Regulations 1992.

James, P (1993) The European Community: A positive force for UK health and safety law?, London: London Institute of Employment Rights.

Lavelette, M and Wright, C (1991) The Cullen Report - making the North Sea safe? Critical Social Policy, July.

McKay, DI and McKay GA (1975) The Political Economy of North Sea Oil, London: Martin Robertson.

Noreng, O (1985) The Oil Industry and Government Strategy in the North Sea, Boulder Colorado: The International Research Centre for Energy and Economic Development.

OILC (1991) Striking Out: New directions for offshore workers: Aberdeen: Offshore Information Centre

OILC (1994) Safety Auditing from Below, Paper presented to the Conference on Human factors in Offshore safety, Aberdeen, 27-28 September.

Risley, AW (1995) 'The Challenges of the Nineties' in Offshore International, October.

Rutledge I and Wright P (1996) Taxing the Second North Sea Oil Boom: A fair deal or a raw deal? University of Sheffield, unpublished paper.

Spaven, M Ras, H Morrison, A and Wright, C (1993) The effectiveness of Offshore Safety Representatives: A report to the Health and Safety Executive, London: HSE.

The Hon. Lord Cullen (1990) The Public Enquiry Into The Piper Alpha Disaster, Cm 1310, London: HMSO.

Todd, I (1996) Better Business is Better Safety, paper presented to 'CRINE: Learning to Survive', The Queen Elizabeth II Conference Centre, London, 31st January - 1st February 1996.

Tombs, S (1991) Piper Alpha And The Cullen Inquiry. Beyond "Distorted Communication"?, In Cox, R.F. And Walter, M.H., Eds., Offshore Safety And Reliability, London: Elsevier Applied Science.

Tombs, S (1995) 'Law, Resistance and Reform: 'Regulating' safety crimes in the UK', Social and Legal Studies, Vol. 4.

UKOOA (1994a) 'Reduction of Burdens on Business: DTI Deregulation Initiative', letter sent to all members of UKOOA, 7 March 1994.

UKOOA (1994b) Minutes of the Council Meeting, 11 May .

Walters, D (1995) Employee Representation and Occupational Health and Safety: The significance of Europe, Journal of Loss Prevention in the Process Industries, Vol 8, No 6.

Whyte, D Tombs, S and Smith, D (1996) Not Required Back (or Don't Rock the Boat if you Want to Keep Your Job). Occupational Safety and Health, January.

Woolfson, C (1995a) "The Deregulation of the British Continental Shelf - the Hidden Agenda' in Offshore Safety in a Cost Conscious Environment, Aberdeen: Offshore Information Centre.

Woolfson, C and Beck, M (1995b) Seven Years After Piper Alpha: Safety Claims and the New Safety Case Regime, Glasgow: University of Glasgow.

CHANGING PERCEPTIONS OF RISK:

The Implications for Management

EVE COLES & DENIS SMITH
Sheffield University Management School
9 Mappin Street
Sheffield, S1 4DT, UK

STEVE TOMBS
Liverpool John Moores University`
School of Law, Social Work, Social Policy
Josephine Butler House
1 Myrtle Street
Liverpool, L7 4DN, UK

Introduction

"Risk is like beauty - it exists in the eye of the beholder." (Pitzer, 1999)

This book has sought to bring together a number of perspectives on the management of risk. Its ultimate aim has always been to raise the significance of risk within key areas of political and organisational activities and with a key focus on the management and control of that risk. There is little doubt that, in spite of our advances in knowledge and management practices, risk is destined to remain an issue of some concern. At the time of writing, for example, issues of risk and safety have once again been firmly catapulted to the top of the public agenda. Whilst there were a number of events that served to raise the public consciousness of risk – including the completion of the BSE inquiry, the anniversary of the Paddington rail crash, the impending inquiry into the sinking of the Marchioness and the sinking of a number of Greek ferries – two events will be highlighted to illustrate some of the key issues discussed in this text. These events are the crash of the Air France Concorde and the Firestone tyre recall.

On July 25 2000, an Air France Concorde, with 109 passengers and crew on board, crashed on take off from Charles DeGaulle Airport, Paris, demolishing part of a hotel and killing a further five people on the ground. The accident occurred only a day after British Airways and Air France had announced that tiny two and a half inch creaks had been found

E. Coles, D. Smith and S. Tombs (eds.), Risk Management and Society, 285–295.
© 2000 *Kluwer Academic Publishers. Printed in the Netherlands.*

in the wings of eleven of the thirteen operational aircraft. Whilst the two events are unconnected, they illustrate the sheer complexity that organisations face in dealing with risk and the potential for multiple modes of failure within complex systems. As the news bulletins and commentators struggled to make sense of the accident (it now seems that debris from a catastrophic tyre burst on take off punctured the starboard wing and caused the fatal fire) issues of risk and safety, quite naturally, dominated and underpinned the media debate. Since the crash British Airways representatives have repeatedly assured the travelling public (or at least that elite sub-set of passengers that could afford to fly on the aircraft) that the safety of passengers was their primary concern and that, ultimately, their aircraft were safe. As a result of the initial air accident investigation, the air-worthiness certificates of all British Concordes were withdrawn some three weeks after the crash and the fleet was grounded (the French Concordes did not fly during that period). During the period of this crisis, the public were informed, perhaps for the first time, of a number of potentially dangerous incidents involving the aircraft that have occurred over the thirty years it has been in service. Perhaps not surprisingly, the debates and flows of information surrounding the safety of the aircraft have been shaped by the accident. From being generally considered as safe aircraft, Concorde is now faced with an uncertain future. Concorde is not, of course, the first aircraft to be involved in a fatal accident and, unfortunately, neither will it be the last. What is interesting about this accident is the manner in which perceptions of safety can be shaped by a single event and how different approaches to regulation are employed within different industries. The accident also raises considerable questions concerning the semantics of risk – what does 'safe' and 'potentially dangerous' mean to those people who are exposed to the hazard.

The passengers that perished on the Concorde were embarking on the holiday of a life-time, they probably (as far as we can know) attached very little risk to the journey they were going to make and there is little doubt that, in their minds, the potential benefits of the trip outweighed the possible hazards. As far as they were concerned, the aircraft had never had a fatal crash or been involved in any serious (life threatening incidents) and the crew were highly trained professionals who were well versed in safety issues. Had the passengers known about the 'potentially dangerous incidents' involving the aircraft over its years in operation, one wonders if they would have still been prepared to make the flight. By comparison, the crew flew in the aircraft on a regular basis and almost certainly felt relatively safe even though they would have had more detailed knowledge of the aircraft's safety record.

The questions raised here are ultimately ones of knowledge, communication and perception. Firstly, at what point does the amount of information available to the public enable them to make an informed judgement regarding the risk and secondly, how much of that information will be disregarded in order to achieve the goals that have been set? To what extent do organisations ensure that the tacit knowledge held by individual members is codified and made more explicit. Risk inevitably brings with it discussions of victims and those who are culpable in the causation of the event. It surfaces questions concerning the role of systems operators in the triggering of the accident sequence but it also raises questions about the role of the managing organisation in setting the environment and culture within which such active error can occur. A key question, and one which has received some attention within this volume, is the extent to which the operators of the technology are ever fully aware of the impact that their actions can have on the performance of the system without being informed of the latent potential for error that is embedded within organisational norms, assumptions, culture and protocols.

A second recent example of risk, this time in the USA, serves to illustrate these issues. Somewhere in the order of 2226 complaints were made to the National Highway Traffic Safety Administration (NHTSA), mainly during the period 1998-2000[1], concerning the vulnerability of certain Firestone tyres, which had been produced for the sports utility vehicle (SUV) market in the USA. (National Highway Traffic Safety Administration, 2000). Many of these tyres failed catastrophically whilst the vehicle was travelling a speed and they appear to have resulted in 103 reported fatalities in the USA over the time period (National Highway Traffic Safety Administration, 2000), although it is possible that some under reporting may have occurred. The tyres were factory fitted to Ford Explorers, although they were also used on other vehicles, and it appears that the tyres failed under warm weather conditions and most of the deaths occurred in the southern states of the USA, where the ambient temperature may have been a major factor in the accidents. Of these accidents, Ninety-two of the fatalities were reported initially to the National Highway Traffic Safety Administration (NHTSA) and a further 11 fatalities were identified by Firestone. Perhaps as a result of the possible litigation associated with the events, Firestone made a request to the NHTSA asking that the data relating to the fatalities and associated claims be treated as confidential information (see National Highway Traffic Safety Administration, 2000). This was rejected by the NHTSA. Whilst these deaths achieved high profile coverage within the USA, the 'crisis' that came with them emerged over a longer period of time. Both Ford and Firestone mounted an extensive media campaign in an attempt to reassure consumers about the safety of the product. At the height of the crisis, the two companies took out full page advertisements in the press in an attempt to put forward the companies perspectives on the crisis and the actions that they were taking. At the height of the crisis, the CEO of Ford appeared on US television in an attempt, again, to reassure consumers.

Exploring the elements of risk management

What these cases illustrate are the organisational dynamics of risk. In both cases, some evidence exists to suggest that early warnings of impending problems may have been ignored. In both cases, the deaths triggered a much broader crisis for the organisations concerned and they also impacted upon their strategic partners. The events also required that those responsible for managing within those organisations attempted to provide consumers (and other stakeholders) with reassurances of safety and, by implication, confirm management's control of the situation.

Any systematic review of crisis events over a period of time would probably highlight many of the issues raised by the brief examination of these two cases. Indeed, Turner (1976) identified a number of core elements in his seminal work on disasters. Turner's work (1976; 1978) clearly has a relevance to the work presented in this volume. In the first instance, the manner in which warnings are ignored by organisations can be seen as a reflection of the core beliefs and assumptions held by senior managers and the associated precautionary norms that they put into place (Turner, 1978). The extent of these precautionary norms – or the patterns of control within organisational systems – has been the subject of considerable discussion within this volume.

The first of Turner's elements, centred on the rigidities of core beliefs and perceptions within organisations and the associated climate that these factors generated. Smallman has described the difficulties of building conceptual models of risk and so setting decision making rules within a climate of scientific uncertainty. He emphasises the need to include all stakeholders within the risk management process and look beyond the traditional towards the research and development of a collaborative paradigm of risk management. Glendon et al, presented a case that takes due consideration of the need to look beyond the confines of quantitative analysis towards "the risk perceptions of decision makers and factors and dimensions that affect their decisions on risk" (this volume p53). They also suggest that a continuing regime of undertaking case studies within their industry and other sectors including the private, corporatised and public sectors will allow for more generalised models of risk assessment and risk management to be developed thus allowing for political, social, cultural and economic influences and their impact on risk management to be more fully understood.(p52)

Turner's second element considers the role of decoy phenomenon in distracting managerial attention away from the main problem. In his early work Turner argued that,

"......action taken to deal with that problem distracted attention from the problems which eventually caused trouble. In other words, a contributory factor to the disasters was the attention paid to some well-defined problem or source of danger which was dealt with, but which distracted attention from another dangerous but ill-structured problem in the background" (Turner, 1976, p. 388).

Stahel outlines the growing discrepancy between insured and economic losses in catastrophic events over the last century. He points out that risk is much more a moral issue than it is technological one. He further points to the fact that 'worst cases' are ruled out in studies on national risk analysis thus creating a danger of the biggest catastrophes being accepted without any preventative measures and that it is taken for granted by business that the State will pick up the pieces. Only by moving towards a more sustainable economic approach through a series of measures that range from a re-evaluation of the way time is used through prudent use of cash resources to the professional qualification of personnel responsible for operations and maintenance can a 'Precautionary Principle' be adopted. In short as O'Riordon notes,

> "precaution requires action ahead of scientific proof, when cost effectiveness of response cannot be guaranteed, and when costs to the public, in terms of behaviour change, new taxation or policy shift, could cause hardship or resentment". (O'Riordon, 1996)

The chapter by Smith and McCloskey in this volume provides a useful link between the sustainability and precautionary principle arguments raised by Stahel and the sunsequent discussions of de-regulation provided by Pearce and Tombs, Tombs, Gifford, and Whyte.

Turner's third element concerns the disregard of non-members (ie outsiders to the dominant elite within the organisation) where those who offer warnings of impending catastrophe are dismissed by those powerful interests within organisations. Again the chapter by Smith and McCloskey illustrates the dynamics of this process. Utilising the British cases of BSE and GMO they explore the nature of failure, the limitations of expertise, organisational learning from failure and the current implementation of the precautionary principle by government and other organisations. They further enhance the debate surrounding the political nature of risk highlighting some of the problems presented by the development of the so-called new knowledge elites and their central position within the risk policy debate. Whilst they argue, like Stahel, that safety and security is a moral issue properly belonging to government they also expose the limitations of governments to effectively implement the precautionary principle in the current climate of self regulation and erosion of the state.

The fourth element outlined by Turner concerns the problems surrounding information difficulties. Clearly, issues of information difficulties abound within risk debates. The brief discussions of the Concorde crash illustrate some of the problems that face an organisation in the aftermath of a catastrophic failure. Although there is obviously a discernable amplification of risk issues following a major accident like the Concorde crash the contributions to this volume clearly illustrate that the issues of risk perception, risk communication, regulation, and the social acceptability (or unacceptability) of risk have become increasingly important within society in general (see also for example Hadden, 1991, Jaffe, 1991, The Royal Society 1992, Smith and McCloskey, 1998, Pitzer 1999). Where these issues are important, and where further work is needed, is in terms of the problems of information and knowledge capture within the incubation of risk potential within organisations. In particular, the role of manager's cognitive schema in helping them to shape their decision making around risk and the discourses that come with that are worthy of much greater study. Turner raises two additional elements that have a relevance here. The first is the involvement of outsiders within these decision making process and the second is the minimising of emergent danger. A greater understanding of the cognitive processes around risk will help to provide us with greater insights into these issues and the interactions that take place between them.

Finally, Turner points to the importance of regulation and the lack of compliance with it as a means of incubating the potential for failure. It is here that this text has made a significant contribution to debates on risk. Notions of regulation, as a form of control, have achieved prominence within this volume. Gifford notes in the introduction to his contribution to this volume that "Statutory regulation is part of the government's role in risk management" (p207). This being said, it is clear that, as Gifford himself states, the 1980s witnessed a clear impetus towards deregulation in the UK. However, we need to be careful in speaking simplistically of deregulation, since there are a myriad of ways in which the states role in risk management can be, and has been, undermined. Thus, alongside deregulation we need to recognise

> "non-regulation, weak enforcement of extant law, hostility to the notion of employee representation, cuts in the budgets of the Health And safety Executive and government research establishments. The privatisation of government science, the denigration of public service and the demoralisation of scientists and HMO inspectors" (Gifford, this volume, p208; see also Tombs, Whyte, and Beck and Woolfson, this volume).

If it is clear that we cannot refer simplistically to a unilinear trend towards deregulation, it is clear that prescriptive, external state regulation has come under sustained political and academic attack since the early 1980s, much of this attack centring around the role of regulation in relation to risk reduction. Whether arguing that external regulation is ineffective, inflexible, bureaucratic, counter-productive, or costly (see Slapper and Tombs, chapter 8), the overall effect of this attack has been to provide momentum for a generalised shift away from prescriptive regulation to performance standards or goal-setting regulation. The relative merits of these approaches are considered at length in the contributions to this volume by Whyte, Beck and Woolfson, and Gifford, (and see also Smallman, pages 73 - 78), and need not be entered into here. Indeed, there is also a substantial literature on regulation generally, and in relation to risk reduction in particular. Suffice to say that there is an urgent need to move beyond ideologically-motivated prescriptions regarding regulation and risk and to undertake detailed, and indeed comparative, studies of various forms of regulation in practice, and to attempt to generate data on the effects of these differing forms of regulation. In these respects, the work contained in this volume on the regulatory shift in the UK offshore oil industry is instructive and important.

Thus, for our considerations, the key implication of the hegemony of discourses of globalisation is that Governments can exert less political control over economies - economic management is relegated to the task of over-seeing the operation of 'free' markets - and over the key actors in these economies, namely corporations and, most significantly, multi- or transnational corporations. Having accepted - more or less willingly - the dictats laid down through the discourses of globalisation, governments self-regulate to the extent that they exercise supreme caution in imposing additional "burdens" (that is, regulation, or more specifically the costs arising from so-called "social" regulation and the effects of corporate competitiveness) upon business. Such discourses have found a perfect complement in the deregulatory initiatives of recent British governments: this was clearly the case in the case of successive Conservative Governments who, as key architects in an international neo-liberal project (see Pearce and Tombs, and Tombs, this volume), played a crucial role in promoting the discourses of globalisation, both materially and ideologically; but it must be recognised that the current Labour Government remains committed to this legacy, portraying itself as subject to the whims of an international market within which Britain can be a key player only by ensuring the continued development of deregulated labour and capital markets (see Hay, 1999).

A critique of the discursive and material processes whereby regulation is organised off political agendas retains implicitly a commitment to the argument, developed in various contributions to this volume, that regulation is a necessary though not sufficient condition of effective risk management and reduction. We must emphasise that this is *not* equivalent to arguing that state regulation represents a panacea in the context of effective risk management. A critical examination of deregulation does not entail a denial of the fact that there must always be a balance to be struck between 'over-regulation' and under-regulation (see Beck and Woolfson, this volume). However, what is also crucial is that the terms of the debate are set clearly when we engage in such considerations. This is why understanding the ideological bases of some techniques utilised within risk management (see above, and Beck and Woolfson, this volume) - notably forms of cost-benefit analyses - is important, since such techniques commonly over-estimate costs and under-estimate benefits associated with risk reduction in general, and state regulation in particular (and on this, see European Agency for Safety and Health at Work, 1999).

These considerations around regulation lead us to a decidedly macro-level set of considerations. For running through the contributions to this volume - and indeed through almost all considerations of risk and risk management - is a fundamental, if often implicit, question: just what level of risk is socially acceptable. It should be clear, however, that to ask this question is to raise complex political issues. Whyte (this volume) points to the limits of technological approaches to risk management, while Stahel (this volume, p101) emphasises that risk acceptability is a moral not a technical issue; and if it is a moral issue, once this moves beyond the realm or level of the private individual, as the considerations in the volume almost all do, then this becomes a political issue. Indeed, for us, to ask what level of risk is socially acceptable is to imply - and thus it should be to address explicitly - a whole series of other, related questions: what kind of society is this; what kind of society do we want; third, who decides or determines the answers to each of these questions. These questions remain to be answered and will inevitably be part of a wider research agenda as risk remains high on the political agenda throughout the 21ˢᵗ Century. What is clear, is that Turner's early work in this area still have a relevance within current debates on risk. What is also clear, is that organisations and managers within them still show a reluctance or an inability to learn effective lessons concerning the manner in which the processes of incubation can be addressed. It is hoped that the contributions to this volume will serve to jog the corporate memory a little in this regard. If it can achieve that then it has been successful.

Conclusion: towards an integrated understanding of risk management and society

The discussion of the previous section seems, at first sight, to take us a long way from the technical elements of risk and risk management to which we referred in the earliest sections of this chapter. Yet they are a perfectly consistent, yet for us logically prior, set of considerations when approaching questions of 'risk management and society'. In conclusion, we wish to make explicit two guiding principles on the basis of which we have developed this introductory chapter, and which we wish to propose are integral to the development of effective considerations and actions around risk management.

First, risk management, both as an intellectual enterprise and as a set of actions must proceed on the basis of a sensitivity to theory and practice. Understanding risk management must entail a range of concrete examinations of the nature and origins of particular risk management techniques and strategies, and implementation of these via risk management programmes at organisational (and, indeed, governmental) level. Social theory around 'risk', 'risk management' and 'society' must be utilised as a source of frameworks for understanding and assessing the nature of risk management strategies and programmes; these understandings and assessments are then used as a means of interrogating the validity and utility of these frameworks drawn from, or developed within, social theory. Thus understanding risk management and society requires a commitment to theory and theory development, but this must not be treated in any abstracted sense. It must be used to understand, and then to be interrogated by, practice. Similarly, risk practices should not be treated in abstract - case studies, for example, have real utility, but this utility is under-developed if they are not then interrogated within broader theoretical frameworks All too often, risk management - as many other spheres of intellectual enquiry and practical activity - has suffered from an over-emphasis upon 'theory' or 'practice' to the detriment of the other.

Second, risk management must be addressed at both a micro- and macro-level; and, at some point, a framework or frameworks are required whereby analyses conducted at these levels can be integrated. In this respect, the use of case studies is instructive. Thus we endorse the exigency with which Glendon et al conclude in their contribution to this volume:

"By undertaking further case studies both within the same industry and in other industry sectors and countries, more generalised models of risk assessment and management could be developed. For example, organisations in private, corporatised and public sectors can be compared in respect of their risk management practices. By undertaking further studies of this type, political, social, cultural and economic influences - increasingly recognised as critical to risk management - can be evaluated for their impact upon risk management practices in organisations" (Glendon et al., this volume, 52).

Thus we must engage in, but move beyond, discrete case-studies, while particular case studies cannot be treated in abstraction. And there is an important general point here also: while it is useful to engage in considerations of dimensions of risk, management and society as set out in this chapter, beyond a certain point of analysis - let alone in any practical sense - these *cannot* adequately be treated as discreet areas or issues.

Recently, in a review of the sociology of industrial injury, Nichols has set out, albeit in a different context, precisely the range of levels that an adequate explanatory framework must encompass or incorporate. While his concern is with industrial injury and not risk *per se*, the elements to which he points, and the levels at which these elements are to be found or operate, are clearly pertinent for any discussion of risk and risk management. While Nichols does not dismiss what he call micro-level studies of the production of industrial injury, adequate analysis cannot stop there, since only a partial understanding can be gained from such study; one must constantly bear in mind that "if men make accidents they do not do so under conditions of their own choosing" (Nichols, 1997: 117). Thus, a micro-level social science, which is limited in its ability to stretch "upwards .. to a more macro-level of analysis" (p.93), obscures any "serious regard to the processes of capital accumulation and the role of the State" (ibid.). And so Nichols argues the case for a political economy of industrial injury (and see Woolfson, this volume). This approach would of course treat safety (and in particular its complex relationship to profitability) "at the level of the enterprise" (Nichols, 1997: 103), but at the same time recognise that "a broader set of conditions always informs the process of production and, within this, the social production of injury" (ibid.: 198), so that analysis must move beyond a fixation with "questions about the final motivation of conduct of actors" (ibid.). A political economy approach, then, will attempt to take account of:

"growth and stagnation in the world economy, of the problems and nature of particular regimes of accumulation, including their resort to labour intensification and corner cutting, and the extent to which through legislation and unemployment there is weakening of labour's capacity to resist. Account must also be taken of the reshaping of national and international industrial composition and of the relation between big and small capital and the rise and fall of different categories of labour with different relations to their employer" (Nichols, 1997: ?03).

The dimensions and scale of such an approach, reiterated by Beck and Woolfson in this volume (39-40), points to a daunting, and largely ignored, research agenda.. Indeed, we would add to this range of factors, for Nichols' focus is very much on the material; there is also a need to address ideological and cultural factors, and these must be addressed from the level of the work group and the organisation through to some attempt to grasp dominant ideas and ideologies at both national (Gifford and Tombs, this volume) and international levels (Pearce and Tombs, this volume).

Finally, we should emphasise what is centrally entailed in such an approach. Risk management must encompass an understanding of the operation of power at all levels of analysis, while power itself can only be fully captured by such an integrated approach. Risk management is defined by power at every level. As Hopfl has stated,

"the capacity to construct statements regarding risk rests on a power relationship within organisation and between organisations and the outside world. In this way, the definition of a situation may carry more significance than another" (Hopfl, this volume, 134).

With some notable exceptions, an attempt to grasp power has been largely absent from concerns with risk management. This volume does, we hope, offer a glimpse of the urgency, contours, and promise of this task.

NOTES

[1] There were some incidents in the early 1990s but most of the events occurred during 1998-2000. The first fatal accident included in the NHTSA data appears to be in 1993 with the first complaint in 1990 (see National Highway Traffic Safety Administration, 2000).

REFERENCES

Associated Press (2000) Ozone hole over a city for first time. MSNBC News. Accessed 6th October 2000. http://www.msnbc.com/news/452791.asp?bt=nm&btu=http://www.msnbc.com/tools/newstools/d/news_menu.asp

European Agency for Safety and Health at Work (1999) *Economic Impact of Occupational Safety and Health in the Member States of the European Union*, Bilbao: European Agency for Safety and Health at Work

Ford (2000) Ford CEO Jac Nasser Announces Safety Initiatives During Congressional Testimony (accessed 5th October 2000) http://www.ford.com/default.asp?pageid=106&storyid=945

Hadden, S G (1991), 'Public Perception of Hazardous Waste' in *Risk Analysis: an International Journal*, Peleum Press, New York, Vol.11, No. 1, pp47-58

Hay, C. (1999) *The Political Economy of New Labor*, London: Unwin Hyman

Hood, C, & Jones, D, (eds) (1996), *Accident and Design, contemporary debates in risk management*, UCL Press, London

Jaffe, M (1991)' "Comments on "Public Perception of Hazardous Waste", in *Risk Analysis: an International Journal*, Vol.11, No. 1, pp58-60

Nasser, J. (2000) Statement of Jac Nasser, Chief Executive Officer of Ford Motor Company, September 6, 2000 (accessed 5th October 2000). http://www.ford.com/default.asp?pageid=106&storyid=946

National Highway Traffic Safety Administration (2000) Database Correction September 21, 2000 (includes September 19, 2000; 15th, 22nd and 31st August, 2000). http://www.nhtsa.dot.gov/hot/firestone/Update.html

Nichols, T (1997) *The Sociology of Industrial Injury (Employment and Work Relations in Context Series)*, Mansell

O'Riordon, T, (1996) 'Exploring the Role if Civic Science in Risk Management' in Hood C, and Jones D, (eds) *Accident and Design, contemporary debates in risk management*, UCL Press, London

Pitzer, CJ, (1999), 'New Thinking on Disasters; The Link Between Safety Culture and Risk-taking', in *The Australian Journal of Emergency Management*, Emergency Management Australia, Vol 14, No. 3, pp41-50

Rosenthal, I (1991), "Comments on "Public Perception of Hazardous Waste", in *Risk Analysis: an International Journal*, Peleum Press, New York, Vol.11, No. 1, pp61-62

The Royal Society (1983), *Risk Assessment*, A Study Group Report, London

The Royal Society (1992) *Risk: Analysis, Perception and Management. Report of a Royal Society Study Group* London: The Royal Society. Chapter 6

Turner, B A (1976) The organizational and interorganizational development of disasters, *Administrative Science Quarterly*, 21, pp. 378-397.

Turner, B A (1978) *Man-Made Disasters*. London: Wykeham

Slapper, G. and Tombs, S. (1999) *Corporate Crime*, London: Longman.

Smith, D, & McClosky, J, (1998), 'Risk Communication and the Social Amplification of Public Sector Risk', in *Public Money and Management*, CIPFA, London, Vol , No , pp41-50

INDEX

Advances in Natural and Technological Hazards Research

Series Editor: Prof. Dr. Mohammed I. El-Sabh, *Département d'Océanographie, Université du Québec à Rimouski, 310 Allée des Ursulines, Rimouski, Québec, Canada G5L 3A1*

Publications

1. S. Tinti (ed.): *Tsunamis in the World.* Fifteenth International Tsunami Symposium (1991). 1993   ISBN 0-7923-2316-5

2. J. Nemec, J.M. Nigg and F. Siccardi (eds.): *Prediction and Perception of Natural Hazards.* Symposium Perugia, Italy (1990). 1993
                                                  ISBN 0-7923-2355-6

3. M.I. El-Sabh, T.S. Murty, S. Venkatesh, F. Siccardi and K. Andah (eds.): *Recent Studies in Geophysical Hazards.* 1994   ISBN 0-7923-2972-4

4. Y. Tsuchiya and N. Shuto (eds.): *Tsunami: Progress in Prediction, Disaster Prevention and Warning.* 1995   ISBN 0-7923-3483-3

5. A. Carrara and F. Guzzetti (eds.): *Geographical Information Systems in Assessing Natural Hazards.* 1995   ISBN 0-7923-3502-3

6. V. Schenk (ed.): *Earthquake Hazard and Risk.* 1996   ISBN 0-7923-4008-6

7. M.I. El-Sabh, S. Venkatesh, H. Denis and T.S. Murty (eds.): *Land-based and Marine Hazards.* Scientific and Management Issues. 1996
                                                  ISBN 0-7923-4064-7

8. J.M. Gutteling and O. Wiegman: *Exploring Risk Communication.* 1996
                                                  ISBN 0-7923-4065-5

9. G. Hebenstreit (ed.): *Perspectives on Tsunami Hazard Reduction.* Observations, Theory and Planning. 1997   ISBN 0-7923-4811-7

10. C. Emdad Haque: *Hazards in a Fickle Environment: Bangladesh.* 1998
                                                  ISBN 0-7923-4869-9

11. F. Wenzel, D. Lungu and O. Novak (eds.): *Vrancea Earthquakes: Tectonics, Hazard and Risk Mitigation.* 1999   ISBN 0-7923-5283-1

12. S. Balassanian, A. Cisternas and M. Melkumyan (eds.): *Earthquake Hazard and Seismic Risk Reduction.* 2000   ISBN 0-7923-6390-6

13. S.L. Soloviev, O.N. Solovieva, C.N. Go, K.S. Sim and N.A. Shchetnikov: *Tsunamis in the Mediterranean Sea 2000 B.C. – 2000 A.D.* 2000
                                                  ISBN 0-7923-6548-8

14. J.V. Vogt and F. Somma (eds.): *Drought and Drought Mitigation in Europe.* 2000   ISBN 0-7923-6589-5]

Advances in Natural and Technological Hazards Research

Kluwer Academic Publishers – Dordrecht / Boston / London